D0581523

TEACHING AND APPLYING MATHEMATICAL MODELLING

ELLIS HORWOOD SERIES IN MATHEMATICS AND ITS APPLICATIONS

Series Editor: Professor G. M. BELL, Chelsea College, University of London

Statistics and Operational Research

Editor: Professor B. W. CONOLLY, Chelsea College, University of London

Baldock, G. R. & Bridgeman, T. — **Mathematical Theory of Wave Motion**
de Barra, G. — **Measure Theory and Integration**
Berry, J. S., Burghes, D. N., Huntley, I. D., James, D. J. G. & Moscardini, A. O. — **Teaching and Applying Mathematical Modelling**
Burghes, D. N. & Borrie, M. — **Modelling with Differential Equations**
Burghes, D. N. & Downs, A. M. — **Modern Introduction to Classical Mechanics and Control**
Burghes, D. N. & Graham, A. — **Introduction to Control Theory, including Optimal Control**
Burghes, D. N., Huntley, I. & McDonald, J. — **Applying Mathematics**
Burghes, D. N. & Wood, A. D. — **Mathematical Models in the Social, Management and Life Sciences**
Butkovskiy, A. G. — **Green's Functions and Transfer Functions Handbook**
Butkovskiy, A. G. — **Structure of Distributed Systems**
Chorlton, F. — **Textbook of Dynamics, 2nd Edition**
Chorlton, F. — **Vector and Tensor Methods**
Dunning-Davies, J. — **Mathematical Methods for Mathematicians, Physical Scientists and Engineers**
Eason, G., Coles, C. W. & Gettinby, G. — **Mathematics and Statistics for the Bio-sciences**
Exton, H. — **Handbook of Hypergeometric Integrals**
Exton, H. — **Multiple Hypergeometric Functions and Applications**
Exton, H. — **q-Hypergeometric Functions and Applications**
Faux, I. D. & Pratt, M. J. — **Computational Geometry for Design and Manufacture**
Firby, P. A. & Gardiner, C. F. — **Surface Topology**
Gardiner, C. F. — **Modern Algebra**
Gasson, P. C. — **Geometry of Spatial Forms**
Goodbody, A. M. — **Cartesian Tensors**
Goult, R. J. — **Applied Linear Algebra**
Graham, A. — **Kronecker Products and Matrix Calculus: with Applications**
Graham, A. — **Matrix Theory and Applications for Engineers and Mathematicians**
Griffel, D. H. — **Applied Functional Analysis**
Hanyga, A. — **Mathematical Theory of Non-linear Elasticity**
Hoskins, R. F. — **Generalised Functions**
Hunter, S. C. — **Mechanics of Continuous Media, 2nd (Revised) Edition**
Huntley, I. & Johnson, R. M. — **Linear and Nonlinear Differential Equations**
Jaswon, M. A. & Rose, M. A. — **Crystal Symmetry: The Theory of Colour Crystallography**
Johnson, R. M. — **Linear Differential Equations and Difference Equations: A Systems Approach**
Kim, K. H. & Roush, F. W. — **Applied Abstract Algebra**
Kosinski, W. — **Field Singularities and Wave Analysis in Continuum Mechanics**
Marichev, O. I. — **Integral Transforms of Higher Transcendental Functions**
Meek, B. L. & Fairthorne, S. — **Using Computers**
Muller-Pfeiffer, E. — **Spectral Theory of Ordinary Differential Operators**
Nonweiler, T. R. F. — **Computational Mathematics: An Introduction to Numerical Analysis**
Oldknow, A. & Smith, D. — **Learning Mathematics with Micros**
Ogden, R. W. — **Non-linear Elastic Deformations**
Rankin, R. — **Modular Forms**
Ratschek, H. & Rokne, Jon — **Computer Methods for the Range of Functions**
Scorer, R. S. — **Environmental Aerodynamics**
Smith, D. K. — **Network Optimisation Practice: A Computational Guide**
Srivastava, H. M. & Karlsson, P. W. — **Multiple Gaussian Hypergeometric Series**
Srivastava, H. M. & Manocha, H. L. — **A Treatise on Generating Functions**
Sweet, M. V. — **Algebra, Geometry and Trigonometry in Science, Engineering and Mathematics**
Temperley, H. N. V. & Trevena, D. H. — **Liquids and Their Properties**
Temperley, H. N. V. — **Graph Theory and Applications**
Thom, R. — **Mathematical Models of Morphogenesis**
Thomas, L. C. — **Games Theory and Applications**
Townend, M. Stewart — **Mathematics in Sport**
Twizell, E. H. — **Computational Methods for Partial Differential Equations**
Wheeler, R. F. — **Rethinking Mathematical Concepts**
Willmore, T. J. — **Total Curvature in Riemannian Geometry**
Willmore, T. J. & Hitchin, N. — **Global Riemannian Geometry**

TEACHING AND APPLYING MATHEMATICAL MODELLING

Editors:

J. S. BERRY, B.Sc., Ph.D.
Lecturer in Applied Mathematics, The Open University

D. N. BURGHES, B.Sc., Ph.D.
School of Education, University of Exeter

I. D. HUNTLEY, B.A., Ph.D.
Department of Mathematics, Sheffield City Polytechnic

D. J. G. JAMES, B.Sc. (Maths), B.Sc.(Chemistry), Ph.D., F.I.M.A.
Department of Mathematics, Coventry (Lanchester) Polytechnic

A. O. MOSCARDINI, B.Sc., M.Sc., Dip.Ed.
Department of Mathematics & Statistics, Sunderland Polytechnic

ELLIS HORWOOD LIMITED
Publishers · Chichester

Halsted Press: a division of
JOHN WILEY & SONS
New York · Chichester · Brisbane · Toronto

First published in 1984 by

ELLIS HORWOOD LIMITED
Market Cross House, Cooper Street, Chichester, West Sussex, PO19 1EB, England

The publisher's colophon is reproduced from James Gillison's drawing of the ancient Market Cross, Chichester.

Distributors:

Australia, New Zealand, South-east Asia:
Jacaranda-Wiley Ltd., Jacaranda Press,
JOHN WILEY & SONS INC.,
G.P.O. Box 859, Brisbane, Queensland 40001, Australia

Canada:
JOHN WILEY & SONS CANADA LIMITED
22 Worcester Road, Rexdale, Ontario, Canada.

Europe, Africa:
JOHN WILEY & SONS LIMITED
Baffins Lane, Chichester, West Sussex, England.

North and South America and the rest of the world:
Halsted Press: a division of
JOHN WILEY & SONS
605 Third Avenue, New York, N.Y. 10016, U.S.A.

British Library Cataloguing in Publication Data
Teaching and applying mathematical modelling. –
(Ellis Horwood series in mathematics and its applications)
1. Mathematical models – Study and teaching
I. Berry, John, 1947-
511'.8'07 QA401

Library of Congress Card No. 84-4561

ISBN 0-85312-728-X (Ellis Horwood Limited)
ISBN 0-470-20079-0 (Halsted Press)

Printed in Great Britain by The Camelot Press, Southampton.

Contents

Foreword

F. J. M. Laver C.B.E.,
Pro-Chancellor, Exeter University

In Britain, the teaching of foreign languages used to concentrate on grammar and literature, with the result that few pupils could apply what they had learnt to the commonplaces of real life. One view of mathematics sees it as a kind of language: certainly, it has often been taught as an abstract system of disembodied relationships—not actually meant to be acted upon. Today, language laboratories cultivate fluency in spoken tongues, and modelling teaches a colloquial mathematics rather than the severely classical formulations that are intoned by a very few, very high, priests.

I hold, most strongly, that mathematical modelling should be taught by requiring students to design and develop their own models for some practical situation, and that they should not just analyse and apply those of others. Of course, this recommendation is much easier to state than to put into practice, for young men and women are limited in the experience and expertise they can use to help them decide which relations are relevant, and what coefficients are appropriate. But, I am bound to say, such considerations have not stopped many of riper age.

I believe that the work of designing a model, and the pleasure of polishing it, are invaluable, in particular because concealed hazards beset the use of ready-made models—including that special variety called packaged programs. I will mention just two of these.

First, designers may fail to make plain (indeed, may not even realize) the limitations and assumptions inherent in their models, so that the unwary use of them in appropriate circumstances, or with inadequate data, may lead to difficulties. Thus, '90, 70, 90' is widely used as a digital model of a fashion model, but it is skimpy even on geometry, and wholly uninformative about kinematical, much less psychodynamic, behaviour.

Second, students—inevitably more expert than wise—may be overly impressed by the evident speed and apparent accuracy of computer-propelled models. Experience teaches (at least, it teaches some) that little value attaches to an answer in 10 milliseconds with 9-digit precision when tomorrow will do, and the input data are liable to errors of +1%. Precision is not the same as accuracy, and neither is identical with significance. Oskar Morgenstern's book (*On the Accuracy of Economic Observations*, Princeton University Press 1950) should be required reading for all who dare to devise econometric models.

Teach skill and economy in the design of models, but care and humility in the application of their results.

Prologue

David Burghes
Exeter University

1. INTRODUCTION

The idea of an *International* Conference on the *teaching* of mathematical modelling was discussed more than two years before it was held. As the last few months before it opened passed, I seriously wondered about the decision taken to hold such a Conference; but having seen so many people taking part—having met new friends and renewed acquaintances and welcomed old friends—I was sure we had a real opportunity to make an impact on the way we teach applications of mathematics.

While we had described to us many interesting and important mathematical models, this was *not* the main theme of the Conference. The theme of the Conference was the teaching of applications of mathematics through mathematical modelling and the teaching of mathematics through its applications. There had already been many international Conferences on the theme of *mathematical models and modelling*, but none in our experience on the theme of the *teaching* of mathematical modelling. We could justly claim to have been the first such Conference. I would also like to stress the international flavour of the Conference. Although we had expected the U.K. component of the Conference to be the dominant one, we were more than delighted to welcome visitors from twenty other countries. This international audience provided all of us with the opportunity to learn what was going on in other parts of the world, and so influence both our own teaching and what was going on in other countries.

2. TRENDS IN TEACHING MATHEMATICS AND ITS APPLICATIONS

In both schools and universities, many of us would say that there is nothing new in the ideas behind mathematical modelling—they have been doing it all the time in the teaching and learning of mechanics. There is some truth in this, for after all, Newton, in the 17th century, was probably the best mathematical modeller of all time. From Halley's observations about the periodic return of his comet, came the deduction from Newton of the inverse square gravitational model. This century has seen an increased use of the Newtonian model in mechanics, but this increased use has not really been reflected in the way we teach mechanics either at school or university. We have lost the roots of applied mathematics in, I suspect, an attempt to make the subject more attractive, by which I probably mean easy, to students.

The influence of the so-called 'modern mathematics' of the 1950s and 1960s, at first moved applied mathematics into a more axiomatic and precise topic—aiming to mirror what appeared to be the attractive qualities of pure mathematics. It is a trend, which I am delighted to see, has stopped. We are moving applied mathematics back to the real world, and we are questioning why we teach applications of mathematics. I am sure that this is a healthy sign, and one of the main reasons why we have here at this Conference a large number of people interested in the teaching of mathematical modelling.

A further reason for the increasing interest in the teaching of applications of mathematics is the present world economic situation. To put it bluntly, we are in a mess, and it is to mathematicians that we look for help in finding acceptable solutions. It is clear that, like in the American project to put a man on the moon, we need to use our mathematical expertise to solve our problems—and this pressing need has its influence on both academics and students. It seems important to firstly know why (or if) mathematical topics have important practical applications and secondly to be competent to use such topics in solving *real* problems.

3. MATHEMATICAL MODELS AND MATHEMATICAL MODELLING

Much of the activity in mathematical modelling can be directly attributable to the SSRC report 'The Training of Mathematicians' by Ron McLone (1973) published exactly ten years ago. In this report, it was found that employers were relatively happy about the basic mathematical skills of their new graduates, but appalled by their ability to translate a real problem into a mathematical one and distressed by their apparent inability to communicate their results to anyone else.

I know that I took these comments seriously and began searching for a remedy. McLone's suggestion for courses in mathematical modelling seemed a good one, but what do you put into such a course? All the books on mathematical modelling that were published in the 1970s (and there were indeed many) were not about modelling but just about other people's models. They were very interesting books, but of little help in teaching a real course in modelling.

I was really put on the spot when in 1979 I was contracted by a large education authority to run an in-service course in mathematical modelling. This is of course not quite the same as running a course for students—no assessment, all are eager volunteers and want to learn. By the third day, it was clear that they didn't want a course in mathematical models (interesting though they were)—what they wanted to do was to actually use their mathematical prowess in trying to solve unseen practical problems. For these participants, it was not too difficult to change some of the models into modelling problems, where real problems requiring an answer were set.

Two developments quickly followed. Firstly the teachers on these courses (and about fifteen such courses were run in the U.K. in 1979 and 1980) wanted their pupils to also experience using mathematics in a practical situation, and this led to the formation of the Spode Group (1981, 1982, 1983), set up initially to develop such material for classroom use. Secondly, I had the chance to expose the participants from higher education at the first mathematical modelling workshop at Paisley, to modelling rather than models. As you can see from the resulting book edited by the organizers, Glyn James and John McDonald (1981) a similar modelling approach was adopted, and the activities of this Group have undoubtedly influenced the way in which mathematical modelling courses have been taught in Polytechnics in the U.K. over the past couple of years. We will see that influence in some of the talks at this Conference.

The basic philosophy behind both the approach of the Spode Group for schools and that of the mathematical modelling workshop for higher education is that to become proficient at modelling, you must fully experience it—it is no good just watching somebody else do it, or repeat what somebody else has done—you must experience it yourself. I would liken it to the activity of swimming. You can watch others swim, you can practice exercises, but to swim, you must be in the water doing it yourself.

So both groups have attempted to find real problems which simulate the real thing, and provide the experiences needed to become a fully trained applied mathematician.

These then are the background issues to this book. There is undoubtedly an increase in the teaching of so-called mathematical modelling courses (although these are very varied in form and what they aim to teach) and it is my hope that this book will not only provide a further stimulus to this trend, but will also identify good practices and provide helpful advice for those who are designing such courses.

REFERENCES

James, D. J. G. and McDonald, J. J. 1981, *Case Studies in Mathematical Modelling*, S. Thornes.

McLone, R., 1973, The Training of Mathematicians, SSRC Report.

The Spode Group, 1981, Solving Real Problems with Mathematics, Volume 1, Cranfield Press.

The Spode Group, 1982, Solving Real Problems with Mathematics, Volume 2, Cranfield Press.

The Spode Group, 1983, Solving Real Problems with C.S.E. Mathematics, Cranfield Press.

Applications and Teaching of Mathematics

Pollak, H. O.
Bell Laboratories, New Jersey

Society provides the time to teach mathematics in our schools every year. Why? Not because mathematics is beautiful—which it is—or because it provides great training for the mind, but because it is so useful.

If usefulness is the fundamental reason for teaching as much mathematics as we do, then we must exhibit, exercise, and emphasize this usefulness at every opportunity. Furthermore, what we teach and how we teach it must change as the needs of society change. The current focus on mathematics and science teaching in the United States will lead to many changes; also the microprocessor as well as the calculator will profoundly affect mathematics. They will lead, for example, to an emphasis on exploratory data analysis and on the teaching of estimation. More deeply, we will see a totally new relation between mathematics and science, and probably social science, in the elementary school.

Change will be more difficult at the secondary level, because the departmental structure is so rigid. Nevertheless, we will have a new subject in the secondary school—technology. By this, I do not mean only the use of technology in teaching—important as that will be—but also the teaching of the uses of science, and of the social and personal problems that such uses can bring. The effect of all of this on mathematics teaching cannot be ignored. The comfortable single strand of pre-calculus mathematics will be diversified through discrete and combinatorial mathematics, probability, simulation, statistics, and elementary computer science. However, traditional subjects will not only be changed externally, but transformed internally as well. For example, secondary-school algebra will change through the availability of the power of symbolic manipulation on the microprocessor, and the emergence of the algorithmic point of view as an organizing principle.

The curriculum materials to realize such changes must, as we have noted, continually emphasize the usefulness of mathematics, and its relation to our daily world and to the sciences. We must prepare new teachers so that they can handle all of this in the schools, and provide the opportunities for present teachers to grow into these responsibilities. Finally, these great new visions of the elementary and secondary level must not be allowed to die as our students enter the university. The teaching of mathematical modelling, and the fruitful co-operation between mathematics and the many disciplines that motivate and use mathematics, have also been major concerns at the conference.

Acknowledgements

The Organizing Committee would like to express their thanks to the following bodies for financial and other aid toward the running of the Conference:

British Aerospace
British Council
Coventry (Lanchester) Polytechnic
Exeter University
Midland Bank
National Westminster Bank
Nuffield Foundation
Open University
Paisley College
Sheffield City Polytechnic
Sunderland Polytechnic

They would also like to put on record their gratitude to Eva Moger, to Jenny and Nigel Weaver, and to the Conference Secretary, Sally Williams.

Topicality: A Personal Plea for Mathematical Modelling Teachers

G. F. Raggett
Sheffield City Polytechnic

It is my personal view that the teacher of students studying mathematical modelling as a major component of a course should achieve the following minimum aims:

(i) motivate the student's interest in the subject
(ii) adequately prepare the student for a possible modelling career in the outside world.

Further, my contention is that (i) is far easier to achieve than is (ii).

The first aim may be achieved, at a superficial level, by illustrating a large variety of fairly simplistic well-posed situations which may be put in a modelling framework. This framework may then be used as a basis for students to solve for themselves other unrelated problems of a similar difficulty. Certainly, the studying of such problems can be beneficial to the student, particularly if they are varied enough to stimulate a fresh approach to each problem; furthermore, many of the good global fundamentals of mathematical modelling should at least have been experienced.

But what of aim (ii)? Who can seriously suggest that such modelling as indicated above, valuable though it may be, can adequately prepare the student for a career in modelling?

Realistically, aim (ii) can never be met ideally within the context of a limited time span course. For to make such a far-reaching claim would require to expose the student to all the world's modelling problems. Taking a small subset of these is one possible approach, but one questions the wisdom of this as both the complexity and technical background required for each such model would become prohibitive.

Thus one strives for a less than perfect situation in which students are encouraged to study models which are both stimulating and which have many of the overall characteristics of real world problems, and yet which do not necessarily require much technical expertise. The paper will indicate the author's belief that such an ideal may be achieved by utilizing topics of current news interest and show, by example, how this ideal was accomplished when teaching a course in mathematical modelling over the period January–June 1982.

1. OVERVIEW OF PAST APPLIED MATHEMATICS DEGREE COURSES

Certainly prior to 1960, the emphasis on the teaching of applied mathematics at all levels had been somewhat introverted. A typical degree course in the subject would involve taught material on traditional well-trodden areas, along with sufficient applicable background mathematics which would enable an average student to solve well-posed problems. These problems generally suffered from the fact that they were unrealistic and too idealized.

As a form of academic training the above approach appeared to satisfy the necessary minimum criterion of applying oneself to the subject and, indeed, the mathematics utilized was generally of good standing. However, as a form of training for solving the problems of the outside world one could perhaps equate such a course with one on Greek Mythology. While this analogue may seem somewhat extreme, it is not meant in a critical sense within the context of the timing of such courses. At the time referred to, it was thought to be the right and proper thing to be teaching this type of material; alternative approaches were only just being conceived.

During the 1960s many of the forward-looking institutions were replacing their traditional applied mathematics courses with potentially more useful subjects e.g. numerical analysis/methods, statistics, operational research and mathematical control. This change was coupled with the emergence of the computer as a useful tool for illustrating many of the methods developed in such courses. While this change was generally welcomed as a step in the right direction, it often proved difficult to retrain existing staff to approach their new subject matter in a way to match the needs of the changing environment. Thus many such courses initially turned out to be too highly theoretically based with concentration once again on specialized ideal cases with little or no use of the computer to solve more realistically posed problems.

This latter criticism appears to have been largely overcome during the 1970s as computer technology (i.e. software, hardware, storage, speed) developed at the same time as staff were making a more conscious effort to apply their material in a more meaningful manner. Mainly during the latter half of this decade, many institutions were also incorporating courses having a major mathematical modelling component.

While the above resume is undoubtedly a gross simplification and generalization, it does at least indicate the changing nature of many applied mathematics degree courses in the recent past. It is generally accepted that the above change in emphasis has been a move in the right direction insofar as the flavour of many newly developed courses should have more realism than their counterparts of twenty or more years ago. Logically, one would think that such a change in philosophy would motivate students more now than in the past, but this conjecture is open to question.

2. PHILOSOPHY OF CURRENT 'MATHEMATICAL MODELLING' COURSES

As implied above, one of the aims in teaching mathematical modelling appears to be to motivate the student's interest in the subject. This aim seems to be met, for the most part, and is achieved in a variety of different ways.

A common approach is one based on a traditional lecture course in which the student is exposed to a variety of different well-known mathematical models which are solvable using known techniques, more usually analytical but possibly numerical too. The student is expected to gain an understanding of such models and then solve similar, related, well-structured problems.

At the other end of the spectrum, another approach is to design a modelling course based on 'the modelling process' which embraces the important concepts of model understanding, model building and validation as well as the model solution itself. This approach may well be carried out at a relatively superficial level by the illustration of a variety of fairly simplistic well-posed situations which may be conveniently posed in a modelling framework. This framework may then be used as a basis for students to solve for themselves other unrelated problems of a similar difficulty. Certainly, the studying of such problems can be most beneficial to the student, particularly if the problems are varied enough to stimulate a fresh approach to each problem; furthermore, many of the good global fundamentals of mathematical modelling should at least have been experienced.

While both of the above course structures appear most likely to achieve general student motivation, they each have some significant drawbacks (in my view at least). The former type of course, although having some relevance, seems essentially to be one on mathematical models rather than on mathematical modelling. As such the student is deprived of the development of his own models, and in this respect is not apparently forced to think for himself. Such courses may well exist due to the teacher's disinclination to expose himself to a possible open-ended situation, much preferring a well-structured approach.

The latter course provokes positive thought from the student but suffers from the fact that the chosen models are likely to be mathematically superficial. It is also quite likely that the teacher may be very familiar with such models having utilised them a number of times on prior occasions.

To be absolutely fair, courses as outlined above probably exist due to limited time considerations. When time becomes less of an inhibiting factor, one would expect modelling courses to develop as some healthy blend of the two above cited approaches. This leads ultimately, perhaps, to students applying the modelling process, either singly or in groups, to more difficult problems which encompass previously taught well-structured models. The approach proposed here seems to be valid educationally as it should keep the student's interest, use non-trivial mathematics and provoke original thought. It should be remembered, however, that the teacher still rightly maintains overall responsibility for the course and in particular for the problems chosen for student participation. There is still then a danger that these problems become as stereotyped as the models part of the course, if the teacher opts to pose problems which are too familiar to himself. In such circumstances, it can be very tempting for the teacher to impose his own views on students attempting to solve such problems using a different, but valid approach.

3. A FURTHER DESIRABLE AIM FOR 'MATHEMATICAL MODELLING' COURSES

While the above discussed aims are considered most desirable there is one further aim which appears to be often overlooked. This is the aim to adequately prepare the student for a possible mathematical modelling career in the outside world.

It is my personal contention that while the previously stated aims are generally achievable, this latter aim is much more difficult to realize. Ideally, of course, this latter aim can never be met; to make such a far-reaching claim would necessitate exposure of the student to all the world's modelling problems.

Perhaps the most natural way to pursue a partial realization of this stated aim is to consider a limited number of real industrial and/or commercial problems. Well meaning though this approach may be, real problems are usually so complex that one would often find it difficult to suitably adapt them without a significant volume of additional specific background information. Skating over these aspects can often lead to a grossly oversimplified problem which the teacher feels competent to handle. Once again his ultimate familarity with the problem can detract from the learning experience of the student. Furthermore such an approach must naturally leave many real problems untouched and one wonders if the student is indeed being done a service by this sort of approach.

In my view, the reason why the above approach does not adequately achieve the stated aim, is that in the 'real world' the mathematical modeller is a real problem solver in the sense that any problem tackled has not yet been completely solved, as far as the modeller is aware. The modeller, thrust into such a situation, would have to get to terms with the problem and this is quite likely to involve thinking back to simplified models within his own experience, a possible literature search, data collection, discussion with other staff etc. Only such an overview is likely to arm the modeller with the necessary information to proceed further and to decide which information is superfluous and what further information is still required. As the modeller's experience develops within an organization a build-up of information will also naturally occur, related to that particular organization.

4. USE OF 'TOPICALITY' TO PREPARE THE STUDENT FOR A MODELLING CAREER

In a real attempt to simulate the above process I have used the theme of topicality for generating modelling examples. The approach adopted has been to set problems taken from the media over the duration of a taught course. One of my roles has then been to search for suitable news items which could utilize some of the structured models previously experienced by the students. Due to my own imposed requirement that the setting of such problems should be done when they are still relatively 'hot', I have not been able to acquaint myself with the problems to such an extent that they become too structured in my own mind. Furthermore, such problems are often short on some data, which thus needs to be interpreted or followed up elsewhere. This activity may involve an extensive newspaper search, access to library material and interviews with experts. I also use myself as a springboard for students' questions and ideas.

Note that the excessive over-specialism, perhaps required in a particular organization, is often lost on such problems. However, this is acceptable in my mind, as the problem may still be relatively complex yet the students will grasp the problem if they are familiar with current affairs, an aspect which they are actively encouraged to pursue. Thus I see this approach as a good preparation for the outside world, even though problems posed may not necessarily be of an industrial type.

5. BACKGROUND OF COURSE AND ASSESSMENT RATIONALE

For illustration, I present in the next section, some topicality assessment examples actually set to students over the period January to June 1982. The situation pertaining to these students was as follows. They were studying for a B.Sc. Systems Modelling degree at Sheffield City Polytechnic and were in their second academic year following a six-month professional experience placement. The time allocation for the mathematical modelling teaching in this year was an average of $2\frac{1}{2}$ hours per week for 22 weeks and an additional five six-hour practical sessions. Within this allocation, the students had a models course mainly based on differential equations with some introductory control. In addition, the students were introduced to the elements of mathematical modelling and a variety of problems were tackled, some involving fairly low level mathematics but others more on a level with taught course material.

The final examination for this course was geared principally towards testing knowledge of well-structured models, so I felt more than justified in setting open-ended modelling exercises for a substantial portion of the required continuous assessment. Due to the reasons already expounded in this paper, these exercises were chosen particularly for their topicality and were further required to illustrate the application of differential equation models to the solution of real problems. Students were generally encouraged to work in groups on such problems, but this was not forced.

Students were expected to suitably choose a model or models from the course as a basis, then modify these appropriately and estimate some salient parameters. Discussion of their solution was required and indications of further work were expected to be outlined as appropriate.

Once the modelling was completed, each student was expected to write up their own report for assessment purposes. They were further required to decide, on a group basis, the percentage work achieved by each group member. Deadlines were strictly adhered to. Both these latter constraints were designed to test aspects of 'pulling one's weight' and punctuality, both necessary ingredients to adequately survive in today's world.

6. GENERATED TOPICALITY PROBLEMS

The following are problems actually posed over the period indicated above. Included after each problem are my initial objectives related to how I thought the students might tackle the problem posed.

Problem 1. Develop a model to simulate the extensive flooding experienced in the City of York during early January.

Initial objectives

A very nasty problem to open with but I confidently expected students to obtain enough relevant data on flood levels as a function of time. This should then lead to a reformulation of the problem in these terms. I envisaged that most students would realize that an input/output model would be required but anticipated that there may be difficulty in modelling these flow rates. In fact using excess input/output flow rates above a natural equilibrium river level leads to a possible excess input representation based on snow thaw (as indicated in the newspaper reports). This may be represented roughly using Newton's law of cooling (heating). An initial excess outflow rate representation could be taken as linear and the river could be treated as a channel with constant cross-sectional area through the city region.

For the high-fliers, an appreciation of river flood level, day/night temperature variations, time delay, rainfall pattern, non-linear outflow, surrounding terrain and receding of the flood level may be expected.

A possible model for this problem is given in Raggett (1983).

Problem 2. On 4*th February* 1982 *the Daily Telegraph reported that Private Terry Bennett of the U.S. Army survived an* 8000 *foot fall without her parachute opening. It was estimated in the report that her terminal velocity would be about* 120 *m.p.h. on impact (full cutting supplied to students).*

Question: Why may it be assumed that the force of wind resistance acting on Terry during her fall was approximately proportional to the square of her velocity? Estimate a value for this proportionality constant.

Initial objectives

This problem is far better structured than the previous one so I expected more stereotyped answers. A discussion and appreciation of the Reynold's number was initially expected to justify the wind resistance representation. This should naturally lead into the formulation of a differential equation.

It is interesting to note that the mass of the girl and her parachute are required to be estimated to obtain a solution. The resulting differential equation, though solvable using separation of variables, leads to an implicit relationship for the required constant which then needs to be evaluated, preferably using an iterative technique.

Other factors which could be mentioned in an attempt to improve the model are that the fall is from a moving object, the variation of air density with altitude, cross winds, variable gravity. Evidently some of these factors are more relevant than others and a discussion of their relative importance could ensue.

Problem 3. In the St. Moritz World Bobsleigh Championship extracts from the television coverage of the four-man event on Grandstand (13.2.82) *and Sportsnight* (17.2.82) *may be summarized as follows:*

The Swiss crew led by Giobellina won the event with a total aggregate time of 4 *mins,* 31.94 *secs. for the four required runs. The event may be thought of (so says the exuberant David Coleman) as the motion of half a ton of steel and men down a run just* 27 *yards short of a mile with a vertical descent of* 420 *feet giving a finishing speed of over* 78 *m.p.h.*

Question: Estimate a value for the coefficient of friction between bobsleigh and ice.

Initial objectives

A simplified model using above information was anticipated to be an examination of a mass on a fixed inclined plane, with air resistance again taken to be proportional to the square of the velocity. Separation of variables solution and input of given information (some not required) enables at best an upper value for the required coefficient.

Other relevant factors which could be included in a more sophisticated model would be the effect of bends and local gradient changes (maximum about 1 in 6 at St. Moritz), the pushed start and the jumping of the men into the bobsleigh, design of bobsleigh, external temperature effects on the run.

Problem 4. In 1981 *The Indian National Institute for Health and Family Welfare, an Indian Government Department, declared the following aims for limiting its population to the end of the century.*

(*i*) *Reduce the yearly birth rate from about* 36 *per thousand in* 1981 *to* 21 *per thousand in* 2000. *This is to be achieved by expenditure on birth control techniques comprising advertising of all forms and the extensive manufacture of contraceptives.*

(*ii*) *Reduce the yearly death rate from about* 14 *per thousand in* 1981 *to* 9 *per thousand in* 2000, *achievable by more effective medical care.*

(*iii*) *Limit the population in* 2000 *to one thousand million from its* 1981 *level of* 684 *million.*

Question: To what extent do you consider the above aims are compatible? (Note that this problem was not quite so immediately topical as the others posed here.)

Initial objectives

A Malthusian population model with fixed birth and death rates (four obvious cases pertain) is valid as a first step in examining limiting situations. This can almost give an immediate answer to the question as posed.

Better students should then look for realistically possible variable time-dependent birth and death rate functions over the applicable period to confirm their initial findings. A discussion of the practical policy effects related to individual choices might also reasonably be expected.

A possible solution to this problem has been written up and will appear shortly, Huntley and Raggett (in press).

Problem 5. With the current interest generated by the START talks and previously generated by the SALT talks, develop a model for the build-up of nuclear arms in the U.S. and the U.S.S.R. over a recent time span, e.g. over the Reagan administration but not necessarily restricted to this period.

Initial objectives

Standard derivation and discussion of coupled arms race equations expected. I anticipated that it would be extremely difficult to obtain reliable data but took the view that enough newspaper reports were available over the years to estimate some data. Once data was estimated an attempt should be made to predict the salient parameters in the model. This could be achieved by examining the phase-plane solution, or by exhibiting various numerical phase-plane graphical solutions on the computer generated with different parameter values. This package was freely available for student use and details are given in Raggett (1979).

Problem 6. Estimate the effectiveness of the British/Argentinian troops in the recent battle for Darwin and Goose Green.

Initial objectives

This was set as an alternative to problem 5 and the objectives were basically the same, a Lanchestrian type conventional combat war model being anticipated with discussion of its phase-plane solution. I further anticipated (correctly as it happened) that the majority of students would opt for this problem rather than for problem 5. This was due to the fact that the battle had only occurred during the previous week and data were readily available from the newspapers.

Possible discussion of extensions including perhaps reinforcement rates, classification of wounded, air support was expected in this case.

7. CONCLUDING REMARKS

If one undertakes to teach in the above fashion then beware, for the following additional strains are put on the teacher.

(i) Energy to keep constantly up to date with the news and to continually assess that which may be potentially useful for assessment within the framework of the taught course.

(ii) Testing the potential usefulness of any piece of news by following up to see if there is sufficient data available. This does not necessarily have to be as complete as you would perhaps like; indeed, this adds to the fun as does the presence of excess irrelevant information.

(iii) Ideally, ensure that the underlying models required have been taught before the assessed work is set. This timing aspect is particularly difficult and requires a much more flexible approach to the suitable choice of examples from a full possible set.

(iv) Conviction that students will be able to achieve something from the problem, but at the same time not going too deeply into the solution oneself initially so as to avoid a constraining trend of thought.

(v) Provision of some sort of 'model' solution for distribution. In view of (iv) this solution will ideally be obtained as quickly as possible after the completion of the marking.

Adoption of the above process was at first speculative, particularly in view of the considerable extra work involved. However, the motivation and response from most of the students in solving these problems exceeded my wildest expectations. Occasionally some very unexpected solutions were presented which surprised me. I feel better for the experience just as I am sure that the students also do. Perhaps surprisingly to many of my colleagues, I did not experience any real difficulty with students using basic models which they had only recently been exposed to.

REFERENCES

Huntley, I. D., and Raggett, G. F., Indian population projections, in Case studies for Mathematical Modelling III (to appear).

Raggett, G. F., 1979, "Interaction computer graphic for the generation of phase plane portraits" *Computers and Graphics*, **4,** 149.

Raggett, G. F., 1983, A simple model of the York floods, in Proc. International AMSE symposium "Modelling and Simulation", Bermuda; Ed. D. W. Russell.

From Consulting to the Classroom

D. H. Lee
South Australian Institute of Technology

This chapter describes some experience in incorporating elements of consulting and applied research projects into a mathematics teaching programme.

Over the last decade the author and colleagues, active in consulting with industry on applied mathematics and operations research problems, have, wherever possible, sought to inject components of this work into the undergraduate teaching programme. The main vehicle for carrying this experience into the classroom has been the case study: the author has developed approximately 20 case studies ranging from physical distribution problems through to thermal system design. This chapter describes briefly the way some case studies have been used in different classes and comments on their classroom use and impact. One case study, exploring fundamental mathematics of storage systems is presented in some detail.

1. INTRODUCTION

'Mathematics' new prominence in the community develops from its success in modelling a rich variety of systems . . . Therefore instead of mathematical models which carry a historical or 'natural' conviction of their validity with them, modern applications of mathematics are increasingly involved with fresh and novel models where the conception and validation of the model is prominent and just as important as the mathematical theory of the model' (Lee, 1972). Furthermore, the effective use of applied mathematics in solving problems for industry requires

- the recognition of the mathematical questions involved
- the characterization of a solution, that is determining what form answers to these questions may take,

- finding out what data is available,
- developing appropriate mathematical models matched to the data availability,
- using mathematical concepts, theorems and techniques to solve the model,
- interpreting the results and communicating them effectively to the client,
- convincing the client that the solutions are correct and maybe even obvious in retrospect.

Within the restrictions of mathematics service teaching schedules, the author has used case studies, typically running as a thread through the subject, to illustrate these aspects. Nor is it sufficient to illustrate the application of mathematics in one or two specific case studies and assume a transfer of knowledge to problem solving. Some specific discussion of the problem-solving framework is necessary. Section 2 of this chapter contains an outline of some of the ideas which have been presented to illustrate the problem-solving process and the nature of mathematical models. In Section 3 some problems which have led to case studies are described. In Section 4 a case study on storage problems is presented, embedded in a commentary on its class use.

2. A MATHEMATICAL MODELLING FRAMEWORK

Without denying the complexity of characterizing problem-solving processes, it does seem important to make explicit some of the concepts of mathematical model building in a class before expecting students to solve problems. It is part of the curriculum of modern applied mathematics.

A framework for structuring early encounters with a problem, adapted from Wilson (1982), is presented in Table 1. The table includes a simple train

Table 1. A problem-solving framework

Objective
Drive train from Station A to Station B in T seconds.

Constraints
Maximum acceleration force is limited.
Maximum braking force is limited.
The train starts from rest at Station A and must be stopped at Station B.
There is track resistance.

Decisions or controls
Motive force (acceleration, braking) as a function of time.

Disturbances
Resistance force variations.

Acceptable solutions
A sequence of acceleration/braking level instructions at, say, 3-second intervals.

control exercise to illustrate the first stage of structuring the problem for model building.

In order to solve a quantitative problem, a mathematical model which describes the state behaviour of the system is needed. Following Aris (1978), a mathematical model is defined as 'a consistent set of mathematical equations which is thought to correspond to some other entity, its prototype'. In Table 2 a mathematical model of the train journey problem described in Table 1 is displayed. Class discussion of the model should note a typical model approximation, wherein the train is considered to be point mass. Similarly the first model can proceed assuming $R = R_0 = \text{constant}$. Subsequently better approximations $R_1 = a_0 + a_1 V$ or $R_2 = a_0 + a_1 V + a_2(V)^2$ with $V = dx/dt$ may be introduced. Since inequalities can always be transformed to equations, the model in Table 2 conforms to the definition but the explicit role of inequalities in many models is worth emphasis.

Table 2. A mathematical model of a train journey

$m\,d^2x/dt^2 = F - R$
$x(0) = 0$
$dx/dt\,(0) = 0$
$x(T) = x$
$dx/dt\,(T) = 0$
$0 \leqslant t \leqslant T$
$B_{max} \leqslant F \leqslant A_{max}$

where m	is mass of the train
x	is the displacement of the train from the origin along the track $x = x(t)$
t	is time
F	is motive force at the wheels $F = F(t)$
R	is resistance to motion $R = R(x, dx/dt)$
X	is distance from A to B along the track
T	is time allowed for the journey
A_{max}	is maximum acceleration force
B_{max}	is maximum braking force

The mathematical modelling process is presented by discussion around the three schemes in Fig. 1. The first two, from Penrose (1978) and Aris (1978) are well-established in the literature. The third schematic can catalyse a cautionary discussion on differing perceptions of problems, the likely changes in problem form or the clients' interests or all of these during the problem-solving process.

3. SOME PROBLEMS USED IN CASE STUDIES

A fruit-packing company in Australia expanded its operations to include processing tomatoes in addition to its traditional stone and other fruit crops.

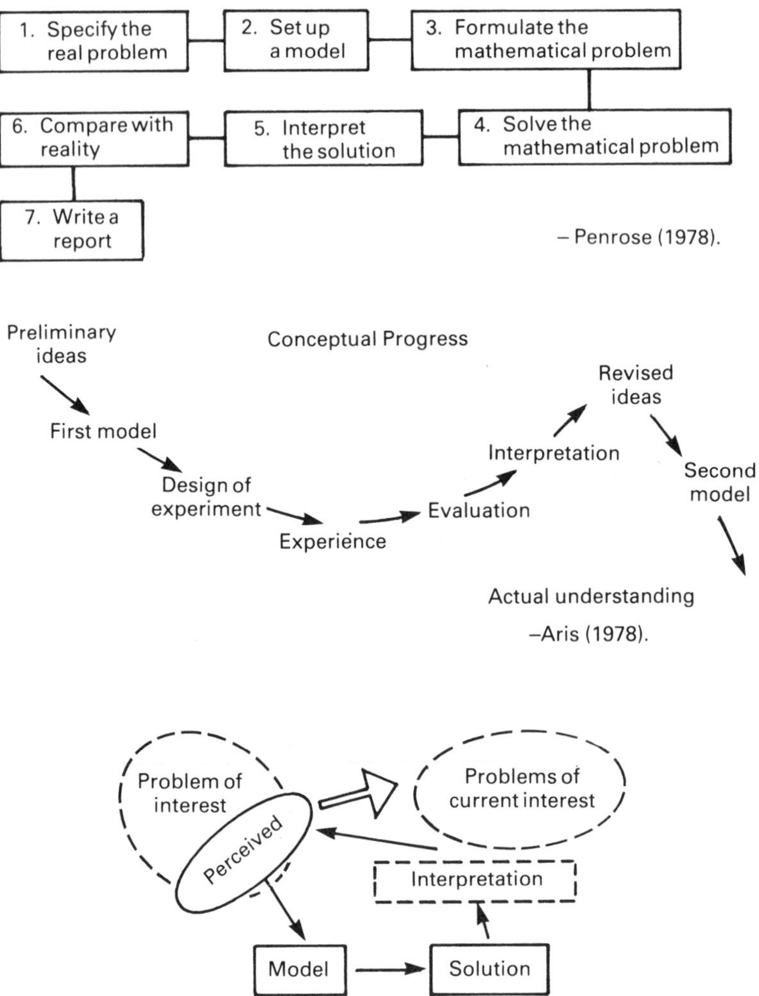

Fig. 1. Three diagrams of the process of mathematical modelling

The tomatoes are grown in several geographically distant regions and transported by lorry in large plastic handling bins. At a management meeting the transport manager raised the question of how many bins were needed to handle the crop; he estimated 4,000 additional bins but the company which had previously held the contract had estimated something closer to 8,000 additional bins. Since bins cost $60 each, the difference is significant at $240,000. If 4,000 bins are too few, some of the crop may be lost, with presumably serious implications for profitability; conversely, borrowing funds to purchase additional bins is a major added cost which needs to be economically justified.

The author and a colleague were consultants to the company on this small problem. The essence of the problem has been decanted into a case study, called the tomato job, used in teaching a management science subject in a business degree programme. The case study starts with the scenario similar to the introduction given above and poses the following question.

You are the factory manager. What is your decision on the number of bins?

No other prompting is given but after a while it is agreed that some further information is needed. After exploring the information needs in an unstructured way, it is a good time to introduce the ideas of Section 2 of this chapter, which may lead to a problem statement as shown in Table 3. The data requirements flow more easily now.

Table 3

Objective

Determine how many bins to handle crop
or Determine how many bins to maximize profit

Constraints

Grower contracts
Trucking schedules
Factory capacity
Store capacity
Handling equipment

Decision variables or controls

Number of bins
Other handling equipment
System changes

Disturbances

Seasonal variations
Equipment malfunctions
Load inspections

Solution specification

Number of bins
System changes, e.g. trucking schedules

Several alternative mathematical models present themselves. Using Fig. 2 as a pictorial model, one can develop a solution via simulation or a queueing model where the time for a bin to complete the cycle is a key variable. Alternatively, by noting bottlenecks in the flow of bins around the network, one can determine how many bins are held in various stages when these processing rate bottlenecks are attained. Hence, estimates of the bins required to sustain the system can be obtained.

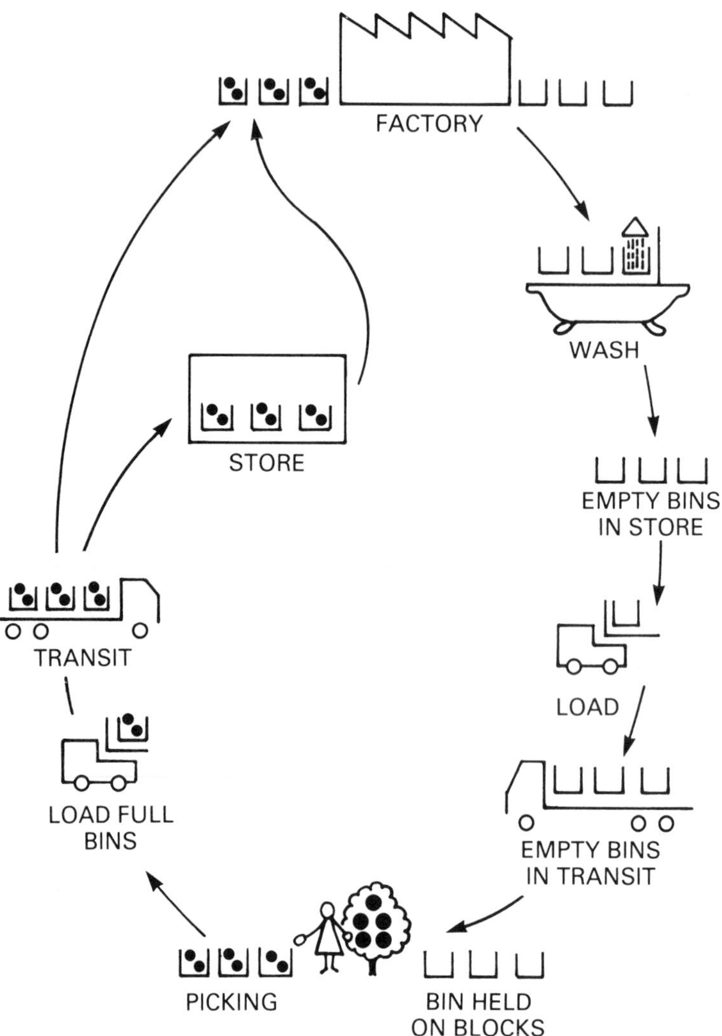

Fig. 2. A schematic model of bin flows

This case study has been quite successful in motivating the need for management science and introducing the model building-solution process.

Other physical distribution consulting exercises have also been very useful in generating understandable case studies. Consulting work for several industries has led to separate case studies in

- loading road tanker vehicles to ensure maximum payload within axle loading limits; solution of this model yields useful exercises in permutations and linear algebra (Lee, 1979),

- designing the geometry of a road tanker for maximum load-carrying capability,
- scheduling packing teams to meet time constrained packing schedules for a bakery,
- many variations of routing and location problems.

A number of engineering design problems have also lent themselves to case studies.

- determining the length of belts to drive pulley systems is one (Lee & Galvin, 1980),
- designing the minimum cost gas network for a gas field, or a compressed air system (Lee & Mills, 1981),
- finding the size of store needed in a gas network or a thermal system (Lee & Mills, 1983).

A basic storage case study, with ingredients inherent in many real problems, is presented in detail in the next section.

4. STORAGE PROBLEMS

Storage of commodities in nature and in man-made systems is a widespread phenomenon. A mathematical model of storage is important for system design. There are many variants of the storage problems permitting opportunities for deterministic and stochastic, discrete and continuous mathematical modelling exercises. A general preamble for a storage case study may take the following form:

> Many commodities are stored in nature and in man-made systems. Dams store water from winter run-off to meet summer consumption demands. Factories hold stocks of materials needed for production and, in turn, generate an inventory of finished products for distribution to consumers. In an industrial complex some processes generate by-product gases which are stored and used subsequently as a fuel.
>
> A store may be needed because at certain times demand for the commodity exceeds supply or because supply is not available when and where needed. Alternatively storage may be introduced into a system to reduce overall costs. A store does not generate supply, but, by cushioning peak loads, or by holding over supply until it is needed, it may allow the size of the generating plant to be reduced. Alternatively, the presence of a store may allow a surplus to be generated when charges or costs are low; by drawing on this surplus in high cost periods, overall operating costs may be reduced.

A specific example could take the form:

> A production process uses varying amounts of gas which is generated in a plant coupled to a store whence the gas is drawn on demand. Gas generators

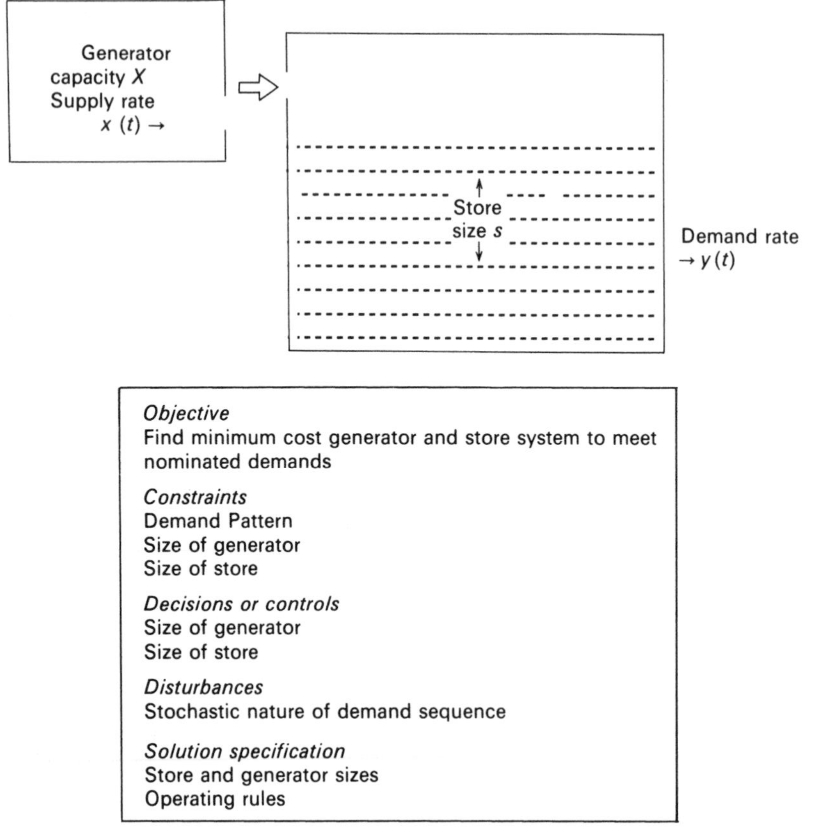

Fig. 3. A schematic of a simple storage system

are only available in discrete sizes. Find the minimum cost system to meet a specified demand pattern.

Almost invariably the modelling process is assisted by drawing a diagram (Fig. 3) and introducing notation.

The system is to operate over a time interval $0 \leqslant t \leqslant T$ and meet some demand which will be specified by $y(t)$, the rate of demand for the commodity. The rate of supply of the commodity is denoted by $x(t)$ and the level of commodity in the store is $s(t)$. Assume, for definiteness, that only non-negative values of the commodity are meaningful. Note carefully that the variables x and y represent rates of supply and demand whereas s is the actual level of the commodity in store. The model could, for example, represent the storage system for the process gas problem, $y(t)$ representing the demand flow rate for the gas in m^3/second (normalised to one standard atmosphere pressure), $x(t)$ the rate of production, again at m^3/second and $s(t)$ the volume of gas in store measured in m^3 reduced again to one atmosphere pressure.

The question as to whether to prefer a discrete or continuous model arises. Suppose a discrete model is chosen.

Let the time interval of interest $0 \leqslant t \leqslant T$ be divided into n equal subintervals of length $\Delta t = T/n$ and introduce the notation

$t_i = i\Delta t, \quad i = 0, 1, 2, \ldots, n$

y_i is the rate of demand for the commodity in the ith subinterval, $t_{i-1} \leqslant t \leqslant t_i$
x_i is the rate of supply generated during the same subinterval, $t_{i-1} \leqslant t \leqslant t_i$
S_i is the level of the store at time $t_i = i\Delta t$, the end of the ith interval or the beginning of the $(i+1)$th interval.
Assume for this article the store is initially empty: $s_0 = 0$.

Since I often use small numerical examples in the formative stages of the model-building process, I encourage the students similarly.

Fig. 4(a) shows a prototypical discrete demand sequence y_i, $i = 1, 2, \ldots, 8$ with subinterval length $\Delta t = 1$. The demand sequence may represent an approximation to the broken line demand curve superimposed on it or it may be the natural or measurable specification of demand.

The basic storage model can now be developed. Determination of a minimum cost configuration can only be pursued when the behaviour of the system can be described.

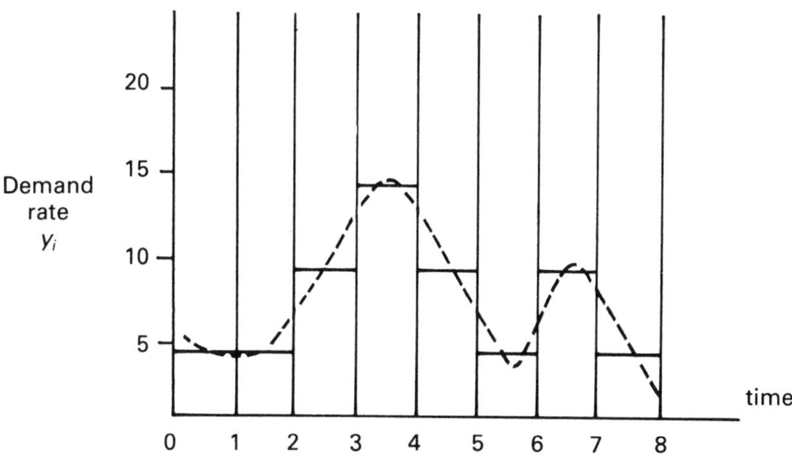

Fig. 4(a). Discrete demand sequence y, $0 \leqslant t \leqslant 8$, $\Delta t = 1$

The fundamental equation of this storage model derives from the conservation condition

$$s_i = s_{i-1} + (x_i - y_i)\Delta t \quad i = 1, 2, \ldots, n \tag{1}$$

subject to the conditions

$$0 \leqslant s_i \leqslant S \tag{2}$$
$$0 \leqslant x_i \leqslant X \tag{3}$$

where S is the capacity of the store and X is the design output rating of the generator. In words, equation (1) requires that the level of commodity in the store at the beginning of any interval equals the level at the beginning of the previous interval augmented by the net or algebraic accumulation of commodity during the interval. The second condition, constraint equation (2), requires that the level of commodity in store cannot be negative nor exceed the capacity of the store for a meaningful solution.

Feasible solutions, the set from which the minimum will be drawn, can now be characterized.

A feasible solution for the discrete storage model is a value of a store size, S, a generator design capacity X, and values of a supply sequence $x_1, x_2, \ldots, x_n$ satisfying equation (1) subject to constraints (2) and (3). For example, one solution can always be obtained by choosing $x_i = y_i$, $i = 1, 2, \ldots, n$ and $S = 0$; if the supply rate can match demand rate everywhere then no store is required. Then $X = \max_i \{x_i\}$. There are, in general, other feasible solutions which use a store. By construction, in Fig. 4(b) a feasible solution of the

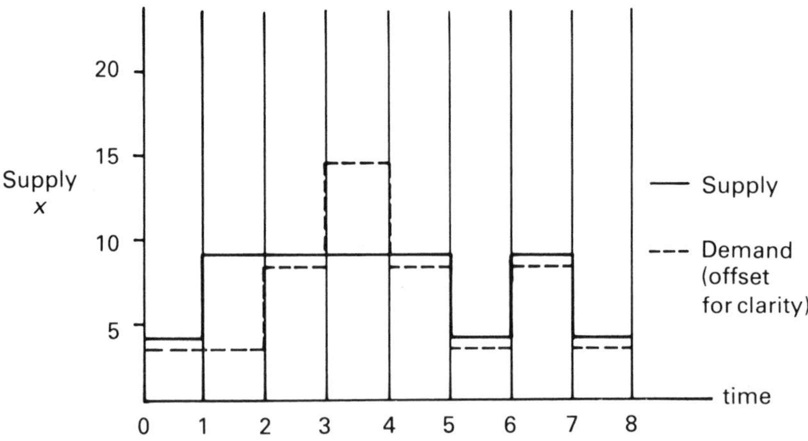

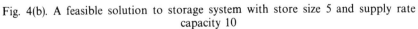

Fig. 4(b). A feasible solution to storage system with store size 5 and supply rate capacity 10

discrete storage model satisfying the demand sequence of Fig. 4(a) via a store of capacity $S = 5$ is displayed. Fig. 4(c) plots the associated behaviour of the store.

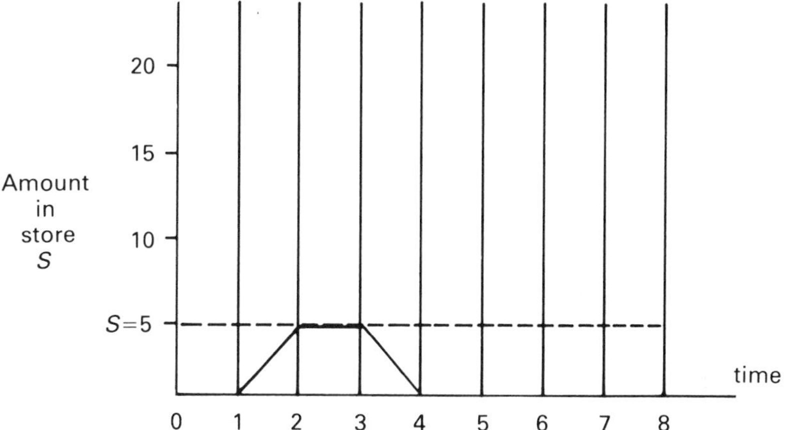

Fig. 4(c). Storage levels for the nominated solution

It is easily verified that the recurrence relation [1] holds within the constraint (2) with $S = 5$.

Some general results about these feasible solutions can be extracted from the conditions (1) to (3). Upper and lower bounds often provide good insight into the system behaviour. An upper bound $X = \max_i \{x_i\}$ has been established. A lower bound on x can be established via a little algebra.

An equivalent statement of equation (1) can be obtained by expansion as follows

$$
\begin{aligned}
s_i &= s_{i-1} + (x_i - y_i)\Delta t \\
&= (s_{i-2} + (x_{i-1} - y_{i-1})\Delta t) + (x_i - y_i)\Delta t \\
&\quad \vdots \\
&= s_0 + (x_1 - y_1)\Delta t + (x_2 - y_2)\Delta t \ldots (x_i - y_i)\Delta t \\
&= s_0 + \sum_{r=1}^{i} (x_r - y_r)\Delta t \quad i = 1, 2, \ldots n. \qquad (4)
\end{aligned}
$$

Since $s_0 = 0$ here

$$s_i = \sum_{r=1}^{i} (x_r - y_r)\Delta t.$$

But the constraint (2) requiring $s_i \geqslant 0$ for all i in turn implies

$$\sum_{r=1}^{i} (x_r - y_r)\Delta t \geqslant 0$$

$$\sum_{r=1}^{i} x_r \Delta t \geqslant \sum_{r=1}^{i} y_r \Delta t$$

and in particular with $i = n$

$$\sum_{r=1}^{n} x_r \Delta t \geqslant \sum_{r=1}^{n} y_r \Delta t$$

which is readily interpreted as the requirement that total commodity supply must equal or exceed aggregate demand for a feasible solution.

But using (3), it follows that

$$X \sum_{r=1}^{i} \Delta t \geqslant \sum_{r=1}^{i} x_r \Delta t \geqslant \sum_{r=1}^{i} y_r \Delta t$$

$$X . t_i \geqslant \sum_{r=1}^{i} y_r \Delta t$$

$$X \geqslant \sum_{r=1}^{i} y_r \Delta t/(i\Delta t) = \bar{y}_i$$

requiring the design capacity of the supply plant to be greater than any moving average value of demand. Below this value of X, no solutions are possible. It is a lower bound on X.

The test data may be used to understand the implications of this result.

For the data shown in Fig. 4a, the mean demand rate over the whole period $0 \leqslant t \leqslant 8$.

$$\begin{aligned}\bar{y}(8) &= \tfrac{1}{8}(5+5+10+15+10+5+10+5) \\ &= 8.125\end{aligned}$$

establishes a necessary lower bound on X the design capacity of the supply plant, namely $X \geqslant 8.125$. But note that there may well be an intermediate moving average value greater than this, imposing an even tighter bound on X, a greater lower bound.

Table 4 shows the calculation of $\bar{y}_i, i = 1, 2, \ldots, 8$ for the data of problem 2. The maximum value of $\bar{y}_i$ occurs in the fifth interval, when $\bar{y}_5 = 9$. This is the greatest lower bound on X, the design capacity of the generator; it is not possible to find solutions with $X < 9$. When $X = 9$, a direct application of the fundamental storage relation with $x_i = X = 9$ everywhere, viz.

$$s_i = s_{i-1} + (9 - y_i) \quad (s_0 = 0) \quad i = 1, 2, \ldots, 8$$

yields the values shown in the second last row of Table 4. Since all $s_i \geq 0$ and the maximum value of s_i is $s_2 = 8$ by choosing $S = 8$, the second condition of storage

$$0 \leqslant s_i \leqslant S \quad i = 1, 2, \ldots, 8$$

Table 4. Moving averages of demand and a trial storage solution for $X = 9$

Interval i	1	2	3	4	5	6	7	8
Demand y_i	5	5	10	15	10	5	10	5
$\bar{y}_i = \sum_{r=1}^{i} y_r/i$	5	5	6.67.	8.75	9	8.33	8.57	8.125
Trial s_i for $X = 9$	4	8	7	6	0	4	3	7
Trial s_i for $X = 10$	5	10	10	5	5	10	10	15
Actual s_i for $X = 10$	0	5	5	0	0	0	0	0

is satisfied and this is a solution for $X = 9$. Note that the store is empty at t_5. If it were not empty anywhere in the solution period $0 \leqslant t \leqslant 8$, then there would be excess capacity in the store and a solution with smaller S must exist.

It has been established, for the example in Fig. 4(a), that a solution to the storage model exists for $9 \leqslant X \leqslant 15$. From this feasible set, it is possible to extract the one minimizing the cost of the system.

If costs of generating plant (discrete sizes X) and storage (continuous sizes S) are given as increasing functions of size, the system cost minimisation problem can be solved. For each discrete generator size available the minimum associated storage can be found directly.

Already it has been shown that for $X = 9$, a store of capacity 8 units is needed. This was obtained by substituting directly in storage equation and verifying that (2) held, hence $S = \max_i \{s_i\} = 8$. But a similar trial approach with x_i fixed at $X = 10$ yields the last row in Table 4.

The solution is indeed feasible but, since $s_i > 0$ everywhere it cannot correspond to the smallest possible store size. There are several approaches to finding the smallest S. Perhaps the simplest is based on the assumption that, when demand y_i is greater than design capacity then the plant is operated at design capacity $x_i = X$ for the duration of that peak and the area between the demand profile and supply capacity value equals the amount of commodity which must be drawn from store for a feasible solution. For $X = 10$, there is only one peak, the one in period 4 where $y_4 > 10$ and hence a store capacity of 5 is needed. This confirms the solution already illustrated in Fig. 4(c) is minimal cost for $X = 10$. The technique can be extended, but note that for a general demand profile each peak has to be checked *and*, equally importantly, the ability of the system to replenish store level in between multiple peaks is a further condition which must be satisfied.

The minimum cost system can then be determined by enumeration on available capacities X by comparing minimum system costs based on the above store sizing methodology.

If costs are dominated by generator size X and store size S and costs are approximately linear in the size range of interest an LP (or integer programme if X remains discrete) arises naturally

$$\min\ c_1 S + c_2 X$$

subject to

$$S_0 = 0$$

$$S_i = S_{i-1} + x_i - y_i \quad i = 1, 2, \ldots, n$$
$$0 \leqslant s_i \leqslant S$$
$$0 \leqslant X_i \leqslant X$$

where c_1 is cost per unit size change in store S and c_2 is cost per unit size change in generator capacity X.

This LP is a good mathematical model of the problem.

Many further mathematical or model variations are possible. For example, should bleeding of surplus gas be allowed in the model; is $s_0 = s_n$ a better end condition than $s_0 = 0$?

In concluding this review of the storage case study, the link from discrete to a continuous model is outlined; it provides a useful opportunity to reinforce basic calculus ideas.

Equation (1) can be written as

$$(s(t_i) - s(t_{i-1}))/\Delta t = x(t_i) - y(t_i)$$

or

$$(s(t_i) - s(t_i - \Delta t))/\Delta t = x(t_i) - y(t_i)$$

and the limiting form as $\Delta t \to 0$, $t_i \to t_{i-1}$ yields

$$\frac{\mathrm{d}s}{\mathrm{d}t} = x(t) - y(t) \tag{5}$$

where $t = t_{i-1} = t_i$ in the limit.

Thus the rate of change of the level of the store at time t is equal to the difference between supply and demand rates at that time. Integrating (5), or more directly using the definition of a definite integral as the limit of a sum in equation (4), the fundamental conditions for this storage model with continuous demands is

$$s(t) = s(0) + \int_0^t [x(r) - y(r)]\,dr \tag{6}$$

subject to the analogous feasibility constraint

$$0 \leqslant s(t) \leqslant S. \tag{7}$$

5. CONCLUSION

Case studies from first-hand consulting experience have introduced a level of vitality and relevance into the curriculum and teaching process.

REFERENCES

Aris, R., 1978, *Mathematical Modelling Techniques*, Pitman.

Lee, D. H., 1972, Mathematics in the Community, *Int. J. Math. Educ. Sci. Technol.*, **3** 160.

Lee, D. H., 1979, Optimal Loading of Tankers, *J. Opl. Res. Soc.*, **30**, 323.

Lee, D. H. & Galvin, S. R., 1980, Belt Drives: A Case Studies for Mathematics Students, *Int. J. Math. Educ. Sci. Technol.*, **11**, 559.

Lee, D. H. & Mills, R. G. J., 1981, Network Design of Compressed Air Services, *Proc. 5th Aust. O.R. Conference.*

Lee, D. H. & Mills, R. G. J., 1983, Gas Storage in an Energy Network, *Proc. 6th Aust. O.R. Conference.*

Penrose, O., 1978, How can we teach mathematical modelling? *J. Math. Mod. for Teachers*, **1**, 31.

Wilson, I., 1982, Analysis of Decision Making Tasks, *ASOR Bull.*, **2**, No. 2, 8.

Modelling, 'successful' and 'unsuccessful'

D. G. Medley
University of Leeds

All mathematical models of the real world are equal in much the same way as all animals are equal in *Animal Farm*; that is, all are in principle conjectures to which no credence can be given until they have been extensively tested and corroborated. In practice, some models are virtually sure of success right from the beginning; others prove to be almost valueless.

A distinction should be made between *modelling* and the part of the modelling process which is commonly called *the mathematical model.* Modelling consists of many steps, beginning with a gathering of miscellaneous ideas, and ending with rigorous testing, and presenting the conclusions whether predictive or decision-making, in a form suitable for use.

Innovative modelling is a time-consuming process. Even when the eventual model is of extreme mathematical simplicity, neither the teacher nor the pupil will find the necessary decisions easy to make. The exercise is extremely valuable, but the model is likely to be, at best, only partially successful; 'failure' is often to be expected.

More success though fewer demands on modelling skill, is likely in fields where there are already many established, well-corroborated laws to appeal to, and the number of significant factors is identifiable and small.

Textbooks can assist either by providing mathematical background, or by providing insight into the modelling process, or by providing interest and motivation. It is important to recognise which of these purposes a particular book is serving.

1. INTRODUCTION

The detailed calculations of orbits, fuel consumption and the like which were made before sending the first men to the moon were almost sure of success—

the launch would never have been undertaken had there been serious doubts about the models used. So what were the characteristics of the model which made it possible to invest so much trust in it?

In the first place the central part of the enterprise was the determination of position and velocity, i.e. it fell clearly inside realms of study (geometry and mechanics) for which *well-corroborated overall models were already available.* Consequently, and this is a second important characteristic, *the class of relevant factors to be considered could be enumerated with some confidence*, namely initial relative positions and velocities of earth, moon and space-craft, and subsequent forces on the craft. True, the detailed careful modelling of forces called for much care, foresight and skill, but the enumeration of expected forces and contingencies was a not impossible task. The tension in the control room and across the world testified to the fact that certainty about the outcome was somewhat less than 100%, but at such a high level as there was of *a priori* expectation of success, it seems appropriate to talk of successful predictive modelling.

I use the word 'successful' of a modelling when there is a high expectation that any predictions of the model (in the range it refers to) have a high probability of being quantitatively correct to the level of accuracy indicated by the modeller. If no level can be indicated or if the main predictions are demonstrably false the model is automatically 'unsuccessful'. (It must be made clear that 'unsuccessful' models are by no means necessarily bad models or a waste of time: unsuccessful models are the usual precursors of successful models, and the process of constructing an unsuccessful model may itself bring with it many useful insights. To avoid the pejorative ring of the word 'unsuccessful' I shall sometimes substitute the word 'unsubstantiated.')

2. MODELS AND MODELLING

Before looking at typical modellings from textbooks, it is worthwhile pointing out the distinction between a *modelling* and a *model.*

Typical stages in the modelling of a real world problem are shown diagrammatically in Fig. 1, which is copied from Burghes and Borrie (1981).

Box 1 in Fig. 1(a) is crucial, and should, by rights, be drawn much larger in relation to the rest of the diagram, it should be seen to contain at least the elements enumerated in Burkhardt (1981) as in Fig. 1(b). The first explicitly mathematical step embraced in boxes 1, 2 is the identification of significant variables.

Step 1 I shall take to include what Burkhardt (1981) refers to as 'Generation of Ideas'. Step 2 must embrace his 'Selection of Ideas'. Every author gives a slightly different diagram or description, but the main implications are similar.

In *well-worked* fields of study, steps 2–5 of Fig. 1 sometimes seem to occupy the centre of the stage. It is all too easy for the mathematician to regard 4 as the

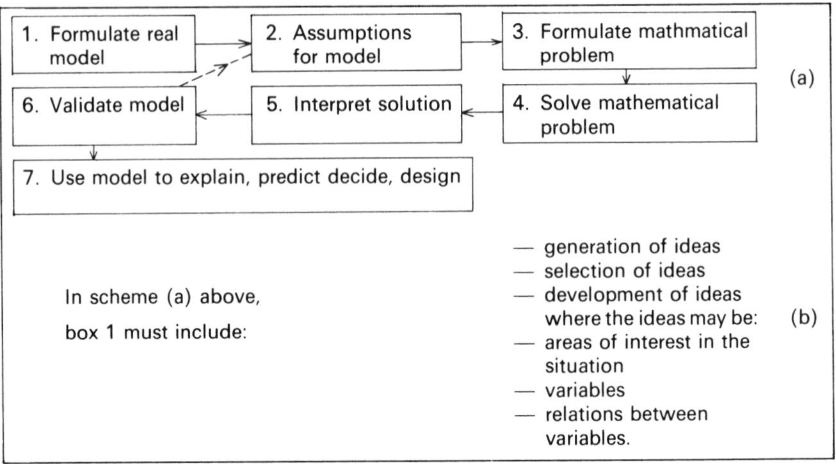

Fig. 1.

nub of the process. Steps 3 and 4 constitute the mathematical model. Every teacher knows that even in routine exercises it is steps 1, 2, 3, 5, 6, 7 which are most difficult to teach by chalk and talk, but it is only if the teacher habitually gives full weight to those steps that the student will be prepared to attempt modelling.

In *innovative* modelling (a phrase I use for modelling in areas where recognized laws of nature are sparse), the essential modelling steps 1, 2, and 6, 7 usually demand *much* more thought and effort than step 4, which may be mathematically trivial. (This is presumably one reason for the apparent triviality of the mathematical syllabus recommended in the Cockcroft Report (1982) at secondary level!)

When assessing the purposes served by any particular textbook it is important to keep the distinction between the modelling process, steps 1–7, and the mathematical model, steps 3 and 4, clearly in mind; for although some books give a good idea of the whole modelling process from start to finish, most home in on the mathematical model and step 4 thereby falsifying and trivializing the modelling process. It is not always easy to tell from the title or even from a superficial glance at the text which kind of material is to be found inside, but books which talk of 'motivation' are particularly dangerous. An introductory description of a problem provided to create interest, seldom adequately serves the purposes of steps 1–3 of the modelling process.

My first example of the modelling process comes from an excellent book called *An Introduction to Mathematical Modeling* by Edward Bender (1978). It is an example of innovative modelling.

3. INNOVATIVE MODELLING

In previously uncharted fields, there can be no quick or certain route to a unique mathematical model. It is not known in advance what factors are important or what methods will be used. The *identification* processes of modelling, that is, the identification of factors and the identification of appropriate relationships (roughly speaking steps 1 and 2 of Fig. 1) are now of over-riding importance. Step 4 may well turn out to be either trivial or formidably difficult.

The chosen example concerns the cost of packaging domestic goods, and is presented on page 19 of Bender's book as follows:

'Consider a product like flour, detergent, or jam, which is packaged in containers of various sizes. You've probably noticed that larger packages of such products usually cost less per ounce. This is often attributed to savings in the cost of packaging and handling. Is this in fact the major cause or are there other important factors? We try to see where this idea leads by constructing a simple model . . . Cost clearly depends on competition and the scale of business. We neglect these factors and concentrate on expenses due to materials and handling. Since we are neglecting some important factors (name some), the resulting predictions will be crude. In addition there are various constants involved which we do not even pretend to evaluate.'

Compared to the exercises set in textbooks and examination papers, Bender's presentation of the problem gives very little lead towards an ultimate model. But compared with a real-world situation a lot of pre-modelling has already gone into the above statement. Real-world modelling in any relatively untrodden field is much more open-ended than any textbook description can convey. The worst service to real-world modelling is to present an innovative modelling as the only, or even the best, approach.

In his presentation of the problem, Bender has already made three points worthy of note:

(1) For an understanding of what actually determines the prices of goods in shops, a far more thorough search for possible influencing factors would be needed.
(2) Before even beginning to construct a model, it is desirable to have some idea of what you are ignoring; both an enumeration of the factors which you are going to consider, and which will appear as variables in your theory, *and* an explicit statement about the general character of the factors you are ignoring are necessary if the status of the resulting model is to be clear.
(3) The presence of unevaluable constants in a theory detracts from the usefulness of the theory (cf. coefficient of restitution, e, in mechanics).

If a teacher is not prepared to tackle a problem such as the above without reference to an approved solution in a textbook, there is little prospect of conveying the flavour of modelling or its methodology, to the pupil.

I can only briefly indicate Bender's subsequent modelling steps. He enumerates five quantities contributing to the final price:

(a) Cost of producing (which I interpret as buying and processing the raw materials).
(b) Cost of the packaging process (excluding packaging material).
(c) Cost of shipping (and other transport).
(d) Cost of the packaging material.
(e) Profit mark-ups by middlemen.

He then (partially) justifies the assumptions relating a, b, c, d to the weight w of commodity concerned,

$$a \propto w$$

$$b \simeq fw + g, \quad \text{constants } f, g \geqslant 0$$

$$c \propto w$$

$$d \simeq hw + kw^{2/3} + m, \quad \text{constants} \quad h, k, m \geqslant 0$$

The profit mark-up of middlemen, he assumes, can be accommodated by attributing a proportional mark-up to any or all of the terms appearing above, i.e. by suitable modification of the unknown constants. The final expression for cost per ounce (sic) is therefore

$$\frac{c}{w} = \frac{a+b+c+d}{w} = n + pw^{-1/3} + qw^{-1}, \qquad \text{(I)}$$

$$\text{constants } n, p, q, \geqslant 0$$

Equation (I) is Bender's mathematical model. It is worth while pointing out that in his preface Bender states that his book is intended as a first course in modelling at undergraduate or postgraduate level—though examples like the above one, would be quite appropriate for school use. The point is that the more open-ended the problem, the greater the proportion of time needed for the essential modelling process 1–3 and 5–7. Even when the eventual model is of extreme mathematical simplicity, neither the teacher nor the pupil will find the necessary modelling decisions easy to make. If it is desired to teach innovative modelling, the simultaneous introduction of difficult or 'interesting' mathematics would be folly. Not only will the pupil be most willing to apply those simple tools of arithmetic etc. of which he/she has already some degree of mastery, but the teacher too will have most confidence if modelling is first attempted in the mathematical context of simple arithmetic, probably in the lower forms of secondary schools. This is in accord with the recommendations of the Cockroft report (Cockcroft, 1982). For more about modelling at that early level I must refer the reader to books like Burkhardt (1981). My own comments will refer mainly to modelling at sixth form or university level.

Now I have not finished the discussion of Bender's example, but have only reached steps 3, 4 (corresponding to boxes 3, 4 in Fig. 1), the mathematical model.

The conclusions which Bender proceeds to draw from his model, i.e. from equation (I) and the equation derived from (I) by differentiation are: that costs of material and handling would, by themselves, ensure the decrease of c/w for increasing w; and that 'doubling the size of the package tends to result in greater savings per ounce when the packages are small than when they are large. We have said "tend to" because the model is crude'. Bender's main text only takes us to step 5 of the modelling process in Figure 1. At this point he refers the reader to a problem set for the reader later in the chapter, where some attempt is made at validation.

Bender gives tables of prices which he himself collected by examining the shelves of a local supermarket, and writing down the prices of packs of various sizes of powdered milk, ketchup and the like. From these tables the reader is invited to test (or 'validate') the model. Alas, for one commodity, flour, even the first quantitative conclusion was false—the medium-sized pack cost less per ounce than the large pack. For only two commodities were as many as four sizes available. In short, validation of the model proved impossible; in my terminology the model was 'unsuccessful' because it could not be recommended for quantitative predictive use.

Nonetheless, this was good modelling: some insights were achievable and, because a genuine validation process was attempted there was no real danger of false conclusions being drawn.

A modelling may be useful even when it is 'unsuccessful', and, indeed, as I have already indicated, successful models usually have unsuccessful models as their forerunners.

Unsuccessful or only *partially* successful models often figure in textbooks of models. It is important to recognize them as such, and not to give them credence which they have not earned, just because some algebra appears on the page.

On the basis of their presentation, some of the examples in Burghes and Borrie's book are unsuccessful models. One, for example, entitled 'Exploited Fish Populations', is presented as follows. After a brief reference to fishing in the North Sea, the mathematical model

$$\mathrm{d}w/\mathrm{d}t = \alpha w^{2/3} - \beta w$$

is written down with minimal explanation. I am sure this cannot be a successful model according to my definition! The emphasis is on the solution of that differential equation. There is no serious pursuit of steps 1–3 or steps 5–7 of the modelling process. Clearly the purpose of the book is to exemplify the use of differential equations, not to teach modelling.

I have now instanced examples at the successful and unsuccessful ends of the modelling spectrum, and have emphasised that the presentation of either a successful or an unsuccessful model may be such as to promote good modelling attitudes, or such as to imply that models are mathematical *faits accomplis* appearing from almost nowhere, and accepted with little questioning. So there is a 2 × 2 classification of modelling presentations, as shown in figure 2.

	Successful	Unsuccessful
Good modelling	e.g. space flight (realistic mechanics) A ✓ B ✓ C ✓ D ✓ E ✓	e.g. Bender packaging O.U. Bicycle lamp A × B × C ✓ D ✓ E (✓)
Inadequate modelling presentation	e.g. bad mechanics teaching of successful models A ✓ B ✓ C ? D ? E ?	e.g. Fish populations A × B × C × D (✓) E ×
	Most modelling for 'motivation' purposes	

Fig. 2.

Before pursuing this and other classifications, let me try to identify and summarize the circumstances which favour successful modelling.

4. CIRCUMSTANCES WHICH FAVOUR SUCCESSFUL MODELLING

The favourable indicators are

(A) that it is possible to identify and enumerate a small number of significant factors, it being virtually certain that all others may be ignored without falsifying the situation;

(B) that a sufficient number of already well-corroborated laws, or of easily testable postulated laws, can be recognized relating those factors.

In the absence of (A), no model by a non-specialist is likely to have more than a tiny range of validity, and even to write down an equation may be an invitation to misuse! Even among specialists, weather modelling and models of the economy (whether those favoured by left-wing or those of right-wing politicians or even those of the Treasury) are notoriously unreliable in application, by reason of (A).

In the absence of (B), very heavy demands are placed on the ingenuity of the modeller. The category of well-corroborated laws starts with the laws of arithmetic—the first mathematical modelling tool to be developed by man. The next most useful laws are probably those of geometry and mechanics, (I regard algebra, calculus, statistics, trigonometry as more mathematically sophisticated formalizations of the laws of arithmetic and geometry.)

To (A) and (B), for successful modelling, should be added three prerequisites for any *good* modelling, successful or unsubstantiable:

(C) that the subject matter is familiar, or is easily appreciated in all significant respects by the modeller;
(D) that the mathematics which will probably be involved lies almost exclusively within the area of competance of the modeller;
(E) that sufficient time and thought are given to ensure that the study is a serious one.

Bad statistical modelling can often be seen to spring from the absence of (C). The result is a modelling of the data instead of a modelling of, or relevant to, the phenomenon.

5. TEXTBOOKS, TEACHING, AND THE SPIRIT OF MODELLING

I do not know if a textbook could be written which would fully serve the needs of a teacher of modelling at sixth form or undergraduate level. I rather doubt it. I certainly believe no such textbook exists at present (not even my own, Medley, 1982!) It is very important to recognize precisely what any textbook fails to do, so that the teacher may not be tempted into offering a course which is a travesty of modelling as well as having only a trivial mathematical content. For mathematics on the one hand and modelling on the other are necessarily competitors for time, notwithstanding the fact that both the *roots of mathematics* and *the fulfilment of mathematics* are to be found in modelling.

I see the relationship between mathematics and modelling not as shown in Fig. 3(a), but rather as in Fig. 3(b). The modelling activities are shown by the left-hand boxes 1, 4, 5 and the large arrows, in Fig. 3(b), in boxes 2 and 3 the mathematics takes off and lives a life all of its own (or, in 'applied mathematics' a life enriched by the modelling arrow $1 \rightarrow 3$).

The shortcoming of which I am most aware in my own book, as a primer of modelling, is that only a few modelling examples are discussed—though at least they are given suitably extended treatment. Most of the space is devoted to the fairly conventional ground in boxes 1 to 4 which is needful for a course in mechanics, impregnated, I hope, by the spirit of modelling, to which I shall return later.

The Schools Council Mathematics Applicable Series of textbooks faces the dilemma of both covering a substantial mathematical syllabus and providing

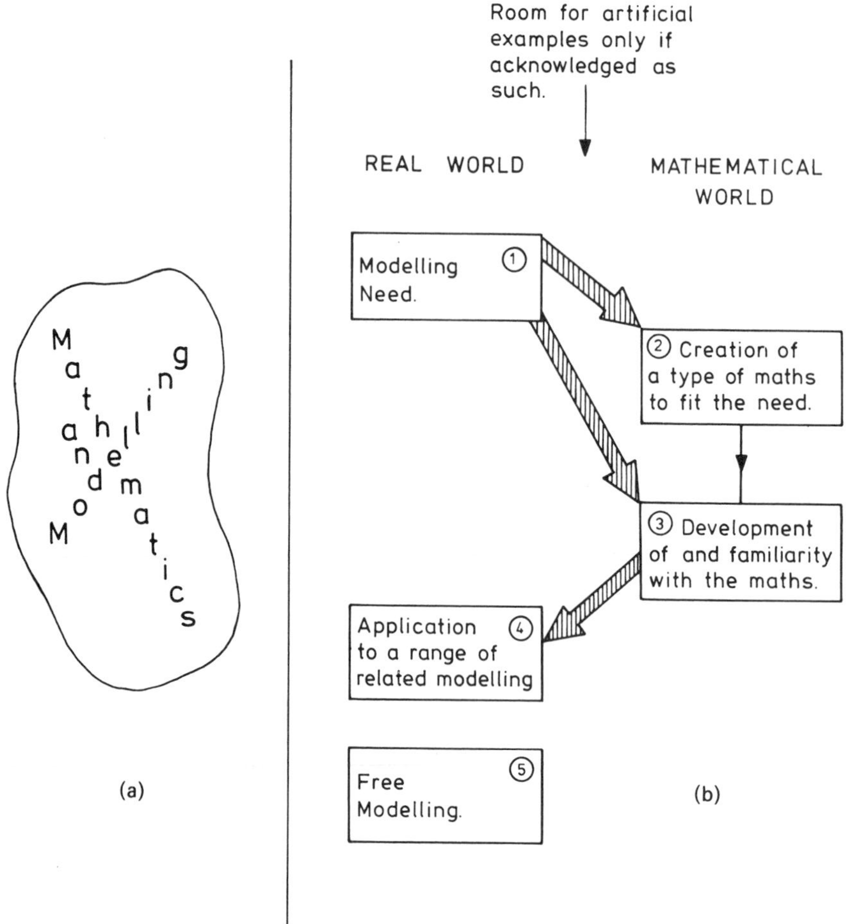

Fig. 3.

an introduction to modelling in a rather different way. Numerous briefly presented, neat versions of real life problems are intermingled with the mathematical development. The books conform to the intimate mixture scheme of Fig. 3(a) rather than to the clearer line of Fig. 3(b). The essential steps represented by boxes 1 and 6 in Fig. 1 are almost completely absent. A very useful series, but needs a strong, clear-headed teacher to develop the modelling into more than a *divertissement*.

On the evidence of only a small sample of Open University study-books and televised programmes, I believe that the modelling arrow 1 → 2 in Fig. 4(b) is badly treated, but that very good material is provided relevant to the modelling

(a)

x $mk\dot{x}^2$ mg

$$m\ddot{x} = mg - mk\dot{x}^2$$

$$\frac{d}{dx}\left(\frac{\dot{x}^2}{2}\right) + k\dot{x}^2 = g$$

$$\frac{d}{dx}\left(\dot{x}^2\right) + 2k\dot{x}^2 = 2g$$

$$\dot{x}^2 = \frac{g}{k}\left(1 - e^{-2kx}\right)$$

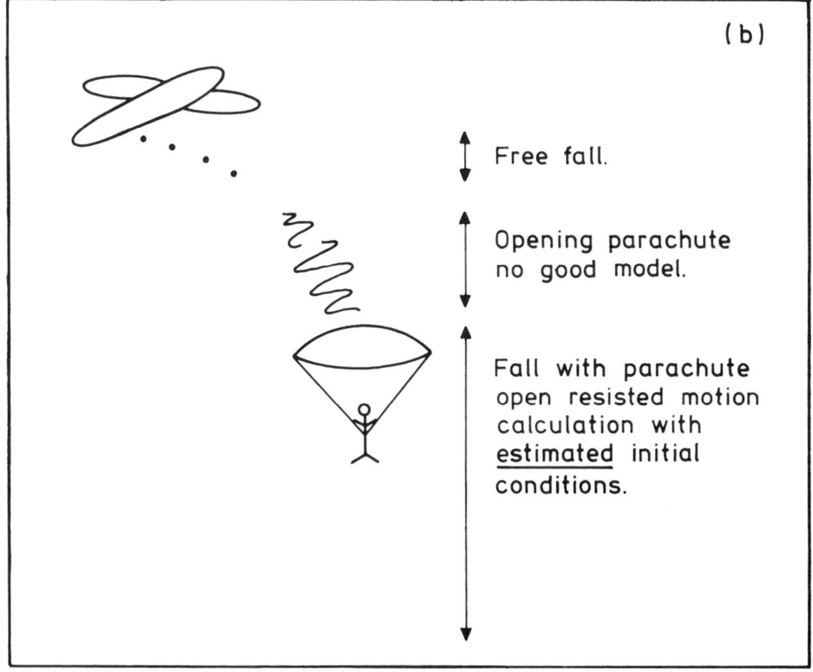

Fig. 4.

arrow 3 → 4, and there are in their televised programmes even more notable contributions to box 5 (free, innovative real-world modelling), much of which would be impossible to parallel in a textbook. I call to mind one superb programme in which two teams were set the problem of deciding on the best position for a bicycle lamp on a given bicycle; after a considerable length of time, and after very properly making a list of possibly relevant factors—like the speed of the bicycle or the width of the light beam—one of the two teams had not even got as far as asking the question 'best for what? This programme (and others) made it abundantly clear that modelling is serious and time-consuming, not slick and quick.

In the Appendix to this paper I shall try to indicate briefly the kinds of services supplied by a selection of other books. I am sure that there are good texts which I have missed out through ignorance. My comments should be taken as personal assessments, not authoritative judgements.

The conclusions from looking at books must be that: (i) original modelling outside conventional, well-authenticated fields is an *activity*, and is better conveyed by the teacher-in-action than by any written account; (ii) the uncritical reproduction by the teacher of unsuccessful or partially successful models serves little educational purpose; on the other hand, active criticism/attempted validation of such models can be a valuable exercise (the recognition that widely accepted models of economics, politics and the like can be *good* but nevertheless to a large extent *unsuccessful* models would be well worth achieving); (iii) the uncritical reproduction by the teacher of successful models leads to dull, stagnant teaching; (iv) in well-authenticated areas of science, successful modelling is often possible, but teaching should always be pervaded by *the spirit of modelling*, i.e. by veracity and the recognition of areas of only partial success.

Let me illustrate the difference between conventional model-out-of-a-hat teaching and 'teaching informed with the spirit of modelling' by reference to a familiar problem in the conventional repertoire of mechanics—a problem of parachute fall.

The model-out-of-a-hat version is represented in Fig. 4(a). The only introduction given is 'Let us model the man plus parachute as a particle, and the air resistance as mkv^2'. By contrast, a modelling of the real world problem would start with a picture such as that of Fig. 4(b). The uncertainties about the speed acquired and the distance fallen before the parachute opened would be recognized and, if possible, estimated to an order of magnitude. From the start it would be clear that any classroom modelling of this kind could be only partially successful. In discussing the motion after the parachute had opened it would be stressed that information about air resistance must be looked for (not guessed, or, still worse, postulated without any reference to experimental data), and sources of information would have to be found, for example one of the CRC handbooks (Tuve and Bolz, 1973) or Medley (1982).

To conclude, it seems to me that:

(1) Traditional subject matter offers higher expectation of success than does novel subject matter, and can be approached in as fresh a spirit;

(2) Innovative modelling by a teacher is very demanding, and following someone else's line of attack serves no substantial purpose when the result, in

any case, is unlikely to be predictively reliable. If the pupil can be encouraged to learn by doing, great educational benefit will accrue. *It is adventurousness which needs to be nourished, not credulity.*

6. APPENDIX. SERVICES RENDERED BY BOOKS TO TEACHERS OF THE MODELLING USES OF MATHEMATICS

The first prerequisite for mathematical modelling is a foundation of essential mathematical techniques. Help is given in laying such foundations not only (at various levels) in modern texts such as Lin and Segel (1974), Medley (1982), Schools Council (1975) and others, but by old-fashioned textbooks like Workman (1904) and Humphrey (1930). Formal exercises can often be transformed into realistic modelling examples by teaching 'informed with the spirit of modelling'. This spirt will be encouraged by explicit recognition of the origins of various branches of mathematics in real world needs, as in Alexandrov *et al.* (1963), Melzack (1976), Medley (1982), and, here and there in Lin and Segel (1974), Noble (1967) and Schools Council (1975).

The second prerequisite for mathematical modelling is a non-trivial appreciation of the subject matter to be pursued. Although some such appreciation may be assumed in familiar realms, e.g. space, motion, shopping, it may not be assumed for problems rooted in chemistry, biology, economics or the like. Necessary semi-specialist background knowledge in some special areas is provided by Alexander (1982a, 1982b), Bender (1978), Daish (1972), Lin and Segel (1974) and Noble (1967).

Non-trivial insights into the methodology of mathematical modelling can be found in Andrews and McLone (1976), Aris (1978), Bender (1978), Burkhardt (1981), Daish (1972), Lin and Segel (1974), Medley (1982) and Ravetz (1961).

Attractive material for the general reader including numerous short accounts of mathematical models (many of which merit critical discussion) are to be found in Alexander (1969), Brebbia (1978), Burghes and Borrie (1981), Lighthill (1978) and Schools Council (1975). The special study in Ravetz (1961) both makes good reading and casts a clear illumination on the craft of modelling.

Advanced undergraduates or postgraduates should be interested in the mathematical models given in Aris (1978), Brebbia (1978), Lin and Segel (1974) and Noble (1967).

The philosophically minded mathematician will enjoy both Hesse (1963) and Ravetz (1961).

REFERENCES

Alexander, R. McN, 1969, *Animal Mechanics*, University of Washington Press.
Alexander, R. McN, 1982a, *Optima for Animals*, Edward Arnold.

Alexander, R. McN, 1982b, *Locomotion of Animals, Blackie.*

Alexsandrov, A. D., Kolmogorov, A. N., Laurent 'ev, M. A., 1963, *Mathematics, its Contents, Methods and Meaning* (3 vols.), MIT Press.

Andrews, J. G. and McLone, R. R., 1976, *Mathematical Modelling*, Butterworth.

Aris, R., 1978, *Mathematical Modelling Techniques*, Pitman.

Bender, E. A., 1978, *An Introduction to Mathematical Modeling*, University of California Press.

Brebbia, C. A. (ed.), 1978, *Applied Numerical Modelling*, Pentech Press.

Burghes, D. N. and Borrie, M. S., 1981, *Modelling with Differential Equations*, Ellis Horwood.

Burkhardt, H., 1981, *The Real World and Mathematics, Blackie.*

Cockcroft, W. H., 1982, *Mathematics Counts*, H.M.S.O.

Daish, C. B., 1972, *The Physics of Ball Games*, E.U.P.

Hesse, M. B., 1963, *Models and Analogies in Science*, Sheed and Ward.

Humphrey, D., 1930, *Intermediate Mechanics*, Longmans.

Lighthill, J., 1978, *Newer Uses of Mathematics*, Penguin Books.

Lin, C. C. and Segel, L. A., 1974, *Mathematics Applied to Deterministic Problems in the Natural Sciences*', Macmillan.

Medley, D. G., 1982, *An Introduction to Mechanics and Modelling*, Heinemann Educational Books.

Melzack, Z. A., 1976, *Mathematical Ideas, Modeling and Applications*, John Wiley.

Noble, B., 1967, *Applications of Undergraduate Mathematics in Engineering*, Macmillan.

Ravetz, J. R., 1961, *Galileo and the Strength of Materials*, School Science Review.

Schools Council, 1975, *Mathematics Applicable* series, Heinemann Educational Books.

Tuve, G. L. and Bolz, R. E. (eds), 1973, *CRC Handbook for Applied Engineering Science*, CRC, Ohio, U.S.A.

Workman, W. P., 1904, *School Arithmetic*, U.T.P.

Modelling in the Classroom—How can we get it to happen?

Hugh Burkhardt
University of Nottingham

The teaching of modelling skills demands a teaching style which includes crucial elements which lie beyond the style elements of nearly every teacher or lecturer. Studies show how hard it seems to be for teachers to change their style in any significant way. Though many 'modelling' courses and books are still really only about *models*, not modelling, demanding of the student only that he learn and use, and not create or modify models, enough has been done in the past twenty years to show that there exist teachers at all levels who can help their students develop their own modelling skills. How can we help all the teachers who should contribute to this work to acquire the skills necessary to get modelling to happen with their students? Recent research will be described which emphasizes the difficulty of this problem, but shows some possibilities for a breakthrough.

1. STIMULATING CURRICULUM CHANGE

In any curriculum innovation that is aimed at a wider public than the innovators themselves, there are four important elements:

- specifying the pupil achievements that you want,
- devising teaching methods that work with the target group of students,
- developing teaching methods that are accessible to the target group of teachers, and
- disseminating the whole system on the scale intended.

These elements might appear to be *stages* of the development process but they are not, because the later elements must feed back to those listed earlier the important constraints and possibilities that they imply; thus development is a cyclic process, like problem solving itself, of which it is a particular type.

Earlier models of curriculum development tended to be of the simple sequential type. They also largely ignored the third element—accessibility to teachers; there has been an assumption that what one teacher can do, any teacher can do as long as they have the 'right attitude' and a bit of in-service training—research suggests that this is quite untrue, and that styles and levels of performance in teaching as in everything else differ enormously, with innovators tending to come from amongst those with the widest range. It is impossible to be sure without further experiment how far this accounts for the failure of so many projects to spread much beyond the small circle of pioneers initially involved. This paper sets out some of the difficulties and promising avenues for tackling them in the context of teaching modelling.

At present it seems fair to say that the first two elements are, provisionally, in relatively good shape.

(1) We know what we *mean* by modelling as a learning activity at school and undergraduate level (though many 'modelling' courses and books are still really only about *models* not modelling, in that they demand of the student only that he learn and use, and not create or modify models).
(2) There exist *teachers* who have shown with groups of students of a wide range of ages and abilities that they can get such work going as a regular part of the curriculum.

The fourth element, dissemination, is under way in that books and materials are being published at various levels. So far, though the authors may have wider ambitions, they are effectively aimed at those teachers who can in practice convert curriculum ideas and suggestions into classroom reality—a small and select group, perhaps more numerous in higher education for various reasons. The attempt to give modelling a role in the curriculum at large is yet to begin—it is here that the third element, accessibility to an appropriately wide range of teachers, must be taken seriously if there is to be much chance of success.

2. PRESSURES ON THE TEACHER

It is worth listing major pressures on teachers which influence their professional behaviour, including their response to suggested innovations.

Workload on teachers in schools is high. Minute-by-minute in the classroom, the teacher's attention, her information processing capacity, is essentially saturated with processing the many stimuli from individual pupils and making appropriate responses (one teacher, observed in detail by members of the School of Education at Nottingham, was found to have about 1200 (twelve hundred!) such transactions a day). Once or twice each hour a new lesson begins—half a dozen new 'arenas' a day. Marking and some preparation is unavoidable. In these circumstances skills at an 'automatic' mastery level must predominate. The time for reflection, for taking on board new ideas and absorbing them to this level is not timetabled—in most schools even departmental meetings are a rarity and are dominated by practicalities—'who is going up?' etc.

Work '*schemes*', imposed or agreed, provide an authoritative delineation of what is expected.

Examinations, through the tasks they present and the way these are marked, delineate with even greater clarity and force what is 'important'—their social importance and the academic authority of their source, the examination boards, together impose a considerable demand of courage and confidence on a teacher who advocates or attempts something outside. The practical importance of modelling skills for, say, low attainers may be persuasive but it is likely to be regarded as a distraction for the 'exam group' even at age 11 or so.

Urgency, as a sense of shortage of time, is generated by the pressures already referred to in an intense form. There is good reason to believe that this is an illusion in that a small increase in the efficiency of learning, which improved grasp or retention, would reduce the need for endless repetition—but that pay-off will have to be made crystal clear if it is to be believed.

Colleagues reinforce all these pressures in most cases—whether they fully accept the basis of the status quo or not, their active acquiescence in it increases pressure on the individual teacher to conform. Even in a school where a few teachers are actively reviewing the curriculum, the isolated nature of most teaching normally allows others to continue their established methods with little need for any evaluation of justification of these.

Public attitudes can play an important role. Although at most times, the teaching professions are left largely alone to determine the curriculum with little specific pressure from parents, politicians or other public voices, at times a clearly articulated public pressure has existed. In recent years it is only at these times, in mathematics at least, that noticeable change in mathematical education has occurred. The two outstanding changes—the content changes of the 'modern maths' of the 1960s and the general introduction of microcomputers into schools in the 1980s both followed intense 'public' discussion and a consensus in favour of change; other equally arguable reforms in between have not had this support and have made no widespread impact. The process of general absorption took about 10 years in the case of 'modern maths' and has hardly started with micros—the vigorous activity in schools in this area is still largely outside the normal curriculum, including the mathematics curriculum.

If modelling is to make progress, generating explicit public support must be a priority—the self-evident importance of these skills will not be enough. All media, from the Cockcroft Report (1982) to peak time television, will be needed if any impact is to succeed; only in this way, it seems, will all elements of the education system (or 'agencies' in Cockcroft's terms) be given a common direction of momentum.

Pupil attitudes are becoming more important though they are still not decisive—except in a negative sense, so rejection must be avoided. On the whole the curriculum is still 'done to' pupils rather than 'built with' them.

Specific problems are recognized by teachers—difficult topics, lower attaining pupils, mixed ability groups etc. These generate their own pressures.

Given all these pressures, anyone who wants to introduce a change is wise to design it so that the teacher's work is made

- easier
- more fun
- in tune with a recognised public demand
- promising help with specific problems already recognized.

These are very difficult conditions to meet, particularly the first two where the innovator starts with the handicap that change in itself is threatening and takes time to absorb. Nonetheless, less severe constraints, such as the simple requirement that change should be 'good for' the pupils and interesting to innovative teachers, almost certainly ensure that the result will in practice only be of interest to that select, but small group. This is the position with modelling today.

3. DIFFICULTIES FOR THE TEACHER

I now want to summarize the problems that teachers must overcome if they are to teach modelling. Understanding of and skill in modelling itself, suitable materials for classroom use, and teaching style all present difficulties—of these I shall argue that the last presents the most serious challenge in this, as in most other suggested innovations.

Some modelling skill and understanding of the processes involved are probably an important ingredient in teaching it successfully. Modelling as a conscious process came new to most of the pioneers, who almost all had the additional advantage of being professional applied mathematicians of one kind or another. Providing parallel experience for teachers is not easy. I have argued Burkhardt (1981) that, since a thorough qualitative understanding of the empirical context is important, everyday practical problems should form the backbone. This allows teachers to learn to a great extent alongside the pupils, and over their shoulders—probably the most effective channel for in-service training. However, they have first to acquire the confidence to try.

Classroom materials for use in modelling courses are gradually becoming more widely available. Though the need for more materials, better tailored for their particular area of the curriculum will always remain, it is my impression that after the world-wide effort over the last two decades there is no longer a real shortage. Max Bell's Review (1983) for ICME4, Reference 2 and this book provide some of the evidence. In making this assertion, I am strongly influenced by the central importance of learning the *processes* of modelling rather than the content and form of specific models of particular phenomena. The emphasis on process makes classroom materials less important than the learning activities they mirror, and than the teaching style that stimulates and supports these activities. In the end, the pupils themselves can provide plenty of problems to tackle and the background knowledge needed to formulate and validate models to analyse and explain these phenomena.

Teaching style seems to be the crux of the problem of getting modelling to happen in the classroom. It is the central purpose of this paper to focus attention on this, giving some evidence for the assertion and some part-proven ideas of what we might do about it. As to evidence, we can start from the list of essential learning activities from paragraph 243 of the Cockcroft Report (1982), summarized and reordered as

(1) exposition by the teacher
(2) practice by the pupils
(3) discussion
(4) practical work
(5) problem solving
(6) open investigation

and the observation, noted by them in tactful terms and much more firmly stated by HMI (1979), that none of the last *four* activities are to be found on any statistically measurable scale in secondary mathematics classrooms. They are not common in higher education either. The nearly universal teaching style consists of explanation by the teacher with illustrative examples of well-defined exercises, followed by copious pupil practice on *closely similar* exercises.

Furthermore, the standard coaching strategy, when pupils are having difficulty with the exercises, consists of re-explanation and of the teacher breaking the problem into smaller more digestible tasks.

The limitations of such a style in helping pupils to learn to tackle *complete, unfamiliar, half-specified, realistic* problems involving a wide range of skills, are clear and well known. The advocacy of more 'open' teaching styles, giving more responsibility to the pupil for defining and tackling tasks with no specific 'right answers', and with the teacher acting as counsellor and even fellow pupil, has been a central theme of the professional associations of mathematics teachers, particularly the ATM, for at least 25 years. It has been advocated by others interested in education for several thousand years. Why has there been so little progress?

We have evidence that open teaching is harder. Standard expository teaching ('tell-'em-and-test-'em') is *single track*. The teacher understands and explains a single line of argument, uses it for the illustrative examples, and where pupils deviate from it in practice, simply notices this fact and returns them to the standard line. (There is evidence that this is a rather ineffective teaching method for those who have difficulties, even for learning standard mathematical techniques; such pupils, and they are a majority, have already misunderstood or misremembered the explanation at least once and will do so again. Even successful pupils usually use their own minor variants of the taught procedure; their success is linked to their being able to tell when they are right—'debug' their own algorithms—and not to more accurate memorizing.)

Open teaching on the other hand is *multi-track*. It is more demanding on the teacher

(a) mathematically because the teacher must be able to perceive to some extent at least the consequences of the different approach on which the pupils are embarked,
(b) pedagogically because the variety of approaches going on in the classroom at any one time will be much wider, and
(c) personally, because the teacher must have the confidence to accept that he will not know all the answers.

There is some encouragement in that I know of no case in which a teacher who has got used to working in an open way, has later abandoned this approach. Everybody who has succeeded agrees that it is more satisfactory and more satisfying. However this is not, of course, a random sample of mathematics teachers. The vast majority have not brought such open elements into their style range.

A research study of about 200 mathematics lessons Burkhardt *et al.* (1983) using a structured framework for the analysis of teaching style has provided some harder evidence on these matters. An analysis in terms of the pupil learning activities and the classroom roles Fraser *et al.* (1983) of the teacher produced the classification shown in Fig. 1. In normal teaching, the roles in the first 3 groups (Manager, Explainer, Task Setter) predominate, along with the Cockcroft activities (1) and (2) above. The study also showed that a microcomputer, programmed appropriately to act as a 'teaching assistant', led teachers to broaden their style to encompass the 'open' roles (Counsellor, Fellow Pupil, Resource) in a natural way that introduced the 'missing activities' (3), (5), and (6). We shall return to this remarkable result.

4. HELP FOR THE TEACHER

If teaching modelling produces heavy new demands on the teacher, particularly in the difficult matter of style, what prospects are there for helping a substantial proportion of teachers to rise to the challenge? We finish with a few current hopes and possibilities—all these suggestions are at best only partly proven.

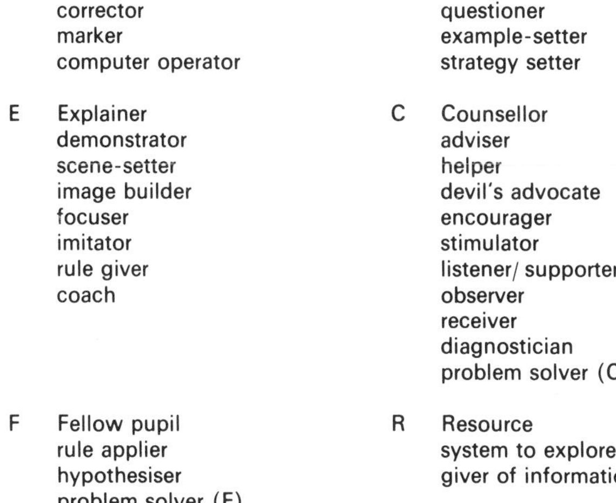

M Manager (tactical)
corrector
marker
computer operator
T Task setter
questioner
example-setter
strategy setter
E Explainer
demonstrator
scene-setter
image builder
focuser
imitator
rule giver
coach
C Counsellor
adviser
helper
devil's advocate
encourager
stimulator
listener/ supporter
observer
receiver
diagnostician
problem solver (C)
F Fellow pupil
rule applier
hypothesiser
problem solver (F)
R Resource
system to explore
giver of information

Fig. 1. Classroom roles.

In-service support is the obvious avenue for promoting the professional development of teachers. However, it must be recognized that the present situation is as it is in spite of 20 years of vigorous in-service effort. Many factors reduce the potential impact of INSET:

- the small number of advisers (less than one per 100 schools)
- the small exposure of each teacher (less than 10 hours per year)
- the difficulty of relating and transferring INSET ideas to the classroom
- suggestions inappropriate for the teachers concerned

We can do little about the first two factors, which are primarily economic—though a *redistribution* of resources (e.g. larger classes for more development time) should be thought about. The latter two problems are a different matter and progress should be possible using an essentially empirical approach to find out what works. Our recent experience suggests that for example, INSET

(1) should be activity based, with the teachers actively involved in trying out what is intended,
(2) should integrally involve work with pupils, in school or on the courses,
(3) should involve more than one teacher from each school working together,
(4) should include feedback on the changes actually produced in classroom practice as a result.

One should mention also the idea of 'cascading', where the 'learners' become 'teachers' at the next stage of the dissemination. Though it is unclear how far, or even if this approach works, it does have a number of positive aspects particularly in promoting an active approach to learning through the realization that you are going to have to teach the stuff yourself very soon. It does imply carefully prepared and specially designed INSET material. Our experiential base for these assertions is, again, the preparation of material concerned with microcomputers in education MEP (1982) and Longman Software (1982), where the demand is so heavy that more normal (and rational?) approaches to INSET are inadequate; the positive feedback on such materials suggests this approach is worth trying in other arenas.

When we look for effective sources of support for the teacher *in the classroom* a number of possibilities, some involving high technology and some much older aids, need to be considered:

Written advice seems to be almost entirely ineffective in promoting style change—this is not surprising where anything is 'total' as teaching style is involved. We should not expect to be able to help somebody, say, to learn to ride a bicycle or to compère a 'TV chat show' by written instructions alone. Teachers' notes, therefore, should not be expected to carry too much of a burden; in any case, they must be very brief if they are in practice to be read.

Video undoubtedly has something to contribute in showing the kind of pattern of classroom activities and teacher behaviour that is being encouraged. It is also our experience that many teachers feel starved of any knowledge of how other teachers work. On in-service courses they clearly enjoy watching videos of other's teaching. This is particularly true if some framework for discussion and analysis of teaching style is provided. SCAN Beeby *et al.* (1979) has proved useful in this regard.

Personal contact has proved extremely effective in a few schools in promoting substantial development including style change. Team teaching, regular staff meetings, and co-operative curriculum development are all important elements. It may also be true that the school departments which have been most successful have unusually well qualified staff. It is certainly true that this approach has not had a large-scale impact over several decades.

Microcomputers can be programmed to act essentially as 'teaching assistant', in a way that has been outlined above. The vast majority of material for computer-assisted learning has been aimed directly at the individual pupil, not involving the teacher in any essential way. Investigations on Teaching with Microcomputers as an Aid (the ITMA Collaboration) has, on the other hand, been concerned with using the microcomputer to enhance the effectiveness of teachers and teaching. The research referred to above has shown the power of the device in this most difficult matter of promoting style change. The potential of this approach has yet to be fully explored but the fact that teachers with a standard approach were led, almost without their noticing, to work easily in open ways seems promising.

5. AN INTEGRATED APPROACH

None of the individually helpful possibilities listed above does anything about the total pattern of pressures with which we began. In a feasibility study with the Joint Matriculation Board, the Shell Centre is trying an integrated approach to gradual curriculum change in mathematics—modelling is one of the activities on the agenda. The essence is to use to examination system to promote step-by-step curriculum changes, each of which has been shown empirically to be accessible to the target groups of pupils and of teachers that the Board serves. Each 'module' in this Teaching Strategic Skills project corresponds to about 5% of the syllabus and comprises

(1) tasks—specimen examination questions, marking schemes and sample answers
(2) sample teaching materials
(3) support materials to help teachers realize the changes implied in classroom activities and behaviour

All the aspects listed above are involved in one way or another. There is a parallel initiative to extend this approach to the area of 'numeracy', in the sense of full functional mathematical ability. How far this approach will succeed remains to be seen. Taking a 'systems view' of the curriculum, however, it seems likely that general change will only be promoted when its full complexity is taken into account, including those 'stabilizing' feedback mechanisms between its different elements which resist change.

REFERENCES

Beeby, T., Burkhardt, H. and Fraser, R., 1979, Systematic Classroom Analysis Notation. Nottingham: Shell Centre for Mathematical Education.

Bell, M., 1983, *Proceedings of the 4th International Congress on Mathematical* Education, Birkhauser, Boston.

Burkhardt, H., 1981, *The Real World and Mathematics*, Blackie, Glasgow.

Burkhardt, H., Coupland, J., Fraser, R., Phillips, R., Pimm, D. and Ridgway, J., 1983, Microcomputers in the Mathematics Classroom, Shell Centre for Mathematical Education, Nottingham.

Cockcroft, W., 1982, *Mathematics Counts.* Report of the Committee of Inquiry into the Teaching of Mathematics in Schools, HMSO, London.

Fraser, R., Burkhardt, H., Coupland, J., Phillips, R., Pimm, D. and Ridgway, J., 1983, Learning Activities and Classroom Roles. Shell Centre for Mathematical Education, Nottingham.

HMI, 1979, *Aspects of Secondary Education in England*, A survey by HM Inspectors of Schools. HMSO, London.

Longman Micro Software, 1982, Micros in the Mathematics Classroom, 14 program teaching units.

MEP, 1982, Micros in the Mathematics Classroom—an inservice pack.

O.U. Students do it by Themselves

J. S. Berry
The Open University
and
D. Le Masurier
Brighton Polytechnic

SUMMARY

A report is given on the mathematical modelling project scheme that forms part of the second level applied mathematics course, called Mathematical Models and Methods (MST204).

Part 1. The Walton Hall view by John Berry

In this part an outline of the project scheme is given, identifying its aims and how the modelling component fits in with the rest of the course. The course ran for the first time in 1982 and the author was a member of the course team.

The students of the Open University study by themselves and seldom have the opportunity of working in small groups. It is desirable, and much more stimulating and rewarding, if mathematical modelling is carried out as a group activity. In this respect, therefore, our students are disadvantaged. However, the course team felt that, in addition to learning mathematical techniques and examining standard models, students should be given the opportunity of investigating, in some depth, an extended problem lasting 40–50 hours over 4 distinct weeks of the year. Inevitably students 'do it by themselves' and an outline of the help given to students for their project work and the advice given to tutors in assessing the students' work is described here.

Part 2. The Part-time tutor's view by David Le Masurier

Part (some would suggest all!) of the process of teaching Open University students is undertaken by part-time tutors, most of whom are in full-time employment as academic staff in H.E. institutes throughout the U.K. For this course, there were approximately 50 such tutors appointed, and this has been almost doubled for the current year (1983). The difficulties of the course and problems for the students of learning at a distance are appreciated more by part-time tutors than by most (some would say all!) Walton Hall academics.

In this part one tutor discusses how well the mathematical modelling component of the course went, and whether the aims were achieved, from two viewpoints: firstly, from the position of the tutor, who is himself relatively isolated, to enquire whether the advice received from Walton Hall was sufficient to enable the tutor to provide the student with worthwhile support, and to assess the work fairly and consistently; secondly, looking through students' eyes, the assistance given to students was satisfactory, so that they gained significantly from their modelling experience.

PART 1. THE WALTON HALL VIEW

1. The Open University teaching system

For the benefit of those readers who are not familiar with the teaching system of the (British) Open University a brief outline is given here.

The Open University welcomed its first students in 1971 and was established to provide undergraduate degree studies, (and more recently taught post-graduate degrees), for part-time students. The majority of these students are in paid employment or busy housewives so that normally they are studying in their spare time (normally evenings and weekends).

The University offers a BA degree which can be taken with or without honours. A student is required to pass six credit equivalents to obtain an unclassified degree (i.e. without honours) and a further two credits to qualify for an honours degree. Open University courses are of two types: full credit courses consisting of 32 weeks study and half credit courses consisting of 16 weeks study, one weeks' study consists of about 10–15 hours work.

Most students are required to take two foundation courses (two full credits), which provide some of the basic skills that an undergraduate at a conventional university would be expected to know on leaving school. 'Second level courses' concentrate on particular disciplines within a subject level so that, for example, in the mathematics faculty, there are second level courses in pure mathematics, applied mathematics, computing and statistics. 'Third level' courses concentrate on particular topics within these four disciplines. It is a part of the second level applied mathematics course on which we concentrate in this paper.

The Open University student may be disadvantaged in many ways. Often a period of study of say two hours' duration must be spent each day, before or after a normal day's work in full-time employment. Furthermore there may be domestic demands, and possibly antagonism from their spouse or family, which make study very difficult. Many students are returning to education after a considerable gap from their school or college days. They may lack confidence in their ability to study and the pattern of 'learning by themselves' can lead to little contact with other students who are experiencing the same problems. The fact that most students overcome these difficulties is a measure of their commitment.

The teaching package of nearly all courses is based on correspondence texts (called units), an integrated series of television programmes (and sometimes radio programmes), audio cassettes (and in the future video cassettes) and a regionally organized tutorial and counselling system. Some courses have associated with them a residential summer school. Much of the teaching package is designed to help students study and learn by themselves.

The student responds to this teaching package by completing assessment materials. These may be in the form of a self-assessment questions or essays where possible solutions are supplied or more usually in a written form which are assessed by a tutor (these are called tutor-marked assignments) or by the computer (these are called computer-marked assignments). These assessed assignments often extend the teaching begun in the correspondence texts and related materials. It is through the tutor-marked assignments that a student who is experiencing difficulties should be identified.

The course tutor has the role of providing 'face-to-face' and 'at-a-distance teaching' of a group of students (usually up to a maximum of 25 students). This tuition takes the form of class tutorials with the group and individual telephone tutorials with each student (i.e. 'distance contact').

Sometimes conference telephone tutorials are arranged between a small group of say 6 students and a tutor. This is a particularly helpful form of tutorial work for disabled students who cannot get to study centres for the face-to-face tutorials.

Course tutors also assess the work of the students in the form of tutor-marked assignments. They receive briefing material about the course and 'tutor notes' at regular intervals which not only consist of mark schemes for the assignments but also up-to-date information on the progress of, and any changes to, the course materials.

The course tutor is a very important cog in the Open University's Teaching Wheel. He (or she) is the link between the course team who produced the materials and the student. It is the course tutor who often fills in the gaps and corrects the errors left by the printed correspondence texts! Most of the course tutors are teachers in higher education and, like the students, are also part-timers.

2. An outline of the course

The second level applied mathematics course is called 'Mathematical Models and Methods' and given the code MST204. The M, S, T identify the course as an inter-faculty course with contributions from the three faculties of Mathematics, Science and Technology. It is a full credit course (i.e. 32 weeks of study) and is about the use of mathematics to solve problems in the real world. Half of the course teaches mathematical methods such as the solution of differential equations, vector algebra, matrix analysis and so on; the other half is about how to represent relevant aspects of the real world by means of mathematical models. Eight units are on Newtonian mechanics and four units investigate other models—these are (animal) population models, heat transfer, linear programming and forecasting. The remaining four study weeks are set aside for a mathematical modelling project. The project is an extended problem giving students the opportunity to set up their own models according to some given specification. The project is a compulsory part of the course and occupies the students for about 40 hours of their time and accounts for 12.5% of their total assessment. It is the development and success of this mathematical modelling scheme that is described in the next two sections.

Appendix 1 contains a copy of the broadcast and assignment calendar for 1982. It shows the relationship between the different parts of the course materials and the unit titles.

3. Some of the difficulties in devising the project scheme

Conventional teaching methods, such as lectures, are not suitable for teaching mathematical modelling. It is important for the student to be actively involved in every part of the modelling process, so that they are 'doing' and not 'watching'. Thus it has been found that the most successful method of giving students the experience of modelling is by project work. But project work can put extra demands on the student and the tutor.

How can we expect students to be able to write reports? How do we mark reports? These are just two of the many questions that had to be answered in devising the project scheme. But being a distance teaching institution other difficulties also had to be tackled. Some of these are listed below:

(*a*) *For the student*

I. Project work in other courses has shown that students can spend far too many study hours on their project, unless the project scheme is carefully structured. The problem for the student working alone is to know when to stop, he does not have much contact with other students 'to compare notes'.

II. Mathematical modelling is usually thought of as a group activity. Discussion of a problem within a group can help to provide the starting point and 'blind alleys' can often be avoided (or dismissed very quickly).

Furthermore the initial brainstorming session during which one familiarizes oneself with the features of the problem is a more enjoyable task when done as a group activity.

Open University course materials are devised for students working alone. Many students live in remote areas and their isolation does not lend itself to group work.

III. The assignments associated with project work usually take the form of reports. Experience of mathematical modelling project work reports at other institutions, show that unless some guidance is given to the student, these reports ramble on in an unstructured way and are very difficult to assess.

IV. Students feel that their marks associated with report work will be much lower than for conventional mathematics assignments where there is a 'correct answer'. Accordingly there is a feeling that unless their model is very good a low project mark will result leading to a disappointing course mark. The course team had to devise a mark scheme that did not favour the good model of the expert, but gave marks for the modelling process.

(*b*) *For the course tutor*

The course tutor has the dual role of adviser and assessor.

V. As an adviser the tutor has the difficult task of helping the student satisfactorily complete the project work without suggesting to the student a particular approach or model. There is the problem of giving the persistent student too much help which can lead to an inflated grade.

VI. As an assessor the tutor must mark a report that describes the modelling activity. The usual form of student assessment that is used in the mathematics faculty are fairly standard exercises with well-documented mark schemes. A tutor looks for a particular part of the solution, for example a differential equation, and gives marks for that part. This approach leads to quite consistent marking across all the tutors for a particular course. However, marking reports was an unfamiliar activity to many of our tutors. This is much more of a subjective exercise and there are instances where two tutors award significantly different marks.

4. The MST204 mathematical modelling project scheme

To overcome the difficulties identified in section 3 the course team spent two years developing variations of the project scheme and trying out different mark schemes and testing them on Open University students. The project scheme teaching materials consist of a project guide, audio-cassette activity, a project-related tutorial session, two tutor-marked assignments, three modelling television programmes (described in another chapter in this book) and summer school modelling sessions.

(i) The project guide

This is not a normal teaching unit but is a guide showing the activities to be carried out during the four project study weeks and contains two examples of student work i.e. case studies. The project guide also includes ideas on how to set about projects. The project work is carefully structured so that the student has a specific set of activities for each study week. This provides a starting point and a goal for each week and by attaching average times for each activity it was intended that a student would not spend an excessive amount of time on the project, (see I above).

A planning chart from the guide is reproduced below:

project week	activity	estimated time	
1	(i) Read *Project Guide* (ii) Carry out modelling exercises (iii) Carry out audio-tape activity (iv) Choose a project, find an audience and plan data sources	3 hours 1 hour 2 hours 4 hours	The times given in the table are rough estimates based on student experiences. They are given so that you can plan ahead.
2	(i) Review week 1 and availability of data (ii) Set up model (iii) Formulate mathematical problem (iv) Plan and write TMA 4	1 hour 5 hours 1 hour 3 hours	
3	(i) Read the tutor's comments on your TMA 4 and review work so far (ii) Collect relevant data (iii) Solve the mathematical problem and interpret the solution using your data (iv) Refine your model, if necessary, and repeat step (iii)	1 hour 4 hours 2 hours 3 hours	Sometime between project weeks 2 and 4 you will attend Summer School, where there will be some project-related activities.

project week	activity	estimated time
4	(i) Review progress so far	1 hour
	(ii) Write a rough draft of your report and discuss it with your audience	4 hours
	(iii) Write your final (TMA 8)	5 hours

(ii) The audio-tape activity

One of the hardest tasks (and one that causes some concern to students embarking on the course) is that of getting started on the project work. This consists of choosing one from the four problem specifications and identifying the important features of that problem. In the conventional teaching institution one way of helping students to get started is by a group brainstorming session (see II above).

To provide the student with a similar experience and to help him 'get started' the project guide has an audio-cassette activity in which the presenter and the student tackle three problems. Associated with the tape are 19 frames similar to those shown on page 56.

The student is encouraged to make a list of all the features that he thinks might need to be included in the model. This *feature list* is then pruned so that there are enough features to adequately provide a model that will describe the problem but not too many to make the mathematics intractable. The students are encouraged to 'keep it simple' particularly in their initial model. This 'starting procedure' is summarized in the following frame and using it our students are able to choose a project (normally the one they can say most about) and get started on their modelling by using the pruned feature list.

The audio-tape activity is a reasonable substitute for the group work described in II.

(iii) The project-related tutorial session

This is usually a two-hour session during which a tutors group of students usually share their concerns about the project scheme. The purpose of the project tutorial is twofold: (i) for the tutor to run through the project scheme and to describe its aims etc., and (ii) to run group modelling sessions by dividing his students into small groups of say 3 or 4 students and setting them loose on a problem (not one of the project specifications!). This latter activity allows the tutor to identify and comment on the following points in the project scheme.

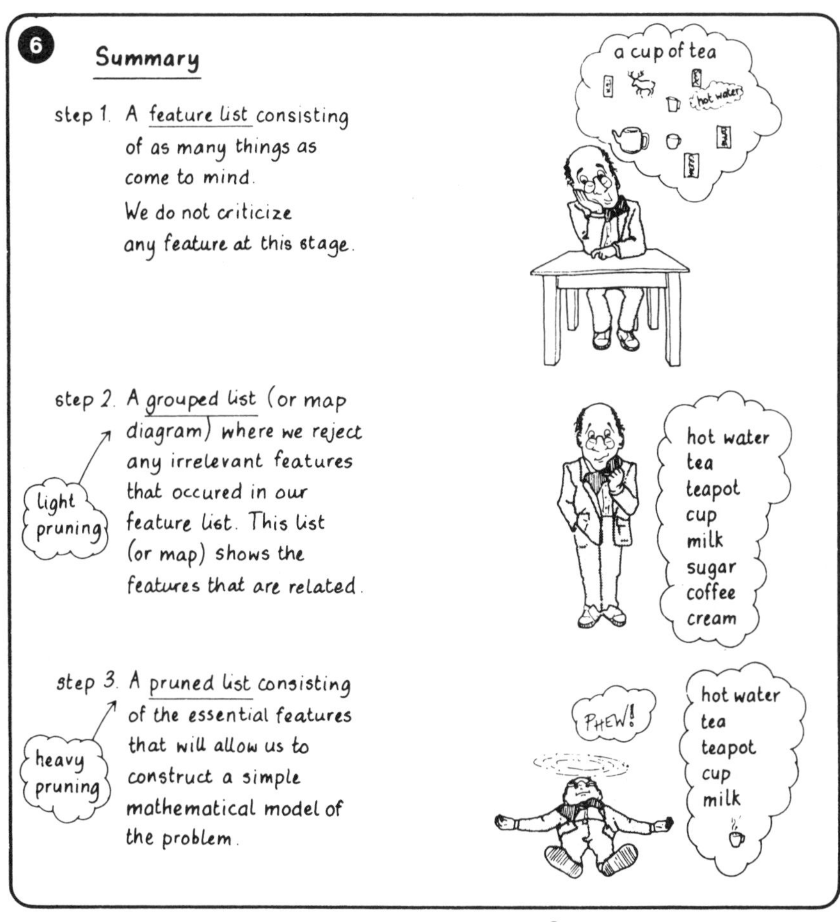

6
Summary
step 1. A feature list consisting of as many things as come to mind. We do not criticize any feature at this stage.
a cup of tea
hot water
step 2. A grouped list (or map diagram) where we reject any irrelevant features that occured in our feature list. This list (or map) shows the features that are related.
light pruning
hot water
tea
teapot
cup
milk
sugar
coffee
cream
step 3. A pruned list consisting of the essential features that will allow us to construct a simple mathematical model of the problem.
heavy pruning
PHEW!
hot water
tea
teapot
cup
milk

(a) what is needed for the assignments
(b) what is meant by 'a single sentence statement of the problem'
(c) where and how one would get data
(d) what is meant by 'assumptions' and that the model must be consistent with these assumptions
(e) it's *modelling* not a fabulous model that is required.

(iv) Tutor-marked assignments

Each student completes two assignments associated with the project work. The first assignment (coded TMA04) is submitted after study week 2 of the scheme and is worth 20% of the project marks. TMA04 can be thought of as an informal contract with the student and tutor. The student identifies the

Problem 2 specification

This problem is about the athletic event of shot putting.
There are two important rules for the shot putter.

(i) the shot must be 'pushed' from the shoulder, not thrown,
(ii) the shot putter must remain within a circle of diameter 7 feet.

The aim of the athlete is to make the shot travel as far as possible before it hits the ground.
How can a shot putter achieve this maximum distance of the shot?

8

My feature list

1. speed of projection of the shot
2. strength of the putter's arm
3. one-handed throw or two-handed
4. size of the shot putter
5. mass of the shot putter
6. type of motion of the putter (i.e. smooth or "hopping" or "turning")
7. material of the shot
8. angle of projection of the shot
9. air resistance on the shot
10. height above the ground of the shot at release
11. size of the shot
12. spin of the shot
13. weather conditions (wind, humidity, temperature)
14. mass of the shot
15. gravity
16. speed of the shot putter

two problems here?

problem to be tackled and shows the start that has been made. The tutor checks that the student has started work and that the approach is likely to lead to a project of reasonable standard. In particular one value of TMA04 is to identify the student in need of help.

The mark scheme for TMA04, reproduced below, shows what is expected of tutor and student.

your task	**tutor action**	marks
1. A statement of the problem under investigation including an outline of the method of obtaining the data (for testing the model) in less than 200 words. (Box 1 of the seven-box diagram)	(i) Will this problem lead to a project of the required standard? (ii) Is the abstract clear? (iii) Is there a single sentence statement of the question?	5
2. State the variables and simplifying assumptions. (Box 2 of the seven-box diagram)	Are the variables and assumptions reasonable and has everything important been considered?	5
3. Outline the model to be used. The relations between variables in the model should be clear. Diagrams should be used if possible. (Box 2 of the seven-box diagram)	(i) Is the model clearly described? (ii) Does it follow from the assumptions? (iii) Is it likely to help solve the original problem?	5
4. Explain the mathematical formulation and how it follows from the model (in less than 200 words). (Box 3 of the seven-box diagram)	(i) Is there an adequate explanation of how the formulation relates to the problem stated in the abstract? (ii) Does it follow from the model? (iii) Is the mathematics achievable by the student?	5

The second assignment (TMA08) is a report of the student's modelling work and carries 80% of the project marks. Again the mark scheme, reproduced below, shows the format of TMA08 and the allocation of marks.

Marking scheme for TMA 8

main section headings	section contents	marks	
Abstract	(i) statement of the problem to include both the starting point and the actual conclusion reached (ii) significance of problem (iii) sources of data		5
Formulation	(i) assumptions (ii) simplifications (iii) important features		5
Initial model	(i) variables defined (ii) model (following on from assumptions) and its solution (iii) interpretation of solution and criticism of initial model	(6) (12) (12)	30
Data	(i) how collected (ii) relevance of data (iii) presentation of data (e.g. diagrams, graphs, etc.)		10
Revisions to the model	(i) revised models based on criticism (ii) interpretation and criticism of revised models (iii) criticism of final model	(8) (8) (4)	20
Conclusions	brief summary of the main results of the modelling		10

The mark schemes are printed in the project guide to help students carefully structure their project reports so as to avoid the muddled essays so often associated with modelling reports (see III). This structuring helps the less confident student break the task of writing a report into a number of manageable sub-tasks. The mark schemes show that it is *the modelling process that is being assessed not the model.* For the tutor, the careful structuring of the project reports and the detailed mark scheme should overcome possible gross inconsistencies in marking (see VI).

(*v*) *Summer school modelling sessions*

One session at Summer School is set aside for discussion of the students' progress with their course project. For many students this is the one opportunity for a group discussion session on their project work.

5. Conclusions

In 1982, the first year of presentation of MST204, the project scheme ran reasonably smoothly. At first the tutors were apprehensive about the project work but they gained in confidence as the year progressed. There were fewer snags with the modelling project scheme than had occurred in other courses with projects and the student workload was about right.

The project guide was satisfactory with the exception of the case studies. These should be replaced by more 'realistic' examples of MST204 student work. Students in 1982 found the guide a help in structuring their project activities and the audio-tape activity did help in getting started.

In the monitoring of the tutors' marking of and commenting on student assignments, the results showed that there were no gross inconsistencies in the interpretation of the mark schemes. The difference between the monitor's mark and the tutor's mark was within the 'error range' for normal mathematics assignments and better than many essay assignments of other faculties. This confirms the results obtained at the pre-course tutor briefing and reported in Berry and O'Shea (1982).

There are still problems to overcome: (i) the summer school modelling sessions are not very successful or helpful to the students course project work; (ii) data collection can be time-consuming and sources of data are often difficult to find; (iii) the workload on tutors is somewhat higher than for the other elements of the course; (iv) parts of the mark scheme are difficult to interpret and some clarification is needed.

Finally the authors of the project guide feel that two of the components of the project scheme, and audio-tape activity and the assignment mark schemes could be easily assimilated into mathematical modelling project schemes in the conventional teaching institution. We firmly believe in the value to the student mathematician of carrying out mathematical modelling and that individual or group *project* work is the ideal method of giving the student the experience to investigate more open-ended problems for which they have to formulate their *own* models.

Acknowledgement

The project scheme discussed in this paper was devised for the course team jointly by the author and Dr. T. O'Shea.

References

Berry and O'Shea, 1982, The MST204 Project Guide, an Open University correspondence text.

Berry and O'Shea, 1982, Assessing Mathematical Modelling, *I.J.M.E.S.T.*, **13**, No. 6, 715–724.

Broadcast and Assignment calendar, O.U. copyright.

Appendix 1. MST 204 Mathematical Models and Methods

Study week	Start date	Course text	Other components/ notes	TELEVISION: Title	TELEVISION: 0705 Mon[1]	TELEVISION: 1100 Sat[2]	UNIT TITLES	ASSIGNMENT: Number	ASSIGNMENT: Cut-off date
	Feb	1 (Computer work; First visit to study centre)	Cassette P (AC302) Cassette 1 (AC303)	1 Modelling a mortgage: living with ill-conditioning	1305 Sat[2] Feb 6	Feb 13	Recurrence relations		
1	Feb 13	2	Cassette 1 (AC303)	2 Direction fields and families of curves	Feb 15	Feb 20	Differential equations I		
2	Feb 20	3		3 Fishing for figures—the logistic equation	Feb 22	Feb 27	Animal Populations: their growth and exploitation.		
3	Feb 27	4	Cassette 1 (AC303)	4 Newton's equation of motion	Mar 1	Mar 6	Newtonian mechanics in one dimension.	CMA 41	Mar 2
4	Mar 6	5	Cassette 2 (AC304)	5 Complex numbers: The exponential form	Mar 8	Mar 13	Complex numbers	TMA 01	Mar 15
5	Mar 13	6	Cassette 2 (AC304)	6 Good vibrations—second order differential equations	Mar 15	Mar 20	Differential equations II	CMA 42	Mar 16
6	Mar 20	7	Cassette 2 (AC304)	7 The fabulous perfect spring	Mar 22	Mar 27	Simple harmonic oscillations		
7	Mar 27	8	Cassette 2 (AC304)	8 Off the record—Resonance and damping	Mar 29	Apr 3	Damped and forced vibrations	CMA 43	Mar 30

Appendix 1. (*contd.*)

Study week	Start date	Course text	Other components/notes	TELEVISION				UNIT TITLES	ASSIGNMENT	
					Title	0705 Mon[1]	1100 Sat[2]		Number	Cut-off date
8	Apr 3	9	Cassette (3 AC305)	9	Ill-conditioning in linear equations	Apr 5	Apr 17	Simultaneous linear algebraic equations	TMA 02	Apr 7
9	Apr 17	10	Computer work / Second visit to study centre (weeks 9–11)	10	BOCM CASE STUDY 1 Linear programming in action	Apr 19	Apr 24	Linear programming	CMA 44	Apr 20
10	Apr 24	11	Cassette 3 (AC305)	11	BOCM CASE STUDY 2 Forecasting in action	Apr 26	May 1	Forecasting		
11	May 1	12		12	Modelling heat transfer—one dimensional steady state heat transfer	May 3	May 8	Heat transfer	CMA 45	May 4
12	May 8	13	Project—week 1	13	Project—1	May 10	May 15		TMA 03	May 12
13	May 15	14	Cassette 4 (AC306)	15	Dots and crosses—product of vectors	May 17	May 22	Vector algebra	CMA 46	May 18
14	May 22	15	Cassette 4 (AC306)	17	Projectiles—motion in more than one direction	May 24	May 29	Newtonian mechanics in three dimensions.		
15	May 29	16	Project—week 2	16	Project—2	May 31	June 5			
16	June 5	17	Cassette 4 (AC306)	30	(Multi-particle systems and Newton's third law)	June 7	June 12	Many-particle systems and Newton's third law.	CMA 47 TMA 04	June 8 June 9

17	June 12	18	Cassette 4 (AC306)	19	Catenaries—numerical approximation	June 14	June 19	Polynomial approximations		
18		19	Cassette 5 (AC307)	18	Integrating by numbers—numerical solutions of differential equations	June 21	July 3	Numerical methods for differential equations.		
19	June	20	Cassette 5 (AC307)	14	Applying matrices—the algebra of re-orientation	July 5	July 10	Matrix algebra and determinants	TMA 05 CMA 48	June 23 June 29
20		21	Cassette 5 (AC307)	21	Special directions—eigenvalues and eigenvectors	July 12	July 17	Eigenvalues and eigenvectors		
21	July 17	22	Cassette 5 (AC307)	22	Two-way motion—simultaneous differential equations	July 19	July 24	Simultaneous differential equations	CMA 49	July 20
22	July 24	23	Project—week 3	23	Project 3	July 26	July 31			
23	July 31	24	Cassette 6 (AC308)	25	Vibration absorbers—normal modes	Aug 2	Aug 7	Normal modes	CMA 50	Aug 3
24	Aug 7	25	Cassette 6 (AC308)	24	Hunting the humps—partial differentiation	Aug 9	Aug 14	Functions of more than one variable	TMA 06	Aug 11
25	Aug 14	26	Cassette 6 (AC308)	26	Line integrals and curl	Aug 16	Aug 21	Vector calculus		
26	Aug 21	27	Cassette 6 (AC308)	29	Multiple integrals in mechanics	Aug 23	Aug 28	Multiple integrals	CMA 51	Aug 24
27	Aug 28	26	Project—week 4	28	Taking the exam.	Aug 30	Sept 4		TMA 07	Sept 1
28	Sept 4	29	Cassette 7 (AC309)	30	Top Gyroscopes	Sept 6	Sept 11	Angular momentum	CMA 52 TMA 08	Sept 7 Sept 8

Appendix 1. *(contd.)*

				TELEVISION			UNIT TITLES	ASSIGNMENT	
Study week	Start date	Course text	Other components/ notes	Title	0705 Mon[1]	1100 Sat[2]		Number	Cut-off date
29	Sept 11	30	Cassette 7 (AC309)	27 From Hasting's to Halley—the orbit of a comet	Sept 13	Sept 18	Motion under gravity		
30	Sept 18	31	Cassette 7 (AC309)	31 Fourier Analysis	Sept 20	Sept 25	Fourier analysis	CMA 53	Sept 21
31	Sept 25	32	Cassette 7 (AC309)	32 Waves	Sept 27	Oct 2	Partial differential equations	CMA 54	Sept 28
32	Oct 2							CMA 55	Oct 5
33	Oct 9							CMA 56	Oct 8
34	Oct 16								

This course will have five components: (i) tutor-marked assignments 01–03, 05–07 (25 %); (ii) computer-marked assignment component (I) 41–52 ($7\frac{1}{2}$ %), (iii) computer-marked assignment component (II) 53–56 (5 %), (iv) project component 04 and 08 ($12\frac{1}{2}$ %), and (v) the final examination (50 %). Substitution, as described in your Student Handbook or Supplement, will apply for up to two TMA's in the TMA component, up to four CMA's in the CMA component (I), and up to one CMA in CMA component (II). No substitution will be allowed for the project component. There will be a threshold of 20 % on the project component. Students not achieving this threshold will be deemed to have failed the course.

Footnotes 1 BBC 1 2 BBC 2 3 Radio 3 VHF 4 Radio 4 VHF

PART 2. THE PART-TIME TUTOR'S VIEW

1. Introduction

This part is a sequel to Part 1. The Walton Hall View, by John Berry. It sets out to answer the issues raised in that paper, and to evaluate some of the advice on project work given by the University to students and tutors. The views expressed are primarily my own, as a part-time course tutor on MST204 working in the South-East region, but I do attempt to reflect the opinions of students and other course tutors that I have met, inside and outside the region. The sources of information are given as references. The project specifications for 1982 are given in Appendix 1.

The main part of the paper investigates the two principal roles of the tutor. Throughout most of the course he must act as adviser, teacher and guide, and this is his prime concern with TMA 04. This is allocated 20% of the total project mark, the remaining 80% being reserved for the final submission (TMA 08) when the tutor must wear his assessment hat. I concentrate on how I interpreted these roles, the dichotomy of the dual role, the conflicts which arose, and how these differences were sometimes amplified by the distance aspect. However, I would suggest that some of the issues raised are not solely related to distance courses, and that some of the lessons learned are applicable to most courses which incorporate modelling and project work.

2. The Open University study year

2.1 *The student*

The student is under a considerable amount of pressure. Each week, from mid-February to the end of September, a new unit must be tackled, apart from the four weeks (2 in May, 1 in July, 1 in August) reserved for project work. Scattered throughout the year would be some half-dozen optional evening tutorials, and about the same number of Saturday schools, all probably many miles away from home. In addition, one complete week in July or August must be devoted to a summer school. If the student is ill, exceptionally busy in his job, or merely wishes to grant himself (and his family) a holiday, the study time is automatically compressed. It is often believed by students and tutors that a number of the academics based in Walton Hall do not appreciate the unrealistic demands that they may place upon them. It is not surprising that questionnaire analysis suggests that there was a tendency for students to do about half the work in the four allocated project weeks, with desperate bursts before the corresponding submission deadlines. No doubt the same pattern emerges for the submission of the 6 other tutor-marked and 16 computer-marked assignments (see Appendix 1 of Part 1). All students with the University are human. The students were evenly divided between those favouring changing the project week schedule, and those voting for the status quo.

2.2 *The tutor*

The part-time tutor operates on a schedule similar to the student. He has to prepare, and run, a number of the evening tutorials and day schools. (Appendix 2 gives my diary for 1982, which is typical). In addition there are 6 tutor assignments (from some 16 students) to be marked, working to the same deadlines as the students. There is a general University request that all assignments are marked and returned within 7–10 days. Time for reflection and careful consideration of students' work is strictly limited.

3. The role of adviser

As Appendix 2 and the previous section suggests, face-to-face contact with a student may be very infrequent. At best it may amount to half a dozen evening sessions during the year, when the main task is to discuss the material contained in the units, and when there are likely to be several other students seeking attention. Letters in which fruitful project discussions were conducted would be a collector's item. The only practical alternative is the telephone, and this is not conducive to a prolonged and worthwhile explanation of a project. It is not a straightforward matter to gain instant recall of the intricacies of a central heating system, even with the appropriate TMA 04 script as an *aide memoire*. At worst, there may be no effective direct contact at all. This year I have two students living overseas, one in the Canary Islands who has used the radio telephone occasionally, and the other in Sri Lanka.

In many cases the student does not seem to view this lack of contact as a tragic loss. From the questionnaire it appears that only 25 % belonged to self-help groups, and only 35 % had a significant discussion about their project work with anyone else, either at home or at work. From these admittedly small samples, I estimate that half of the students would not have talked about their project work with anyone outside summer school apart from their course tutor. O.U. students have to be able to work on their own in order to survive. Even so, the achievements of some of these students were quite remarkable to those of us who believe that frequent discussion is an essential part of modelling and problem solving. It still seems likely to me that a number of students would benefit enormously and gain considerable encouragement and advice from more frequent contacts with students and tutors.

During the early stages (in May) there was the need to get everyone started. What should I do, as a tutor, to assist and to learn of the difficulties that students were experiencing?

I attended a self-help group of 3 students, each, as it happened, attempting a different project specification. I think I can claim a measure of success in my performance as a fly on the wall, as I was later accused by the students of saying too little. At times I sensed that they hoped for some lead from me, but I interjected only to resolve deadlock, or if there was imminent danger of time-wasting. That evening gave me some ideas and confidence for the following week, when I devoted one complete evening tutorial (of two hours) to a mock project, using the motorway problem as an example, as suggested in the Tutor Notes (Appendix 3).

Generally speaking, I found these notes helpful, although suggestions like 'shall we make a feature list?' come suspiciously close to being ideas in my book. My major criticism is the implication that the tutor works entirely with the complete group. Ten students (of the 17 still allocated to me) attended. This seemed too large a number to allow free discussion, and this was a slightly awkward number if I was to have smaller groups of equal size. I settled for two groups of five, but this proved, as I feared, too large for such an open-ended problem and the limited time available. Feature lists were vast, and opinions so diverse, that pruning became a challenging exercise. I felt that I had to exert too much pressure to achieve some measure of agreement within each group, and take too much of a lead in simplification in order to end with a model and mathematical formulation. (A repeat performance this year with even smaller groups confirms my views.) I suspect that I found the session more exhilarating and rewarding than they did. According to the questionnaire, almost all of the students found it at least of some help as preparation for TMA 04.

The length of the TMA 04 scripts which I received varied from 3 to 9 handwritten pages, with an average of 6 pages (about the maximum recommended length). Not surprisingly, I found great difficulty in dealing with the weakest scripts in a positive manner. Their scripts tended to be either short and lacking in content or, possibly worse, lengthy, vague and woolly (i.e. long and lacking in content). The dilemma one faced was how, restricted to the written word, one could make constructive and helpful comments on their material, without demoralizing a student who was probably struggling. Let us consider Mr B as a typical example. His script contained (his headings):

Abstract		80 words
Assumptions	List of 7	100 words
Model A short list of 'variables' related by one standard heat equation.		
The Mathematical Problem		
	(contained no mathematics)	70 words

I wrote more than 200 words—almost as many as Mr B wrote—of criticism and advice (see Appendix 4). Throughout, I attempted to balance criticism with encouragement. It transpired from the questionnaire that Mr B found my comments 'slightly confusing'. Maybe this is not surprising if one is attempting to achieve so many objectives in relatively few words. I turn to the Tutor Notes for assistance (Appendix 5). Clearly it is not simple to develop 'what is good' if there is very little that is good in the script. Almost all students managed to list a reasonable set of assumptions and features. What the weaker students find difficult is to use these lists to construct a model and to formulate a relation. This is the really significant step, and a few practical examples of how this might be achieved by the tutor would be most welcome. I shall have more to say about 'giving advice to students' in the next section.

As a rule, the scripts of the more able students were much easier to deal with, being more nutritious and easily digestible.

Compliments and encouragement were more readily, and genuinely given. Mr E wished to take account of heat loss from his hot water tank. I hinted at Newton's law of cooling and reaped the reward when I read, in his TMA 08,

> Changing the approach from TMA 4, and following my tutor's prompt of 'how about investigating Newton's Law of Cooling' I proceed as follows . . .

A more satisfying success was gained with the co-operation of Mr S. An extremely able individualist, he had not found the project tutorial helpful or relevant and he had ignored my pleas for a simple model. I ended my comments on the 'rather ambitious model' with

> At least try a few values before committing yourself to what looks like an extensive exercise.

I later wished that I had added those words of advice to at least 80% of the scripts, as it is one of the most important points to stress for any student new to modelling. At the time it seemed such an obvious exercise to undertake that it was not always mentioned—a serious omission from the Project Guide. Emphasis is placed on obtaining a mathematical relationship between variables. The Tutor Notes (Appendix 5) encourage the tutor to seek this in preference to the arithmetic calculation (the back of the envelope job). Let me give just one illustration.

A number of students found it hard to obtain the data they needed—or thought they needed—for modelling their hot water system. For example, to determine a U-value for their insulation, they wished to evaluate the expression

$$\left[\frac{1}{h_1}+\frac{b}{k}+\frac{1}{h_2}\right]$$

where h_1, h_2 are the convective heat transfer coefficients at the inside, and outside surfaces, respectively, and b is the thickness of the material and k is its thermal conductivity.

Difficulties arose in discovering values for h_1, h_2 and k. Two points did not occur to them, apparently. Firstly, although precise values may have been unavailable, it was possible to estimate the order of magnitude. It was not appreciated that such quantities could be treated as parameters, and that some idea of their effect could be gained by taking, say, the least and greatest imaginable values. A related issue was the failure to subject such parameters to a preliminary examination, which might have suggested a simplification to the model. It may well turn out that, for any likely value of k, b/k is very small compared to $1/h_1$ or $1/h_2$ for a first model, and so it may be neglected completely.

I am glad to report that Mr S reacted favourably to my criticisms and went on to produce the best project in the group. These relative successes may well have been assisted by the fact that I had met Mr E and Mr S 3 or 4 times prior to submission of TMA 04. I felt that I knew them sufficiently well to appreciate how I should phrase my comments. The advice which one gives, and the manner in which it is given does depend, to some extent, on the personality of the recipient. This is even more important when dealing with modelling, where there may well be no absolutely correct answer. It may also be significant that I had never met Mr B. It is usually hard to gauge the response of people with whom you are not acquainted.

Although I knew Mr K quite well, I believe I damaged his chances slightly by giving too much praise. It was obvious that he had spent a long time on his script, and he told me that he had become very involved with the problem (the economics associated with an O.U. degree). I regarded his TMA 04 as the best I received.

> This has every sign of being an excellent project . . .
> Your model follows on well . . .
> Modelling the promotion prospects allows various possibilities . . .
> There is plenty of scope for extending/improving your model . . . (2 possibilities given) . . . .

The final version of his project was a disappointment. He had not developed his ideas as I anticipated, nor produced so much more than he gave in TMA 04. In the terms of the project guide and mark scheme, he had done more than enough on TMA 04. Somehow there should have been the means to carry some of this effort over to TMA 08.

I was called upon to give project advice very rarely after I had commented on TMA 04. The pressure of the unrelenting sequence of weekly units, holidays and a summer school would probably account for this. I received a few telephone calls in July and August, and this experience appears to have been shared by colleagues. Questionnaire results indicate that such contact was almost always of some use, but I usually felt that I had done little except provide a degree of encouragement (Regional office did allow me three additional hours 'flexitime', with a suggestion that these might be used on the telephone).

At the back of my mind, throughout this advisory period, was concern at the reactions of the students to my comments, whether they would be misunderstood and so send them off in the wrong directions. I was conscious of the need to locate all the errors and then proceed to give sound advice. My written comments on TMA 04 may be the only direct contact on the project work, there for all the world, and the monitors at Walton Hall in particular, to see. Sometimes I felt more remote and at a distance than the students—they had each other. However, two-thirds of the students found the TMA 04 a 'very useful' preparation for TMA 08, and only one found it useless. It seemed to be much more highly rated, in this regard, than corresponding work at summer school.

4. The role of assessor

Marking to an alien mark scheme is not a popular pastime for many academics. Schemes which we have not devised fail to award marks for essential features, yet, in our expert opinion, marks are frittered away on relative trivia. Such views often come into sharper focus on open-ended problems, which are subject to shades of opinion. How does one assess the modelling process?

As an Open University assessor, my interpretation of any mark scheme is open to inspection. Students are likely to compare notes, particularly if they are working on the same project specification. What may appear to be significant differences to the tutor may be quite insignificant in the eyes of the student. The tutor is also aware that Big Brother, in the guise of members of the course team, is watching the new venture with considerable interest and justifiable concern. My efforts can, and will be directly compared with my distant colleagues.

I shall examine my role as assessor firstly for TMA 04, and secondly for TMA 08. In the latter case I include issues which only came to light after final submission, although they may well be more closely related to shortcomings in my advisory role. There were useful lessons to be learned here, which might also be applied to more conventional undergraduate courses. To indicate these I shall use some of the horrors perpetrated by students which might have been prevented if I had given better advice, or seen the student a few weeks prior to submission.

4.1 *Assessment of TMA* 04

Of the 16 scripts I received, 13 were devoted to the hot water specification. Discounting the tedium which may creep into the reactions of the most dedicated professional when faced with repetitious tasks, it may seem to have been advantageous to have such a large percentage devoted to a single problem. However, it is inevitable that one makes comparisons between the various scripts. One sets up a filtering device which may lead to an expectation of an unjustifiably high standard. This is not necessarily to the advantage of the student who rightly anticipates a fair mark.

The Project Guide states:

> This TMA should be short and clear: its main function is to enable your tutor to advise you . . . it will give your tutor some idea of how far you have got and what kind of help you need. You will then be eligible for a maximum of ten marks instead of twenty, but you will receive some notes from your tutor to help you to continue.

I suspect that there is an underlying assumption by the student that the magnitude of the mark awarded reflects the quality, not only of the TMA 04, but of the final project. In many cases this is perfectly reasonable. However, it is the tutor's responsibility to avoid discouraging the student. Even a modest effort should achieve 50%, so that the average mark I awarded was in the region of 75% (the range was actually 40–93), which was close to the national average. Unfortunately, in September there would be the distinct possibility that one's chickens would come home to roost, in the form of a final version (TMA 08) based and revised on my earlier comments, with the thought, in the student's mind, that a good TMA 04 mark implies a comparative TMA 08 mark. The questionnaire asked students how they would have felt if the assessment for TMA 04 had been formative, i.e. commented on by the tutor, but not marked and graded as part of continuous assessment. Students were evenly split over this question, while 60% thought that they would have submitted it for comment even if it was not assessed. Setting out to mark the TMA 04 scripts, I refreshed my memory on the 'Guidelines for marking TMA 04' of the Tutor Notes (Appendix 6). Let us proceed through the assessment scheme (Appendix 7) selecting only those parts which may be considered contentious.

Task 1 (*Statement of the problem*)
(i) Deciding whether the student's effort is 'likely to lead to a project of the required standard' is often difficult. To make a sound judgement merely by reading the abstract is virtually impossible. The question is misplaced.

(iii) This is the first occasion when one may make unfavourable comparisons between different scripts. If a student uses two or three concise sentences in preference to a single, lengthy, complicated effort, should he be penalized? The more ambitious soul may find the constriction painful.

(iv) This question does not go far enough. What is the data required for? Is it to provide values for parameters, or is it to be used to validate the model? I believe the answers to these questions give one a fairly reliable guide to the quality of the final project.

Task 2 (*Variables and assumptions*)
It is a brave or foolish tutor who can confidently assert that everything important has been considered. Having waded through a large number of

assignments, so that I had collected the 'comprehensive' features/assumptions list, it was tempting to refuse to give full marks for what, in practice, were truly praiseworthy efforts. In this respect, those projects marked in the latter batch may have been unfairly penalized, thereby balancing the advantage they may have received from criticisms and advice based on increased tutor experience. To combat this, there are a few techniques the assessor can try, e.g.

(1) marking all projects relating to a particular specification as a single batch;
(2) ranking a batch of projects in order of merit (this gives no absolute values).

Although arrival of scripts was spread over a month (May 24–June 26), the vast majority arrived on two consecutive days (June 8, 9) so such procedures were practicable, to some extent. However, in view of the severe bias towards the hot water problem, the restriction of the number of projects (about 6) which can be squeezed into a single envelope (provided by O.U.) and the natural desire to maintain a steady turnover, the techniques were of limited value.

Tasks 3 *and* 4 (*Model outline and Mathematical Formulation*)
Many students (and tutors) found it an unfortunate and, perhaps impossible constraint to relate the variables without describing the mathematical formulation. In consequence these two areas were often combined in a lumped parameter system in the mind of this assessor. Once this had been accepted, assessment was reasonably straightforward, although I was never happy with awarding marks for correct formulation from unlikely assumptions. Most students answering the questionnaire sportingly accepted that we got their marks 'about right'.

4.2 *Assessment of TMA* 08

The mark scheme for TMA 08 is reproduced as Appendix 8. Advice concentrated, at great length, on convincing tutors that a large discrepancy in raw marks would not be too crucial, and would yield only a small difference in 'final course mark'. General advice was confined to a few short paragraphs, which finished:

> The important thing is to *be consistent* on your marking of each student in your group . . . . Try and be *generous* in your interpretation of the mark schemes . . . .

That is what I set out to do, and I shall comment on three aspects of this process. Firstly I highlight what I believe to be a striking omission from the scheme, secondly I refer to some scripts which made me unsure of the advice (TMA 04) I had been given earlier in the year, and finally I offer a few examples of typical howlers which I confess should never have been committed.

No direct mention is made in the mark scheme of initiative or a number of other characteristics which should be demonstrated in modelling exercises, such as ingenuity, determination, etc. Some of the activities in the area of data collection were meritorious. A few marks are available under 'data' heading— (Appendix 8) but this is not generous. It was fascinating to see what some students did in order to gather information.

For Mr E the procedure was

> to note the gas meter reading at night prior to running a bath, and read it again first thing in the morning, timing both readings. Water temperature was noted using a bath thermometer, sometimes held under the running hot tap, sometimes in contact with the copper cylinder under the insulating jacket . . . . Air temperatures were measured with a greenhouse max/min thermometer, positioned on the floor about 2 m from the airing cupboard . . . . Measurements were made with the insulating jacket in position, with the jacket removed, and then with the timer switching off overnight.

Ms V also spent many pleasurable hours in her bathroom.

> In a preliminary run the following times were observed by listening to hear when the (immersion) heater switched on/off. This involved spending a lengthy period in the bathroom (but proved a useful time to read the Sunday paper!) . . . . As far as possible no other electrical appliances were used during the experimental period. When the desire for a cup of coffee became too great an attempt was made to assess the effect on the reading and an adjustment was made. It is recognised that the results are bound to be very approximate.

Mr N nobly switched off all electrical equipment, including the refrigerator, during the working day, in the middle of a relatively hot summer. Mr C cut up pounds of tomatoes, with able assistance from his wife, so that he could count up the number of seeds per fruit. Such endeavour is typical of the dedication of many O.U. students, to whom I pay tribute, and I think their Churchillian spirit worthy of more than a miserly 2 or 3%.

On the other hand it would be wrong for me to omit this piece of heresy from Mr L

> I suggest that future projects could avoid the requirements for practical work. I know of students who spent hours and hours measuring water temperatures rather than concentrating on producing sensible models! I could also confess that my experimental observations were pure fabrication (!) since I did not consider real measurements to be of any significant value in the modelling process.

This supports my argument—full marks for initiative! Finally, spare a thought for Mr I

> I landed up in the wrong project. I did the hot water project, which was readily solvable by experiment, when I should have been on the seed project, which was only solvable, if at all, by the modelling process.

Of a more embarrassing nature, from the tutor's standpoint, is the TMA 08 which may have suffered from unsatisfactory advice given on TMA 04. I have already mentioned the problem of Mr K in section 3, who may have produced too much at an early stage. He puts his own case rather well:

> I think (I may be wrong) that I tended to be penalised for doing a good TMA 04 (at least I thought it was good) and therefore I appeared to lose too many marks for not changing the basic concepts. While I bow to your higher knowledge, I still disagree with your comments.

He scored 93 % on TMA 04, 61 % on TMA 08, giving 67 % (weighted) overall for his project work.

Even worse is the realization that the advice given on the TMA 04 was either ill-conceived or has been misinterpreted. At that time I criticized the assumptions made by Mr W who was also investigating the economics of an O.U. degree. I considered his model unrealistic, and there was no proper mathematical formulation. He appeared to be frightened away from improving and extending his initial attempts, so that he produced a brief and scrappy essay on a new and, unfortunately, equally unacceptable set of assumptions. I had to search diligently in order to bring him up to the pass threshold. Of his project I wrote:

> It doesn't seem to bear any relationship with TMA 04 which was much nearer to requirements. At least that had some recognisable form. Perhaps I was too critical and put you off.

I have already referred (the hot water system problem in section 3) to the reluctance of students to try out a few numbers at an early stage, to assess the importance of certain parameters. In practice, many students showed a marked reluctance to vary any parameters at all, and sometimes picked the wrong one. Insulating his hot water cylinder, Mr R found that each 12.5 mm thickness cost £4.78½ (note the exactitude). A standard 80 mm jacket cost £8.99. In order to produce an impressive table of values, he steadfastly considered the steady increase in cost for each 12.5 mm thickness. Thus 75 mm insulation cost £28.71 (£4.78½ × 6). The purchase of a standard 80 mm jacket yielded a sharp decrease in cost. Perhaps one should not cavil at this discontinuity when one observes its effect, for Mr R calculated that heating his uninsulated tank, with no time switch, would cost £5,119 over a period of 6 months, whereas the insulated tank cost a mere £720. Moreover, the introduction of a time switch was crucial, since this reduced the heating costs further to £130.134 (precisely).

Could there be another use for the quick sum? If the results are surprising (Mr R did not reveal his true feelings), analysis of a simple (alternative) model might provide a suitable test for these results. Ms G included the insulating properties of copper in her model. She concluded that these were not good, but fortunately discovered that her insulated tank took 6 days to cool from 40°C to 30°C 'and this figure (of 6 days) is too small as I approximated the exponential (law of cooling) to a straight line'.

My gold medal award for misplaced logic, and compensating errors, came from Mr P. He calculated the mass of the water in his tank to be 1.17×10^5 kg. Fortunately he was able to cope with this huge mass by obtaining gas at a special discount rate of 2.72 pence per therm. Even so, it cost him £7.58 per day to run his colossal system, with a time switch installed. He confessed

> The model shown is obviously an over-simplified one and is unlikely to show the true costs of the various systems. It is open to large errors . . . .

Wreathed in smug smiles the tutor is faced with a dilemma when attempting to assess those disclaimers. Are these questionable figures due to unrealistic assumptions, faulty modelling, arithmetic ineptitude or just bad luck? Perhaps this was a smart move on the part of the student, playing the project game. An unlikely result offers wide scope for criticism, and a fairly easy path for a second trip round the modelling circuit (Project Guide, page 6). Almost one-third of TMA 08 marks are available for 'interpretation of solution and criticism of initial model' and 'revisions to the model'. Mind you, these have to be *earned*.

5. Conclusions

Assessing projects in accordance with an imported mark scheme is never a straightforward job. The difficulties are amplified by the awareness that this is a national scheme with details accessible to all, so that students and Walton Hall can make comparisons between tutors. Nevertheless, I believe that this course makes us think a little more deeply about what we do and say. The escape route of the 'impression mark' has been sealed off. We must re-examine our approach to the teaching of modelling, the setting and assessing of projects, and how one guides and communicates advice and criticism. In particular this course has demonstrated that it is possible to construct a national mark scheme for assessing projects although I am convinced some changes should be made. These are suggested in Appendix 9.

Pressures also arise in the offering of advice to students. Distance 'tuition' implies use of the written (or spoken) word so that the student and the tutor miss the visual signals which are so helpful in appreciating how criticism is being given and received. In this context, as in most others, coping with the less able students is more demanding, although, possibly, more rewarding. Such students may be easily discouraged or misdirected, and probably lack the conviction to come back to the tutor when they are stuck. The tutor walks a tightrope, balancing between the roles of adviser and assessor, knowing how much the rewards from the latter role are a function of his success in the former. I am attempting to resolve this dilemma by maintaining closer contact this year.

The average mark for TMA 04 was 70–75 %, with 43 % gaining a mark above 80 %. For project work these may seem to be very high marks. Many students expect to achieve at least 90 % for their other coursework, and the less able would be disappointed with less than 50 %.

When the threads of communication between students and their tutor are weak, it is vital that they are not demoralized. It is understandable that students with a high TMA 04 mark will anticipate a similar award for their TMA 08. This will often prove to be the case, but occasionally the student will not fulfil this early potential, or may not interpret correctly the well-meaning advice given by the tutor, and be disappointed by his final award. The University should warn students (and tutors) of this possibility in the Project Guide and the Tutor Notes.

From an organizational standpoint, there may be problems in waiting to receive a sufficient number of scripts corresponding to a particular specification whereby the tutor can make comparisons and collect his ideas, without appearing to delay their return. There are risks in this collective approach, since it is desirable that the tutor should remain open-minded and receptive to new ideas. As time passes he tends to become more critical, and therefore mark down, which may be unfair to those students whose projects are marked later. On the other hand these same students may benefit from the increased breadth and depth of the tutor's experience gained from marking earlier scripts.

Students are likely to enrol with the O.U. because they feel that they can cope on their own, and, in many cases, they have no choice. In consequence they may not make great efforts to discuss their projects with anyone. However, it is still my conviction that modelling is well suited to group activity, provided that the groups are small—not more than 4—and the problem is of interest and relevance to all members. The Project Guide should place greater emphasis on the desirability and advantages for students to discuss their projects with each other.

Despite their experience and maturity, O.U. students perpetrate errors of the same disastrous scale and frequency as full time H.E. students of more tender years. They have no clearer idea of how to get a feel for a practical situation, nor how to test for the really important features (parameters) of a problem, for example by feeding in some extreme values into a very simple model. The Project Guide should lay much greater stress on the need to test simple models at an early stage by using a few trial values for the variables. This helps to identify those parameters which are likely to have a significant effect. It does not seem to occur to students that they should compare their predicted results with reality. The dichotomy between mathematics and the real world is still very marked for some students. Nevertheless, numerous students showed considerable initiative with their project work. The assessment scheme for TMA 08 is insufficiently flexible to take this into account.

I am sure that most students did gain from their modelling experiences, as the questionnaire analysis endorses—one-third being 'much more confident' about tackling a similar exercise again. Generally speaking, the projects were of an impressively high standard. Of the 15 I finally received, I would rate 3 as excellent, 6 as really good and only 3 as poor. Bearing in mind the novelty of the situation, all concerned should feel more than satisfied with the outcome, and I face 1983 with eager anticipation and more confidence of the scheme than in 1982.

REFERENCES: SOURCES OF INFORMATION

1. MST 204 PG Project Guide (Walton Hall).
2. Tutor Notes (Walton Hall).
3. Report of 1982 Debriefing for MST 204.
4. MST 204 (1982) Student Feedback—Project Week Reports (Survey Research Dept)
5. Questionnaire issued on an informal basis to students allocated to myself and two colleagues. 31 returned.
6. Scripts from my own group of students.
7. Conversations with numerous students and colleagues.

Appendix 1

Specification 1

Investigate the comparative economics of
(i) buying and installing a time-switch for your domestic hot water,
(ii) buying additional insulation for the hot-water tank,
(iii) just leaving the hot water on all the time.
For the purpose of this project: (a) assume that you do not already have a time-switch, (b) otherwise use your own system; (c) solid fuel systems with 'energy to waste' do *not* qualify.

Specification 2

The winter Olympics never seem very far away, with the hour-by-hour television coverage of Downhills, Slaloms and the like. It was while watching one of these presentations—and finding that my dictionary couldn't help to distinguish between a luge and a toboggan—that the question first occurred: would a four-man bobsleigh come down a measured run faster than a two-man bobsleigh? After a little thought, further questions began to occur: What effect does friction have? Should riders lie or sit? What's the best strategy to adopt when pushing?

Investigate the motion of a bobsleigh.

Specification 3 Discuss whether taking an OU degree is a financially worthwhile activity in the long term. Does the award of the degree enhance your salary and prospects sufficiently to make up for the total cost of taking the degree?

How does the economics of an OU degree compare with that of a conventional university degree taken (i) as a mature student, (ii) when you are 18?

Specification 4 When a new variety of plant has been developed, there is initially a shortage of seed. Once the variety has been established as a big seller, a small proportion of the seed produced is retained for producing more seed. In the early stages of development, however, there is a choice between either retaining a larger proportion and getting a small return on sales or retaining a smaller proportion and inhibiting future production. In this project you should set a long-term goal, and then determine a retention policy which will enable you to attain this goal
(i) in minimum time *and/or*
(ii) with maximum profit *and/or*
(iii) according to some other criteria.

Appendix 2. Tutorial contacts and diary of events

Jan 30	Tutor Briefing Session (London)
Mar 2	Evening Tutorial (Brighton)
Mar 10	Due date submission of TMA 01
Apr 7	Due date submission of TMA 02
Apr 20	Evening Tutorial (Crawley)
May 5	*Project Week 1 starts*
May 12	Due date submission of TMA 03
May 18	*Evening Self-help Group*

May 24	*First TMA 04 received*
May 25	*Evening Tutorial (Projects) (Brighton)*
May 26	*Project Week 2 starts*
June 8	Evening Tutorial (Crawley)
June 9	*Due date submission of TMA 04*
June 23	Due date submission of TMA 05
June 26	*Last TMA 04 received*
	Saturday Day School (Croydon)
July 14	*Project Week 3 starts*
Aug 11	Due date submission of TMA 06
Aug 18	*Project Week 4 starts*
Late Aug	*Last-minute advice to students—a few telephone calls*
Early Sept	
Sept 1	Due date submission of TMA 07
Sept 4	Saturday Day School (Canterbury)
Sept 7	Evening Tutorial (Brighton)
Sept 8	Due date submission of TMA 08
Oct 19	Examination

N.B. Dates and events in italics relate to project work.

Appendix 3. Taking a project related tutorial class

Start off by talking through the project scheme and what it's about. Take the 'motorway problem' described on page 19 of the Project Guide. The golden rule is that the tutor should not provide ideas himself; all ideas should come from the modellers. The tutor can act as 'scribe' and see that students are following a sensible process by suggestions like 'shall we make a feature list?'

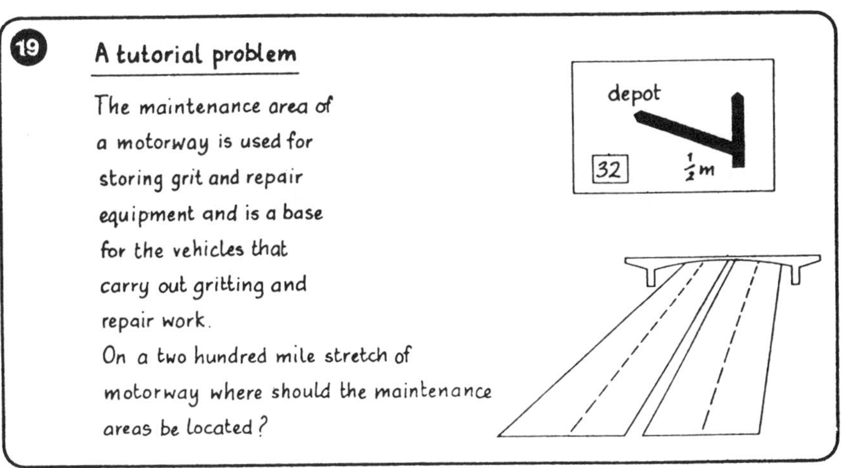

19
A tutorial problem
The maintenance area of a motorway is used for storing grit and repair equipment and is a base for the vehicles that carry out gritting and repair work.
On a two hundred mile stretch of motorway where should the maintenance areas be located?
depot
32
½m

These differ from the M101 summer school investigations in important respects:

1. This is modelling, not puzzle solving
2. No introspection
3. The tutor is not to give 'the answer' at the end, but to drag some answer out of the rest of the group.

The purpose of these group modelling sessions is to show students how to use feature lists as an aid in constructing models.

With the motorway problem you can illustrate to the students the amount of work that is needed to get to the TMA 04 stage. Take the feature list, assumptions and initial model formulated *by the students in your tutorial group* and write an outline TMA 04 based on them. Point out what *you* will be looking for in marking a real TMA 04. Stress the following:

(i) a good statement of the problem (not a copy of the project specification)
(ii) where would you get data to test the model? (look on a map or drive up the M1)
(iii) does the model follow on from the assumptions?
(iv) give an explanation of the maths needed.
(v) it's modelling not a fabulous model that you want.

Let your group of students do the work. Do not try to get to a solution. This is a long, '40 hour' type project, not one for completion in one hour (or so) tutorial.

Appendix 4. Comments on TMA 04 of Mr B

What you have written is clear and concise and you have the outline of a good project, but I don't think you have yet thought through the process sufficiently.

The abstract is O.K., perhaps a brief outline of your model could have been included. Main omission (from TMA) is no information about obtaining data.

List of assumptions is reasonable, but does not include all those you actually make—some significant, e.g. drawing off hot water, ignored, perfect mixing within tank. Others are identified later (e.g. about tank). Also I suspect (iv) and (v) are relatively unimportant at initial stage.

Your model is sound, based on assumptions. Diagram not a bad idea? Do you intend to ignore the *ends* of your tank (not a pipe)? There are no details on modelling the time switch situation—will you neglect cooling effects while heater is off?

I am afraid your mathematics formulation is vague, and really adds very little. The times on/off should have been considered earlier. What is wanted here is a development of the relationship in 'The Model'. How will you actually determine the *costs* and make *comparisons*? Eventually, perhaps how will you present your results? What parameters will you vary, e.g. times on/off, as you suggest, thickness of insulation, etc.?

This is a promising start—keep in touch.

(mark 48 %)

Appendix 5. Tutor notes—giving advice to students

Advice can be given to students either on the PT3 form or on the script (or both) in the usual way. However, the advice given *must not* be a particular model, it is particularly important that students be encouraged to formulate *their own* model. It is the *modelling process* that is being assessed not the model.

TMA 04 is supposed to help tutors identify the struggling students. Consider the TMA 04's of A. R. Smith and M. Bermingham which were sent to all tutors before the briefing meeting, could it have been predicted from these that the TMA 08's would be so poor? The main danger sign seems to be that both students have confined themselves to arithmetical calculations. Such students should be encouraged to use more mathematics: to describe their models in terms of variables and to find relations between them, expressible as equations. Other danger signs: disorganization, no clear statement of problem being tackled, a superfluity of irrelevant data.

Advice should develop what is good in what the student has done so far; tutors should never suggest their own models to students. For example if a student submits only a feature list (mark 20 %) the tutor might suggest to the student some simplifying assumptions for pruning it, or some variables from it to concentrate on, or some help in formulating a relation connecting these variables. The Tutor could encourage the student to telephone him when he has made a little progress. Advice should not be specific to the given problem but should be general modelling advice.

It is part of good project work for students to inform themselves about the real situation being modelled. For example in the bobsleigh project, they should read up a bit about what actually happens in a bobsleigh.

Students should *not* write to any national body asking for help with their project. They should use only data available to the general public, e.g. in their local library; or which they collect themselves.

It is important that the mathematical methods that the student will need is M101 level (or equivalent) or within the first 12 units of MST 204. It is an unwritten rule of the teaching of mathematical modelling that a student should use only mathematical techniques with which he/she is competent and new techniques learnt during the project period weeks (weeks 13–28) are not likely to fit into this category.

A revised model does not necessarily imply harder mathematical methods. It would be including a feature which was assumed not to be important in the initial model.

Students have been asked in a stop press to send two copies of TMA 04 to their tutors. One is for your comments and will be returned via Walton Hall in the usual way. The other copy is for the tutor to refer to in any telephone contact and for the TMA 08 marking.

Appendix 6. Tutor notes. Guide lines for marking TMA 04

The mark scheme for TMA 04 is reproduced as Appendix 2. Marking projects is not easy. There is bound to be *some* variation in the marking of a particular TMA 04 (or TMA 08) by a group of tutors. We expected that this variation would be likely to be greater for project type work than for conventional mathematics TMA's. The marks for the TMA 04 of A. R. Smith given by the tutors at the tutor briefing showed that there was *less* variation than we anticipated, although this particular TMA 04 was *very* difficult to mark.

80 % of the marks given by 28 tutors were within the range 10–15 out of 20, and almost one half (13 of them to be exact) gave a mark of 12, 13, or 14. This variation compares favourably with the marks of the 'more usual' mathematics TMA's.

The following table gives a breakdown of how to award the five marks in each section.

0/5 no attempt, or complete rubbish
1/5 a genuine attempt, of 'threshold' standard
2/5 bare pass
5/5 good work satisfying all the criteria in the mark scheme.

Remember a perfect model is not required, even at TMA 08 stage. Just mark the modelling *process*, whether or not the resulting model is one you think highly of. If the model is bad, this will show up when it is compared with real data. But the student's model must be consistent with his stated assumptions.

Encourage your students to write *their own* statement of the problem not just a rewrite of the problem specification; and award higher marks for those students who do. The 'single sentence statement of the problem' referred to in the mark scheme, is a short precise statement of what the student would say to his 'audience' in describing the problem that is to be solved.

A good test to use a rough guide to the marking of TMA 04 is as follows:

4/20 Feature list only
8/20 Feature list and simplifying assumptions with an attempt at a model
8–14/20 A reasonable attempt at the four headings of TMA 04
15–20/20 A good TMA 04. Likely to lead to a good project.

Remember (i) a difference of 4 marks out of 20 between two tutors' marking of a TMA 04 will only lead to a difference of $\frac{1}{2}$ mark out of the $12\frac{1}{2}$ project marks.

Appendix 7. Marking scheme—TMA MST204 04

You should use the marking scheme given on page 24 of the 'Project Guide' which is reproduced below for your convenience.

your task	tutor action	marks
1. A statement of the problem under investigation including an outline of the method of obtaining the data (for testing the model) in less than 200 words.	(i) Will this problem lead to a project of the required standard? (ii) Is the abstract clear? (iii) Is there a single sentence statement of the question? (iv) Is there a suitable method of collecting data outlined?	5
2. State the variables and simplifying assumptions.	Are the variables and assumptions reasonable and has everything important been considered?	5
3. Outline the model to be used. The relations between variables in the model should be clear. Diagrams should be used if possible.	(i) Is the model clearly described? (ii) Does it follow from the assumptions? (iii) Is it likely to help solve the original problem?	5
4. Explain the mathematical formulation and how it follows from the model (in less than 200 words).	(i) Is there an adequate explanation of how the formulation relates to the problem stated in the abstract? (ii) Does it follow from the model? (iii) Is the mathematics achievable by the student?	5
	total marks	20

You will probably find it useful to look at the sample marked scripts included in the 'Project Guide'.

Appendix 8. Marking scheme—TMA MST204 08 (1982)

You should use the marking scheme given on page 27 of the 'Project Guide' which is reproduced below for your convenience.

main section headings	section contents	marks
Abstract	(i) statement of the problem to include both the starting point and the actual conclusion reached (ii) significance of problem (iii) sources of data	5
Formulation	(i) assumptions (ii) simplifications (iii) important features	5
Initial model	(i) variables defined (ii) model (followed on from assumptions) and its solution (iii) interpretation of solution and criticism of initial model	(6) (12) (12) 30
Data	(i) how collected (ii) relevance of data (iii) presentation of data (e.g. diagrams, graphs etc.)	10
Revisions to the model	(i) revised models based on criticism (ii) interpretation and criticism of revised models (iii) criticism of final model	(8) (8) (4) 20
Conclusions	brief summary of the main results of the modelling	5
Presentation	clarity and layout	5
	total marks	80

You will probably find it useful to look at the sample marked scripts included in the 'Project Guide'.

Appendix 9 Suggested changes to mark scheme TMA 04

your task	tutor action	marks
1. A statement of the problem under investigation in less than 100 words.	(i) are there concise statements of the problems to be solved (in their own words?)	3
2. Unchanged from original (App. 7)	Unchanged from original	4
3. Description of data required, its purpose, and how it is to be collected.	(i) is the data appropriate to the model? (ii) is it satisfactory for its purpose (for parameter testing or validation)? (iii) is method of collection suitable?	5
4. Original 3 & 4 combined (App. 7)	Omit 4 (iii).	8

Additional questions for tutor
(i) Is the mathematics achievable by the student?
(ii) Will the problem lead to a project of the required standard?

Mathematical Modelling—A Major Component in an MSc Course in Mathematical Education

K. H. Oke
Polytechnic of The South Bank

SUMMARY

The MSc course was first offered at the Polytechnic of the South Bank in 1977, and the initial results in developing and running this course was first reported by Oke (1980). The report covered the rationale and objectives of the mathematical modelling component, one of four, that is presented on the course. With the additional experience gained in running such a course, new insights have developed which have led to considerable modifications being made to the structure of the mathematical modelling component. Such modifications on teaching strategy and assessment methods are detailed in this chapter. Some difficulties encountered in finding appropriate problems for modelling exercises, both for the teacher and the students, are identified with illustrative examples on how such difficulties may be overcome.

Some teachers attending the course have opted to carry out an investigation into aspects of teaching mathematical modelling to students at a level with which they are familiar; to date, three investigations have been carried out at secondary level, two at FE level, and one (by a Polytechnic lecturer) at undergraduate level. The results of these investigations are briefly reported on, together with the work of the author—the latter based on work with undergraduates, MSc students (teachers) and various workshops. Emphasis is placed on formulation processes and how they relate to general problem solving strategies. In this context suggestions for further work are made and, in particular, how such endeavours are closely linked with the development of Intelligent Knowledge Based Systems (IKBS) in fifth generation computer studies.

1. BACKGROUND

Mathematical modelling is a major component of the two-year part-time MSc course in Mathematical Education. The MSc was first offered in the Polytechnic in 1977, and the initial results on the development and teaching of mathematical modelling at this level were first reported in a paper by Oke (1980). The paper outlines the rational and objectives of the course, and for convenience, the main objectives are listed again here:

(i) For the teacher to be able to formulate and solve models corresponding to certain classes of physical and organizational situations.
(ii) To enable the teacher to evaluate critically the validity and utility of such models.
(iii) To enable the teacher to develop teaching material relating to simple model building for classroom use.

Although these objectives relate to graduate teachers of mathematics in secondary schools and colleges of FE, by replacing 'teacher' by 'student', the first two objectives could well suit any modelling course.

A number of different teaching styles, classroom activities, and assessment methods have been experimented with in the past six years of development. In spite of these experiments, some of which are detailed in later sections of this paper, the basic curriculum structure has remained the same (see Table 1). Chief amongst the reasons for the choice of curriculum in Year 1 is to

Table 1. Structure of Curriculum

Year 1	Term 1	Term 2	Term 3
	Social and organisational sciences (SOS)	Physical sciences and technology (PST)	Life Sciences (LS)
		Development of simple modelling activities in these areas.	
Year 2	**Term 1**	**Term 2**	**Term 3**
		Teacher-led seminars on general methodology and the teaching of mathematical modelling + lectures on topics of special interest (e.g. simple catastrophe theory).	

demonstrate the breadth of applications of mathematics. This has posed problems, particularly in the physical sciences, in view of the technical command of fundamental principles that is usually required. This criticism also applies, although to a lesser extent, in the other areas. Consequently careful choice of problems to be modelled, both for teachers and their students/pupils, has had to be made. Careful consideration is also given to the mathematical expertise that is needed at the solution stage of each activity; the content and problem solving ability of first year undergraduates (and often much less!) is expected when teachers tackle modelling problems—for secondary school pupils the problem is more acute. Current articles on modelling and attempts at teaching modelling are presented in teacher-led seminars in Year 2.

At South Bank, the modelling course is presented by a team of three staff—one for each of the areas SOS, PST, and LS. The author of this paper concentrates on PST, but the modelling team work closely together on general teaching approaches including the joint running of a combined workshop (for Years 1 and 2) that is held once towards the end of each academic year.

The extent to which the main objectives of the course have been met in the continuing development of the course is now investigated.

2. TEACHING METHODS

In this section, attention is concentrated on Year 1 of the course where modelling is introduced. It should perhaps be pointed out straight away that the experiences gained in developing this course over the past six years have much in common with the experiences of others who have taught modelling at a variety of different levels—for example, see Burghes and Huntley (1982), Oke and Bajpai (1982), James and Wilson (1982), Burkhardt (1979). Inevitably, there is a different emphasis placed on the various teaching and learning styles employed; some would argue that group work, where no hints are offered, is the best way of introducing modelling to students—others (including the present author) have recommended interactive teaching where the whole class and teacher/lecturer work together. At South Bank, we are now using a combination of methods (see Table 2).

It is still felt that interactive teaching is an effective method of introducing modelling to the inexperienced. Teachers have tried it out in schools, see for example Sheridan (1980) and O'Hare (1980), and report favourably on pupil/student responses. Similarly, interactive teaching has been experimented with in FE and with undergraduates in HE. Like all teaching styles however, the care, skill, and sensitive handling by the lecturer/teacher are paramount. Irrespective of level of student, experiences are indicating that the following basic approach is valuable.

Table 2. Teaching and Learning Styles

Year	Style	Brief description
1	Interactive	Lecturer working *with* class, models from scratch from problem statement.
1	Group work	Students (teachers) split into groups of three/four and work from problem statement (1 hour duration).
1 and 2	Workshop	Several problem statements presented, students (teachers) choose problem and work in groups ($2\frac{1}{2}$–3 hours duration). More experienced (Year 2) help less experienced (Year 1) teachers.
2	Seminar	Teacher-led discussion on published paper. General modelling strategies analysed as well as pedagogic implications.

Year 1, Term 2		1 hour per week classroom sessions interspersed with individual 'homeworks'
SESSION 1		Introduction to modelling—any differences between SOS, PST, LS? Differences between professional modeller and classroom work (30 mins). Interactive teaching: U-tube accelerometer—see Oke and Bajpai (1982) for details (30 mins).
	Homework:	Individual attempts to extend model.
SESSION 2		Discussion on individual attempts and presentation of lecturer's approach. How could material be used in secondary school/FE?
SESSION 3		Interactive teaching: Sound-distortion in record player—see Oke (1981) for details (15 mins). Group of 3/4 working (30 mins). Brief discussion of group ideas (15 mins).
	Homework:	Individual attempts to develop simple model.
SESSION 4		Discussion on individual attempts (level of maths required approx GCE O Level). How could material be adapted for use in secondary school/FE? (40 mins). Interactive teaching or extensions of model (20 mins).
	Homework:	Individual attempts at extending model.

SESSION 5 — Discussion on individual extensions.
Lecturer's attempts (30 mins). Discussion on finding *relevant* material for modelling—see The Spode Group (1982), for example, for secondary schools/FE.

Homework: Try to find six problem areas appropriate for a modelling approach in secondary school/FE. Illustrate a possible development for one of them for interactive teaching.

SESSIONS 6–10 — More modelling problems developed on interactive/group lines as in sessions 1–5.

SESSION 11 — Major modelling workshop ($2\frac{1}{2}$–3 hours) with Years 1 and 2 combined. Year 1 teachers work with Year 2 teachers in each group. Each group chooses one from three problem areas to model. Lecturers act as consultants on background information—*no hints* given on modelling approach.

SESSION 12 — Discussion of workshop group efforts—approaches compared with lecturers', brainstorming tactics, individual follow-up at solution stages.

The above is a fairly typically styled one-term programme for the introduction of mathematical modelling to students (teachers) on the MSc as well as undergraduates. However, there is a difference in emphasis—teachers are additionally asked to consider the appropriateness of modelling material for their students. Of course, for teachers to be able to teach modelling, they must first of all themselves learn to model. Teacher feedback from the MSc inevitably changes as their modelling experiences progress. In the summer vacation following Year 1, they have each to prepare a course-work (which is assessed) based on a problem area they have identified. The course-work must state aims and objectives, and indicate clearly how the material might be developed for students at a level with which they are familiar. Teachers must also design self-assessment or follow-up questions for their students.

Teachers and staff are all agreed that group-work, workshops, and individual endeavour are the most instructive methods of learning to model. However, teachers also appreciate the value of interactive teaching and discussion periods, especially at the early stages of a modelling course. As pointed out in the last section, some teachers have tried out, with favourable results, the interactive teaching technique in their own schools and colleges of FE as work for their MSc dissertation.

In choosing modelling material (problem areas) suitable for teachers, two prime considerations that are taken into account are:

(a) The need to 'stretch' the teachers in building up their modelling expertise to a level considerably above that expected of their students.
(b) The background interests and mathematical maturity of teachers and their students.

The reasons for (a) are analogous to those for teachers having a degree in mathematics as part of their preparation for teaching at GCE O Level and A Level. The importance of paying careful attention to the general background of students and the level of mathematics that is expected from them in modelling has been discussed elsewhere—see Oke (1980), Burghes and Huntley (1982), Burkhardt (1979).

It is important not to attempt to introduce a new topic in mathematics by means of a modelling exercise since:

(i) It 'forces' the modelling in a pre-set direction in order to encompass the new mathematical topic.
(ii) It places too great a burden on the student. It has now been generally experienced that it is far better to keep the mathematics down to a minimum—in an introductory course in modelling, the level of mathematics assumed should be several years below that of current achievement.

The question of relevance of modelling material appears to be more contentious. For example, O'Hare (1980), tried out the geometrical design of a pick-up arm to minimize sound distortion in a record player with an average ability class (potential CSE grade 3). This modelling exercise is carried out by MSc teachers, and was also tried out with industrial graduate mathematicians in a recent workshop, see Oke (1981). All found the problem interesting, but the distinguishing feature was that attempts at modelling reflected background knowledge as well as mathematical problem-solving ability. The CSE group drew tangents from an external point to a circle (recording groove), and the industrialists attempted some Fourier analysis on a distorted sine wave.

3. FORMULATION STUDIES

In Year 2 of the MSc course, attention is drawn to some general issues on modelling and how they relate to the teaching of modelling. Teachers are provided with a small selection of research papers and other articles and are required to lead seminar groups in discussing each paper. One of the themes that we are attempting to develop in this way is a study of the relationship between mathematical modelling and problem solving. This is a difficult task, although all teachers agree that it has important implications for the classroom. It is a difficult task for two chief reasons:

(a) Problem solving has concentrated, in mathematical education, on the strategies used in solving well-posed, close-structured mathematical problems. Mathematical modelling concentrates on 'real' problems which are consequently more open and often not well posed.
(b) Both problem solving and mathematical modelling involve concept formation, particularly in the initial or formulation stages. There appears to be little known about concept formation of a *universal* kind which can be put to practical use.

These difficulties are not confined to mathematical modelling and problem solving in mathematical education of course, but are also experienced in the design of any large and complex system. For example, one of the chief considerations in the development of fifth generation computers is the need to develop research into the internal capabilities of an Intelligent Knowledge Based System (IKBS)—see the Alvey Report (1982). In this connection, the report identifies the following areas that urgently need basic research:

(i) Classification, concept formation.
(ii) Summarizing, abstracting.
(iii) Selection, retrieval filtering.
(iv) Reasoning, use of heuristics.
(v) Planning, modelling.
(vi) Learning, memorizing.

The above list has been discussed in a number of teacher-led seminar groups, and under the guidance of the author of this paper, teachers have started to develop a classification of concepts, heading (i), appropriate to the formulation stages of modelling. This work, necessarily, is still in its infancy and ideas to date are now briefly reported.

4. CLASSIFICATION OF IDEAS

The way in which a model initially develops depends to a large extent on the problem being investigated and on how well-posed the problem statement is. Consequently, any attempt to identify ideas and activities which have a general or universal value must be made by studying a wide range of problems. We are currently experimenting with the following crude classification of ideas for the pre-formulation and formulation stages of modelling. (see Fig. 1).

The classification arose from investigations with the MSc teachers, covering several individual formulations for each of four widely differing problems. A common factor was sought for, and it was decided that the information content or 'specificity' level of each concept, question posed, assumption made, etc.,

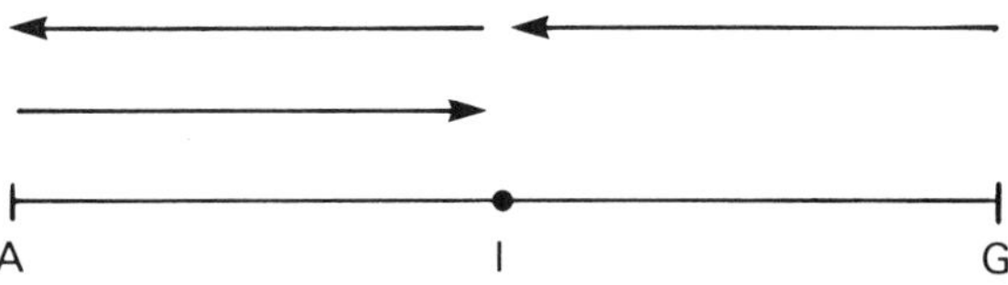

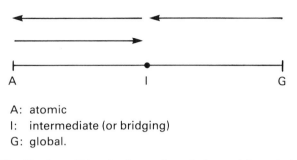

A: atomic
I: intermediate (or bridging)
G: global.

Fig. 1. Classification of ideas in the pre-formulation and formulation stages

would provide useful insights. The most broadly encompassing ideas (e.g. what are the environmental considerations in the design of a large-scale windmill?) are labelled *global* (G); more specific, or sub-problem suggestions (e.g., power required from a windmill assuming fixed-pitch blades) are denoted *intermediate* (I); the most specific variables, relationships between them (e.g. wind-speed required for power calculations on a windmill) are labelled *atomic* (A). Early, or pre-formulation, stages tend to be dominated by ideas and suggestions of G type—especially in brain-storming workshop conditions, although as mentioned earlier this very much depends on how simple and well-structured the original problem is. Some modellers prefer, having identified a sub-problem at the outset which interests them, to develop a model as rapidly as possible, and consequently ideas concentrate largely in the I and especially A categories.

As formulation progresses, re-thinking often takes place leading back to I (and occasionally to G). The arrows in Fig. 1 are merely suggestive of this moving forwards and backwards in information level. It is often quite difficult to decide whether or not a particular idea is of the A or I type (although it is usually easily agreed on whether or not it is G). Consequently, the diagram in Fig. 1 is drawn as a straight line linking A, I and G suggesting a spectrum of ideas (if not a continuous spread).

Attempts, such as this one, need further investigation to see if they lead to any practical value. It is fair to say, however, that by being involved with such work, the teachers in their second year of the MSc increase their awareness of the activities involved in modelling and adopt more flexible approaches to problem-solving in general.

5. ASSESSMENT

The following forms of assessment are used on this course:

(i) Self-assessment questions.

(ii) Formally assessed course-work assignment.
(iii) Written examination paper.

Additionally, if teachers so choose, a substantial project on an aspect of teaching modelling is undertaken and presented as a dissertation (the presentation of a dissertation is an essential part of the MSc course, but teachers may choose another area, e.g. History of Mathematics, in which to work).

Although details are provided in Oke (1980), some changes have taken place over the past three years. Firstly, only one course-work assignment is set towards the end of Year 1; this is partly the result of policy to reduce the overall number of assessments in all subjects. This assignment consists of each teacher finding their own problem in the SOS, LS, or PST areas, and developing a mathematical model relating to this problem. Teachers are expected to define the learning aims appropriate to a level of student with which they are familiar, and to provide self-assessment questions for their students—these questions may test understanding of the developed model as well as test ability to extend or model a similar situation. This course-work assignment is of the same type as detailed for Year 1 in Oke (1980), where attempts at awarding marks for each activity are outlined. We are tending now to give an overall mark (A, B, etc.) for an assignment, without giving a detailed breakdown of how that mark (letter) was arrived at. This is because teachers have considerable choice in how they present their work, and because of the completely free choice they have in the problem (which they find) to model. Ingenuity in modelling (and problem choice), coupled with realistic teaching notes gain most credit.

The second change in our approach to assessment concerns the written examination paper which is taken at the end of Year 2. The paper, which is of three hours duration, is split into two sections: A and B. As reported in Oke (1980), where examples of questions are also given, Section A consists of three (was four) questions—one in each of the SOS, PST and LS areas. Teachers are required to formulate as far as possible, in one hour, a model or models of the problem posed. Section B consists of five or six questions (was four) based on general modelling and pedagogic issues; teachers are asked to refer to Year 2 seminars and published papers presented at these seminars as far as possible in their written answers. The main change that has taken place is that Section A questions are handed out one week in advance of the examination. The reason for this change is to provide teachers with an opportunity to study any background material they think may be helpful in modelling the problem (out of three) of their choice. By this method, teachers also have more time in which to think out their possible formulation approaches.

It is, of course, arguable as to whether or not mathematical modelling should be examined by a (semi-) conventional examination paper. Whilst the general issues of Section B can be so examined, and are perhaps best examined this way since examination preparation 'helps to focus the mind', assessment of creative abilities of modelling itself are probably best left to course-works and projects. We are still reviewing the situation!

6. CONCLUSIONS

Both staff and students (teachers) have benefited considerably from the mathematical modelling component of this MSc course. On the staff team side, we are continuing to review the preparation of modelling; we are still experimenting with teaching techniques—interactive teaching, guided and completely 'open' workshop groups, and with learning styles. Assessment methods are also being carefully monitored. Staff research and development interests, often backed-up with consultancy activities in industry and commerce, are providing valuable insights and ideas for tackling new problems. Teachers are experimenting with teaching modelling in their own classrooms and, coupled with a general problem-solving approach, are finding beneficial changes in their perspective and understanding of the applications of mathematics. Teacher 'feedback' is positively encouraged, both in the 'lecture room' and on course-boards; this feedback is making a significant contribution, not only to the general running of the course, but also to helping staff to identify appropriate problems and their presentation to students in secondary schools and colleges of FE.

REFERENCES

A Programme for Advanced Information Technology, The Report of the Alvey Committee, 1982.

Burghes D. N., Huntley I., 1982, *Int. J. Math. Educ. Sci. Technol.*, **13**, 6, 735–754.

Burkhardt H., 1979, *Bull. IMA*, **15**, 238–243.

James D. J. G., Wilson M. A., 1982, *Int. J. Math. Educ. Sci. Technol.*, **13**, 6, 789–796.

O'Hare J., 1980, Mathematical Modelling for the Average Ability Pupil Aged 14–16, MSc Dissertation, South Bank Polytechnic.

Oke K. H., 1980, *Int. J. Math. Educ. Sci. Technol.*, **11**, 3, 361–369.

Oke K. H., 1981, Minimisation of Sound Distortion in a record player, in *Case Studies in Mathematical Modelling*, ed. R. Bradley, R. D. Gibson and M. Cross, Pentech Press, pp. 31–55.

Oke K. H., Bajpai A. C., 1982 *Int. J. Math. Educ. Sci. Technol.*, **13**, 6, 797–814.

Sheridan M. F., 1980, Mathematical Modelling: A Group Activity for Sixth Form Applied Mathematicians, MSc Dissertation, South Bank Polytechnic.

The Spode Group, 1982, Solving Real Problems with Mathematics, Vols 1 and 2, CIT Press.

Issues involved in the Design of a Modelling Course

A. O. Moscardini, D. A. S. Curran, R. Saunders, B. A. Lewis and D. E. Prior

Sunderland Polytechnic

1. INTRODUCTION

Rapid developments in technology over the last few years have made high power computing facilities more readily accessible. As a result, an increasing number of organizations are beginning to use mathematical modelling with computer simulations as a cheap and reliable way of solving very complex problems. In the past such problems could often only be tackled by very costly and often unreliable small-scale experiments. The inevitable decrease in the cost of computing equipment and software will add further impetus to the use of mathematical modelling by industry and commerce.

Further evidence for the growing importance of mathematical modelling is provided by the number of highly attended conferences and the increasing number of journals which have recently been devoted to the topic. Locally the North East Polytechnic modelling group POLYMODEL has attracted a great deal of industrial support and there has been a considerable demand for the proceedings of the five annual conferences it has organized. (see Cross *et al.*, 1978; O'Carroll *et al.*, 1979; Bradley *et al.*, 1980; Caldwell and Moscardini, 1981; O'Carroll and Hudson, 1982).

The increasing use of mathematical modelling by industry and commerce has led to a movement to change school, university and polytechnic curriculae to include more modelling. Many higher education establishments now provide courses at undergraduate level often as a part of a mathematics degree programme. For example, at Sunderland, modules are included in the BSc Combined Science degree, BSc. Environmental Studies and the HND in Nautical Studies. These courses, however, only provide an introduction to mathematical modelling and are not designed to produce the specialist mathematical modellers required by industry and commerce. Indeed, a course aimed at providing such specialists would make such demands or a student's mathematical sophistication and critical judgement that it is doubtful if it could be successfully offered at undergraduate level. Unfortunately provision of postgraduate courses has not matched that at first degree level. Oxford University does run an MSc in applied mathematics and modelling but the course is heavily biased towards the former. Other universities such as Brunel provide training in modelling but only as part of an MSc course in a related area.

For these reasons Sunderland Polytechnic decided in November 1979 to introduce a full time MSc in mathematical modelling. Because of political pressures, both internal and external to the Polytechnic the course has taken 3 years to develop and has emerged as a two-year part-time MSc. In this paper we review the problems that occurred in the design of the course in the hope that this will be of benefit to other institutes that are planning to introduce a similar course.

2. ORIGINAL CONCEPTION OF THE COURSE

The original course team consisted in the main of applied mathematicians, some of whom had practical modelling experience. This was mainly limited to physical systems modelling, for example the modelling of blast furnaces (Cross *et al.* 1980) and problems arising in the area of electromagnetic fields (Caldwell 1981). Other members of the team had a more academic experience of modelling acquired in teaching modelling to undergraduates and attendance at conferences on modelling.

There are many definitions of mathematical modelling, almost as many as there are practitioners of the art. Although the team had not formally added to this collection in the initial design stages of the course, the concensus seemed to be that modelling consisted of three stages (see Fig. 1):

(a) the translation of a real problem into a set of mathematical equations,
(b) the solution of the equations,
(c) the interpretation of the solution in terms of the real problem.

Although with hindsight it is easy to criticize this view as naive and idealistic, it is hardly surprising in the light of the background and experience of the team.

The content of any course is bound to reflect the interests of the staff involved in its design. It was envisaged that the modelling activity in the course would be confined mainlv to the type of physical systems modelling with which the team members were familiar. This implied that the students would need a solid background in the major 'classical models' found in hydrodynamics, electromagnetism, elasticity and heat transfer. The team quickly realized that few potential recruits would possess such knowledge. This led to

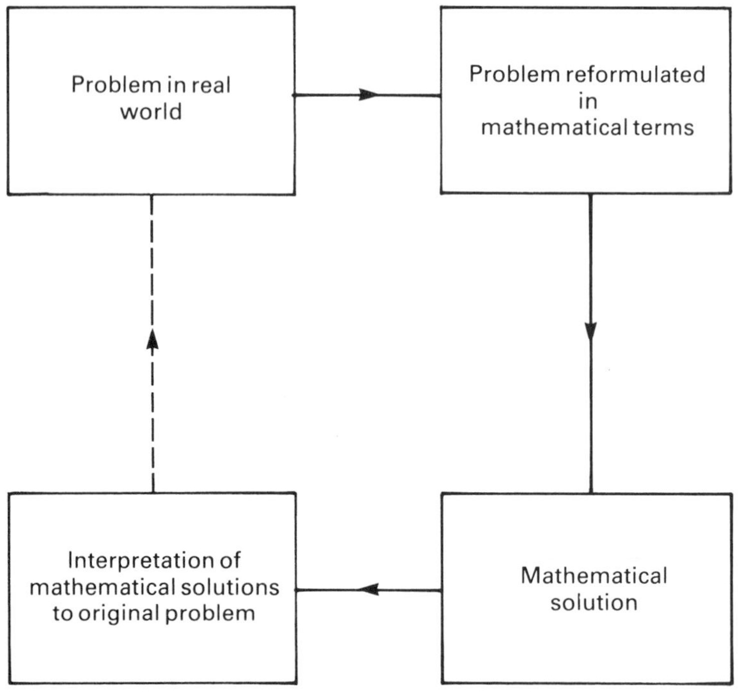

Fig. 1.

the idea of a 'physical systems' course in which the students would be subjected to a crash course in these topics. Some team members doubted the practicability of covering so much ground in the limited time available (60 hours). Others argued that all students would be familiar with some of the subject matter and that at postgraduate level more rapid progress could be expected than in a first-degree course.

In addition, the course would have to include mathematical techniques for the solution of the model equations. Thus modules in the analytic and numerical solution of ordinary and partial differential equations, linear algebra and optimization and control were included. Since students would be expected to develop some of their own computer packages a substantial module in software development was included.

Subsequent discussion brought out several defects of the course. The modelling component was peripheral occupying less than one-fifth of the total contact time. The course was too 'applied maths orientated'. The physical systems requirement would lead to severe restrictions in student entry limiting the intake to engineers, physicists and mathematicians with a strong traditional applied mathematical background. Software development was covered in too much detail and was poorly integrated into the course.

3. DESIGN OF THE PRESENT COURSE

Because of policy decisions outside the control of the design team, the full-time course was resurrected as a part-time course and took about eighteen months preparation for submission to C.N.A.A. During this period, it is possible to identify five major areas of contention where decisions had to be made. These are now discussed in detail.

3.1 Definition of modelling

Although this seems an obvious starting point it is in fact difficult to establish a common definition that will satisfy a team of mathematicians. Many definitions of modelling exist (see Collection of papers, 1977, 1979, 1982) the literature abounds with them and the choice of definition will affect the flavour of the course. Two definitions were discussed (and discarded) during the design period before the present definition was accepted. The progression of these definitions mirrored the development of our modelling ideas and thus the content of the MSc.

The first definition, as shown in Fig. 1, was naive and simple, as was mentioned in section 2. It reflected the views of a team that saw the mathematical techniques as the main thrust of the degree with the modelling a peg on which to hang them. As it became apparent that there was a little more to modelling than this, our definition progressed to something of the type shown in Fig. 2. This definition suggests two important aspects of modelling.

(a) that a methodology of modelling can be developed,
(b) it is a highly iterative technique.

The disadvantages of this definition are that the labels on the boxes are not very instructive and would need a lot of explanation. It also presents too mechanistic a view of modelling but this definition did allow the range of models to be extended to subjects such as Biology, Medicine, Chemistry and Economics. Two major aspects of modelling that are not fully dealt with by this definition are how to establish and how to validate the model. The integration of these aspects into definition 2 lead to our final definition of modelling shown in Fig. 3.

The importance and value of conferences should not be underestimated as it was a conference on validation given by the Institute of Control Theory that provided the impetus for the methodology shown in Fig. 3. The modelling process is now divided into five stages, each of which is represented by a box and the various substages in each box are linked by arrows forming a circle. These circles indicate that there is in general no specific starting point in each stage and that these substages may be investigated several times. One proceeds to the next stage when sufficient understanding of the previous stage has been achieved. The general advancement of understanding is from left to right but

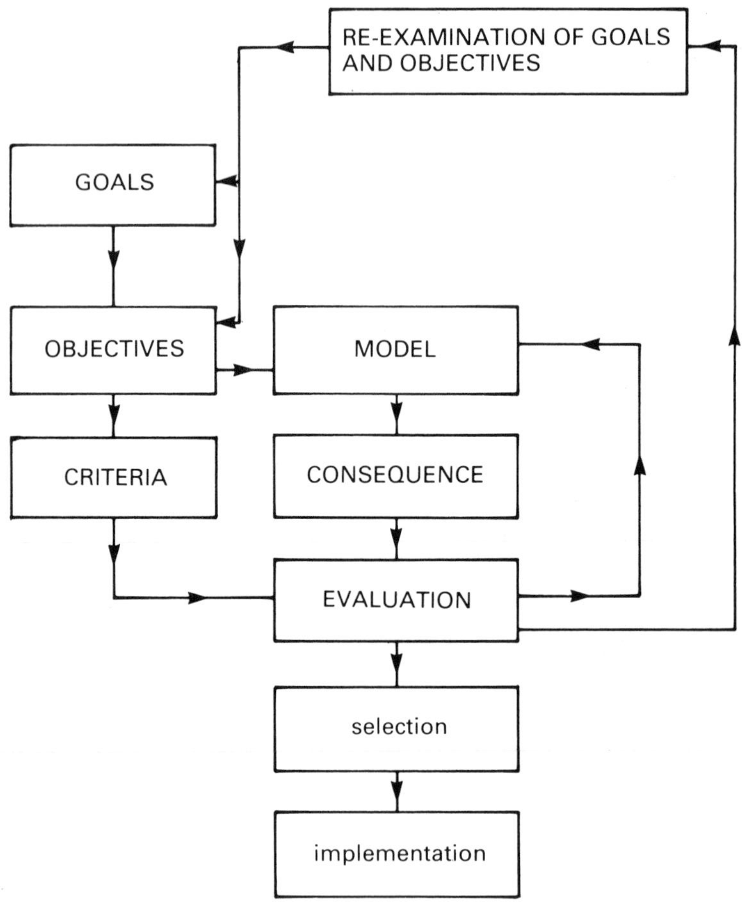

Fig. 2.

the double-headed arrows signify that the modeller may return to a previous stage to re-interpret, redefine or change inputs. When the 'solution of the model' stage is satisfactory, then a validation procedure is followed. The modeller may then implement his solution or return to a previous stage.

The difference between definitions 1 and 3 and their effect on the MSc was tremendous. In definition 3, the mathematical techniques play a still important but smaller role in the process and the course reflects this.

3.2 Entrance qualifications

The mathematical content of a modelling degree will depend to some extent on the assumed background of the students. If highly restrictive entrance

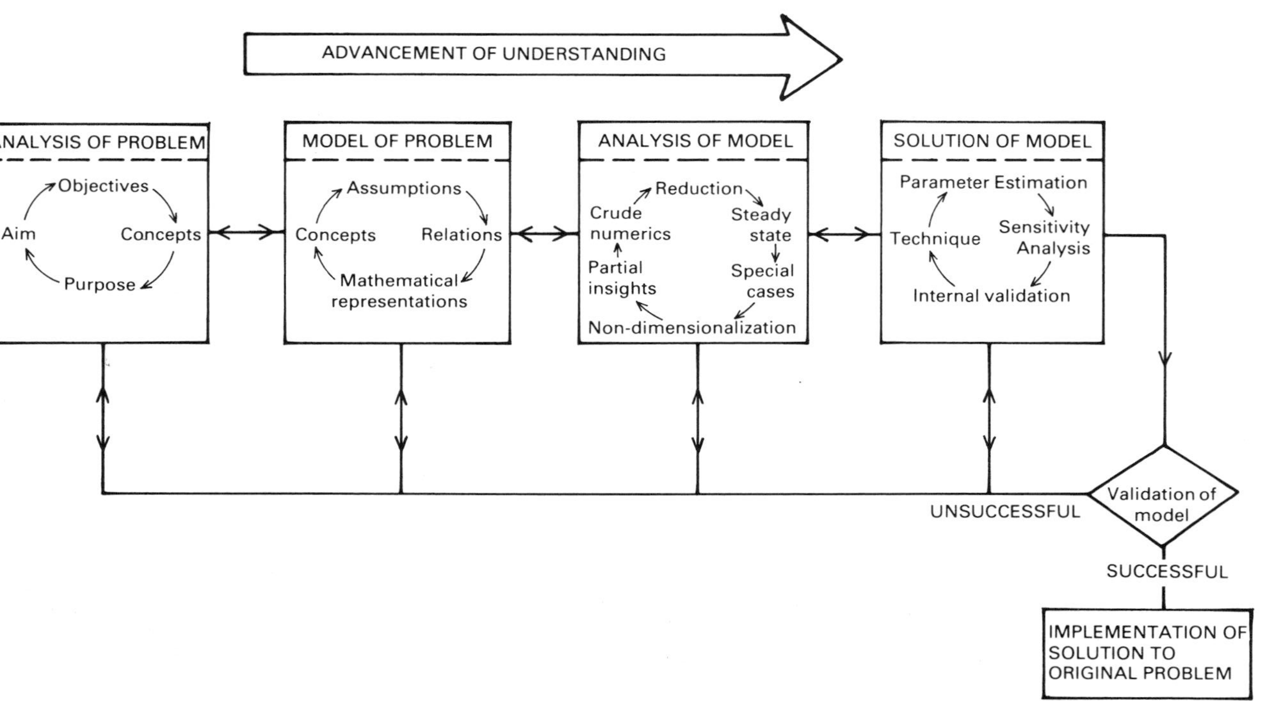

Fig. 3. The modelling process

qualifications are set, then more time is available for construction of sophisticated models but the enrolment may be poor, which in these times would be fatal. If the entrance qualifications are relaxed then complicated systems of balancing studies have to be arranged and the range of models that can be tackled is restricted. In undergraduate modelling courses, modelling situations can be presented together with extensive background material. Also time and help can be given to stages 1 and 2 in the modelling process and thus the range of models need not be restricted. It could be that at the MSc level, models would be expected to be more sophisticated and this approach would not work. But it must be remembered that at this level, students are more mature and that individual reading and research are expected from them. Therefore, there is no reason why models cannot be taken from a variety of subjects not just the specialization of the individual students. (Fresh minds can produce some very interesting models.)

One exception is the case of Physics, which promoted much discussion. Many models involve physical concepts such as gravity, masss, acceleration, viscosity, radiation and the basic laws of thermodynamics, electromagnetic field theory and mechanics. Nowadays, even at MSc level, one cannot assume this background and there is too much here for a student to read up. Even a course devoted entirely to physical systems could not cope unless it was of tremendous length. This is certainly a problem. Our solution was to delete the physical systems course and to provide a choice of models for the students to tackle. Some of these models would involve physical concepts but only students with a suitable background would be allowed to attempt these models. Models discussed in lectures would not involve advanced physical concepts that would need a lot of explanation.

3.3 Types of models and techniques to be studied

There are various types of models—continuous, stochastic, statistical, commercial, economic and decision. It would be difficult to cover all types and our course tended to concentrate on continuous systems. A course of ordinary differential equations which contains phase planes, numerical methods, stiff systems and perturbation methods is given in year one. Many models have parameters that are difficult to estimate. Statistical and optimization techniques are provided as an aid to parameter estimation and a course in finite difference and finite element methods for solving partial differential equations is also provided. These courses should provide enough mathematical techniques to model most continuous systems. The lack of full stochastic and decision models is regrettable but the existence of this type of modelling is discussed with their role and limitations.

As well as for reasons of lack of time, this type of modelling was partially omitted because expertise and interests of the staff were towards continuous modelling. This is obviously an important factor in determining what can be taught, but the other extreme must be avoided. The boundary element technique is a very modern and useful technique for solving partial differential equations and there is considerable expertise in this field in the Sunderland Mathematics Department. This topic was inserted into the MSc several times during the eighteen months for the above valid reasons but was finally rejected because the number of models to which it could be applied was too limited to justify the number of hours needed to adequately cover the topic.

3.4 Role of computing

The title of the MSc is 'Mathematical Modelling and Computer Simulation' and so the computer plays a major role in the MSc. Fortunately, Sunderland Polytechnic has extensive computing resources and there are hardly any hardware restrictions on anything we plan. Computer Simulation will either use a software package or a simulation language, and both aspects are covered.

It takes time to develop sufficient numerical techniques and the associated technical computer programming to be able to generate solutions to model equations. Also undue concentration on this will detract from other equally, important, features of the modelling process. A solution to this problem is to use simulation packages. Providing the simulation package is easy to use and sufficiently 'user-friendly', students are essentially relieved of the burdens associated with acquiring complex computing skills. Thus students can concentrate on the skills associated with analysing systems, formulating the model, generating results and assessing their validity. Both discrete and continuous simulation packages exist and typical of these are two packages developed at Sunderland Polytechnic called IPSODE (Moscardini *et al.* 1980 and Thorp *et al.* 1982 Interactive Programming with Systems of Ordinary Differential Equations) and APHIDS (Cross *et al.* 1983 A Program to Help in Discrete Simulation). Commercial packages exist but there are often problems in modifying them to meet the needs of a particular course. It is much better for the course team to develop their own software and extensive literature exists in this field. A module called Software Package Development explores these aspects in the MSc.

Many languages are specifically written for simulation purposes one of which is DYNAMO (Richardson and Pugh 1981). This language is associated with a modelling technique called System Dynamics developed by Jay Forrester at MIT and is widely used in economic and business modelling, especially those involving feedback. The advantages of this method are that a student can build complicated models from simple concepts and the inter-relationships of the various parts can be clearly shown on the system dynamics diagram. The first two stages of the modelling process are difficult to teach but this approach provides an excellent framework for this type of analysis.

DYNAMO itself does not require a long time to master (which is usually a major disadvantage of simulation languages) and can be mounted on most machines. A course on System Dynamics in both years completes the computing aspect of the MSc.

3.5 Teaching and assessment

Teaching mathematical modelling to students or new graduates in industry is difficult. No definitive pedagogy exists but much research is being conducted in this area, some of it by members of this department. For the purposes of this degree, two principal stages have been identified:

(a) the representation of open-ended problems by mathematical relationships and the interpretation of mathematical solutions in the context of the original problem,
(b) the application of techniques to analyse and solve equations.

These stages require different methods of teaching and assessments. It is intended to cover the first stage by workshop sessions. Here the lecturer will play the role of the problem provider and consultant. His function will be to provide sufficient background material for the group and either to answer students' queries directly or point them towards the appropriate sources. The onus in these sessions lies on the student to apply methodology that has been taught in the course to problems from a variety of fields. Students will obviously favour problems that are related to their own area of knowledge, but by a correct choice of problem and the provision of adequate background material or resources, this method will allow students to tackle problems in areas unrelated to their previous disciplines. The emphasis in these sessions is not primarily in obtaining a solution to the equations but in the movement from the problem to mathematics and back again.

The second stage will be covered by more 'normal' lecturing sessions. It is envisaged that an initial workship will produce the need for the knowledge of certain techniques. These techniques will then be provided by lectures and further workshops can combine both stages to produce completed models.

The two different forms of teaching require different methods of assessment. The knowledge and abilities developed in the workshop sessions could not be adequately tested by formal examinations as this would be imposing artificial constraints of time and availability of information that do not exist in reality. This was therefore, thought best examined by continuous assessment. The techniques will be examined by formal examinations but even here the flavour of the examination will be more towards modelling than applied mathematics.

The continuous assessment will consist of one mini-project per term. This will be an open-ended problem that can be modelled and solved using the techniques that have already been taught in the course. The students will be expected to produce complete results in a fully documented form. Marks will be awarded for creativity, presentation and use of techniques taught in the course.

A major project will be the culmination of the course. This will be a substantial piece of work which unifies the principles and methods of the taught components and the modelling methodology. The mini-projects are seen as ideal preparation for the final project.

This paper presents the final decisions taken for each of the five areas discussed above. It must be emphasized that these pathways were not smooth and seven different courses were devised and rejected during the eighteen-month period.

4. RECOMMENDATIONS TO FUTURE DESIGNERS

With the benefit and wisdom of hindsight, it is now clear how the design of such a course could have been expedited. The hints parallel, as might be expected, the major areas of contention.

The definition of the modelling process must be hammered out and agreed by all the staff. It is essential for all the staff to be committed to the same methodology as differing views of methodology are the cause of most other arguments.

An average profile of the typical student to be attracted to the course should be constructed. This profile will be all-embracing and so it should be agreed as to which deficiencies would be compensated for by balancing studies and which knowledge is to be deemed essential.

The strength and weaknesses of present staff should be assessed. This could be a potentially explosive situation and should be handled with extreme tact but an honest assessment must be made especially for higher degrees. If expertise is lacking in certain areas then this could be remedied by future appointments. If there are no such appointments on the horizon then this will affect the content of the course. There is nothing more disheartening than to construct what is in effect a perfectly sound course and then be told that it can't run because of staff resources.

A typical model should be worked out in great detail and presented to the teaching team. This should be chosen so that it encompasses all the aspects of the methodology. An excellent way of doing this would be to run a staff seminar program where the staff can act as student guinea pigs. Although unpopular, this would unify the ideas of the course team.

None of these ideas is highly sophisticated or original but it is the honest opinion of the course team at Sunderland Polytechnic that had these ideas been implemented, and their conclusion accepted then, a lot of time would have been saved in the design stages.

5. CONCLUSION

This paper describes the construction and final version of an MSc at Sunderland Polytechnic entitled 'Mathematical Modelling and Computer Simulation'. Various 'bottlenecks' in the design process have been identified and the discussions and solutions are stated. Using the benefit of our experiences, advice on how to avoid some of the obvious pitfalls is given. It is hoped that this paper will be of value to anyone involved in the design of a modelling course and the authors would welcome any correspondence from fellow designers.

REFERENCES

Bradley, R., Gibson, R. D., Cross, M. (Eds.), 1980, Case Studies in Mathematical Modelling (Pentech: Plymouth).

Caldwell, J., 1981, Forces on a coil due to an oblate spheroid of magnetic material, *I.E.E.E. Trans. Mayn.*, **17**, 1.

Caldwell, J., Moscardini, A. O. (Eds.), 1981, Numerical Modelling in Diffusion Convection (Pentech: Plymouth).

Collection of papers, 1977, *Math. Gaz.*, **61**, 81; Collection of papers on mathematical modelling.

Collection of papers, 1979, *Bull, I. M. A.*, **15**, 10; Collection of papers presented at the Symposium on 'Mathematical Modelling: its relevance to the teaching of Mathematics in Higher Education', held in January, 1979 at Hatfield Polytechnic.

Collection of papers, 1982, *Int. J. Math. Educ. Sci, Technol.*, **13**, 6; collection of papers on aspects of mathematical modelling.

Cross, M., Gibson, R. D., O'Carroll, M. J. and Wilkinson, T. S. (Eds.), 1978, Modelling and Simulation in Practice (Pentech: Plymouth).

Cross, M., Gibson, R. D., Traice, F. B., 1980, Analysing blast furnace performance, Case Studies in Mathematical Modelling, Eds. R. Bradley, R. D. Gibson and M. Cross (Pentech: Plymouth).

Cross, M., Moscardini, A. O. and Thorp, M., 1983, Interactive computer simulation tools for use in teaching mathematical modelling, *Int. J. Math. Educ. Sci. Techol.*, **13**, 6.

Moscardini, A. O., Thorp, M., Cross, M., 1980, Interactive computer simulation without programming, *Adv. Eng. Software*, **2**, 13.

O'Carroll, M. J., Hudson, P. (Eds.), 1982, Process Modelling (Pentech: Plymouth).

O'Carroll, M. J., Bush, A. W., Cross, M., Gibson, R. D., Wilkinson, T. S. (Eds.), 1979, Modelling and Simulation in Practise 2 (EMJOC: Northallerton).

Richardson, G. P., Pugh, A. L. III, 1981, An Introduction to Systems Dynamics Modelling with DYNAMO (M.I.T., Cambridge, Mass.)

Thorp, M., Moscardini, A. O. and Cross, M., 1982, Interactive discrete event simulation without programming, *Adv. Eng. Software*, **4**, 1.

The Use of Discrete Dynamic Models in the Teaching of Mathematical Modelling

P. H. Landers
Dundee College of Technology

SUMMARY

The chapter considers the digital simulation of mathematical models of dynamic systems and a particular problem which occurs in the necessary discretization of the continuous model. This problem, the generation of spurious transfer function zeros, is a serious one for the control engineer, especially as very often these zeros are non-minimum phase. It is observed that increasing the sampling rate makes matters worse. An engineering problem is considered and the z transform method is shown to be inadequate. Simpler discretization techniques are described which were used to deal with the problem.

1. INTRODUCTION

Mathematical modelling of dynamic systems is strongly linked to computer simulation of such systems, as a mathematical model is required before simulation can be performed and simulation is often used to check the model. For many years analog computers were used for this task. Model responses could be displayed on an oscilloscope or some other recording device, and the results obtained from different models could be easily compared. The usual means of simulating a dynamic model was to write down the differential equation and then use integrators and summers to realize it on the computer. Linear dynamics were particularly easily dealt with and nonlinearity, though a little more difficult, could be incorporated where necessary. The simulation invariably involved the use of negative feedback, with the accompanying risk of instability. If the system to be modelled is stable however, the analog model should also be stable.

Digital computers have now become so common, for other reasons than the requirements of computer modelling, that they have tended to replace their analog counterparts for all but those problems that require great speed of solution. There is then a requirement to be able to simulate continuous systems on digital computers. The obvious procedure is to replace differential equations by difference equations as digital computers cope with delayed versions of variables very easily, provided the delay is an exact multiple of the sample period. Nonlinearity is also easily included in digital models.

2. ANALOG TO DIGITAL TRANSFORMATION

There are two major problems in the transformation of the analog model to a digital one. One is the need to decide on the sampling period. (This can be a difficult problem and it will not be considered here.) The other is the fact that difference equations are analogous to differential equations, but they are not the same. There are several ways of obtaining a difference equation from knowledge of a differential equation. None of these is entirely satisfactory, and in general they produce different difference equations.

Mathematical modelling is used by control engineers to describe systems so that a control strategy may be derived. Once the model is obtained total reliance is placed on it. Usually the system output is sampled and fed to a digital computer, which produces a control signal by operating on the system output according to some control law. This control signal is discrete in nature, being the output of a digital device, and it is this signal which becomes the system input. It is made piecewise continuous by the use of a zero order hold and the method of discretization preferred in this circumstance is an adaptation of the z transformation which takes into account the presence of the hold.

In passing from the s domain to the z domain there is a one–one mapping of the poles of the s domain transfer function to the poles of the z domain transfer function. Any pole at s_i is mapped to a pole at $z_i = \exp(s_i h)$ where h is the sampling period. The left half plane of the s domain is then represented by the unit disc of the z domain. However, the zeros of the z domain transfer function bear no relationship to those of the s domain transfer function. In particular, a transfer function with n poles will have $n-1$ finite z domain zeros regardless of the number of finite s domain zeros. For instance, if the simple transfer function

$$\frac{1}{(s+1)(s+2)}$$

is considered, its analogous z transform, assuming the presence of a zero order hold, is found to be

$$\frac{(1-2e^{-h}+e^{-2h})z^{-1}+(e^{-3h}-2e^{-2h}+e^{-h})z^{-2}}{(1-e^{-h}z^{-1})(1-e^{-2h}z^{-1})}$$

The poles occur, as expected, at e^{-h} and e^{-2h}, but a finite zero occurs at

$$\frac{2e^{-2h}-e^{-3h}-e^{-h}}{1+e^{-2h}-2e^{-h}}$$

This can lie anywhere between -1 and zero on the real axis depending on the value of h.

It may be wondered if the presence of the zero order hold is responsible for this finite zero. However, if the zero order hold is ignored and the usual z transform considered, then we obtain

$$\frac{(e^{-h}-e^{-2h})z^{-1}}{(1-e^{-h}z^{-1})(1-e^{-2h}z^{-1})}$$

This also has a finite zero, although in this case the zero is at the origin of the z plane whatever the value of the sampling period.

3. THE EFFECT OF TRANSFER FUNCTION ZEROS ON SYSTEM BEHAVIOUR

It is relevant then to look at the effect of zeros in general. Zeros which occur in the left half of the s plane are said to be minimum phase, whilst those which occur in the right half of the s plane are non-minimum phase. This is somewhat doubtful terminology, but it is very common. What it means is that a zero at $s=b$ for instance, b a negative real number, will have phase equal to arctan (w/b) for some frequency w, whilst a zero at $s=-b$ will have phase equal to $-$arctan (w/b). The former is regarded as minimum phase, and the latter as non-minimum phase. This certainly seems to be the wrong way round, but the engineer is always concerned with negative phase, of which he invariably has too much. The minimum phase zero (that is, the one with positive phase) reduces the system negative phase, whilst the non-minimum phase zero (the one with negative phase) increases the system's negative phase. The non-minimum phase zeros are in the right half of the s plane. By analogy we may then say that any zero which occurs outside the unit disk in the z plane is non-minimum phase. Clearly such zeros are acceptable; it is the presence of poles in this region which causes instability.

In many instances where mathematical modelling is used to help solve engineering problems negative feedback is applied to the model, and indeed to the system at some stage. The presence of finite non-minimum phase zeros then becomes a serious matter. Consider an open loop transfer function $N(s)/D(s)$, where $N(s)$ represents the numerator of the transfer function, its roots being the finite zeros, whilst $D(s)$ represents the denominator of the transfer function, its roots being the poles. If simple negative feedback is applied, with gain k, the closed loop transfer function becomes

$$\frac{N(s)/D(s)}{1+kN(s)/D(s)}=\frac{N(s)}{D(s)+kN(s)}$$

Clearly the poles of this transfer function are affected by the zeros of the original transfer function (unless k is zero), and for k approaching infinity, the poles approach the roots of $N(s)$, i.e. they approach the zeros of the original function. If the zeros are non-minimum phase then the poles must enter the unstable region of the s plane in order to approach these zeros.

The same occurs in the z plane, but as we have seen, even if there are no finite zeros in the s plane, there usually will be some in the z plane. It will now be shown that these spurious finite zeros may be non-minimum phase. Considering the following s domain transfer functions

$$\frac{(n-1)!}{(s+a)^n}, \text{ a some positive real number, } n = 2, 3, 4.$$

The analogous z domain functions are

$$(-1)^{n-1}\frac{\partial^{n-1}}{\partial a^{n-1}}\left(\frac{z}{z-\mathrm{e}^{-ah}}\right)$$

For $n = 2$ this becomes

$$\frac{zh\mathrm{e}^{-ah}}{(z-\mathrm{e}^{-ah})^2}=\frac{z^{-1}h\mathrm{e}^{-ah}}{(1-\mathrm{e}^{-ah}z^{-1})^2}$$

which has two poles, and also one finite zero, the zero being at the origin of the z plane. For $n = 3$ we have

$$\frac{zh^2\mathrm{e}^{-ah}(z+\mathrm{e}^{-ah})}{(z-\mathrm{e}^{-ah})^3}=\frac{z^{-1}h^2\mathrm{e}^{-ah}(1+\mathrm{e}^{-ah}z^{-1})}{(1-\mathrm{e}^{-ah}z^{-1})^3}$$

which has three poles, and also two finite zeros, one zero being at the origin of the z plane, and the other being within the unit disc. For $n = 4$ we have

$$\frac{zh^3\mathrm{e}^{-ah}(z^2+4\mathrm{e}^{-ah}z+\mathrm{e}^{-2ah})}{(z-\mathrm{e}^{-ah})^4}=\frac{z^{-1}h^3\mathrm{e}^{-ah}(1+4\mathrm{e}^{-ah}z^{-1}+\mathrm{e}^{-2ah}z^{-2})}{(1-\mathrm{e}^{-ah}z^{-1})^4}$$

Here there are four poles and three finite zeros. One of the zeros is again at the origin and the other two are at

$$(-2\pm\sqrt{3})\,\mathrm{e}^{-ah}$$

Clearly as $h \to 0$, one of the zeros approaches the point on the real axis given by $-2-\sqrt{3}$, which is well outside the unit disc.

It might be argued that the sampling period can be raised to such a value that the zeros all occur within the unit disc. But this would mean reducing the sampling frequency and the aliasing problem then becomes apparent. Indeed, the sampling frequency has usually been raised specifically to avoid aliasing, and the engineer's response to problems resulting from discretization is usually to raise the sampling rate, on the assumption, apparently false, that this will bring the discrete system behaviour closer to that of the original continuous system.

Considering Table 1, where the parameter a has been set equal to unity, the effects of varying the sampling period, and therefore the sampling frequency, can be seen. A sampling frequency of 0.75 Hz is just low enough to cause the zero to occur within the unit disc. As the parameter a is effectively the cut-off frequency in radians per second, the cut-off frequency in hertz is 0.159. The minimum practical sampling frequency is about five times this, about 0.8 Hz, and 0.75 Hz may therefore be regarded as just acceptable. But in a practical situation a zero at 0.984 is much too close to the edge of the unit disc. In any real system parameter variation would be such that a significant margin would be required. Zeros at a distance from the origin of 0.753 might be acceptable and for this a sample rate of 0.625 per second is required. This, however, is below the practical minimum for the avoidance of aliasing.

Table 1. The variation in location in the z plane of one zero of the discretized form of the transfer function $3/(s+1)^4$

Sampling period h (seconds)	Sampling frequency $(1/h)$	Zero location
2	0.5	−0.505
1.6	0.625	−0.753
1.3	0.75	−0.984
1.3	0.77	−1.02
1.0	1.0	−1.37
0.5	2.0	−2.26
0.2	5.0	−3.06
0	∞	−3.73

The plant then may be minimum phase, but the discrete model may not be. The controller is designed to control the model, in the hope that it will then control the plant. The controller may then have an unstable pole resulting from an attempt to deal with a spurious non-minimum phase zero. Although the model may not exist in hardware form, the controller based on it does.

4. A REAL PROBLEM

For some years the author was concerned with the modelling of the linear and nonlinear dynamics of a large load driven via a gearbox by an electric motor

(Landers, 1982a,b). The transfer function of the linearized model was of the form

$$\frac{W_L}{V} = \frac{K}{as^4 + bs^3 + cs^2 + ds + e} \tag{1}$$

where W_L is the angular velocity of the load and V is the voltage applied to the motor terminals. Such a system has four poles, and in the z plane, three finite zeros. The position of the zeros varies with the sampling period and this is shown in Table 2. Only for sampling periods greater than about 0.025 seconds are all the zeros within the unit disc. This implies a requirement for a sampling rate of less than 40 per second. But the system is known to have a resonance at about 30 Hz, and this resonance greatly affects system stability. Even direct application of the sampling theorem indicates a requirement for a sampling rate greater than 60 per second, whilst practical considerations indicate a requirement for greater than 300 per second. However, the highest breakpoint in the frequency domain is slightly in excess of 100 Hz. A sampling rate of greater than 500 per second is then required. The digital simulation did, in fact, use a step size of 0.0003 seconds. The z transform model is totally unacceptable at this sampling rate, being unstable for ridiculously low loop gains, and other methods had to be used.

Table 2. The variation in location in the z plane of the three finite zeros of the discretized form of the transfer function (1) (see text)

Sampling period h (seconds)	z_1	z_2	z_3
0.0001	1.045	0.945 $\pm$j0.0407	
0.0003	1.143	0.84 $\pm$j0.109	
0.001	1.566	0.525 $\pm$j0.245	
0.002	2.509	0.217 $\pm$j0.277	
0.003	4.27	0.0369 $\pm$j0.232	
0.004	8.388	−0.0641 $\pm$j0.163	
0.005	27.39	−0.12 $\pm$j0.077	
0.0055	280	−0.18	−0.096
0.006	−39.71	−0.241	−0.062
0.01	−5.591	−0.384	−0.0162
0.012	−4.138	−0.426	−0.0107
0.014	−3.156	−0.486	−0.00765
0.018	−1.215	−1.031	−0.00467
0.02	−0.813 $\pm$j0.704		−0.00393
0.025	−0.0805 $\pm$j0.983		−0.00302
0.03	+0.544 $\pm$j0.739		−0.00278
0.04	+0.59 $\pm$j0.598		−0.00283
0.05	−0.803 $\pm$j0.285		-2.05×10^{-3}
0.1	+0.218 $\pm$j0.643		-9.12×10^{-4}

5. FRACTIONAL DELAY

The presence of fractional delay also tends to produce non-minimum phase zeros. Wellstead *et al.* (1979) give the following example. They consider the transfer function

$$\frac{25\exp(0.16\,s)}{S}$$

with a sampling period of 0.1 seconds. The z transform with zero order hold is obtained as follows. Consider the transfer function

$$\frac{25\exp(0.16\,s)}{s+a}, \quad h = 0.1$$

Letting $\Delta = (n-1)\,h + \delta = 0.16$, n a positive integer, therefore $n = 2$ and $\delta = 0.16 - 0.1 = 0.06$
We then consider

$$\frac{25e^{s(h-\delta)}}{s(s+a)} = \frac{25e^{0.04s}}{s(s+a)}$$

where the s in the denominator results from the presence of the zero order hold. The residue at $s = 0$ is $25/a$ and that at $s = -a$ is $-25e^{-0.04a}/a$. From this is obtained

$$\frac{25}{a(1-z^{-1})} - \frac{25e^{-0.04a}}{a(1-e^{-ah}z^{-1})} = \frac{25(1-e^{-ah}z^{-1} - (1-z^{-1})\,e^{-0.04a})}{a(1-z^{-1})(1-e^{-ah}z^{-1})}$$

Including the two units of delay resulting from the fact that $n = 2$, and taking into account the numerator of the expression for the zero order hold by multiplying by $1 - z^{-1}$ results in

$$\frac{25z^{-2}}{a(1-e^{-ah}z^{-1})}\,[(1-e^{-0.04a}) + z^{-1}\,(e^{-0.04a} - e^{-ah})\,]$$

Allowing a to approach zero, we obtain, via de L'Hopital's rule

$$\frac{25z^{-2}\,(0.04 + 0.06z^{-1})}{1-z^{-1}} = \frac{z^{-2}(1+1.5z^{-1})}{1-z^{-1}}$$

and clearly there is a zero at -1.5 in the z plane. Clarke (1981) has remarked that the zero would be minimum phase if δ were less than $h/2$. Of course, it is not surprising that the presence of pure delay may result in non-minimum phase behaviour, as pure delay is itself non-minimum phase, but this is another example of the problem.

6. THE EFFECT OF ZEROS IN THE *s* DOMAIN ON THE *z* DOMAIN

From the foregoing it is natural to wonder about the effects of the presence of zeros in the s domain. If we consider the following transfer function

$$\frac{s+3}{(s+1)(s+2)}$$

we find that this becomes, in the z domain

$$\frac{z(z-2e^{-2h}+e^{-h})}{(z-e^{-h})(z-e^{-2h})}$$

This has a zero at the origin of the z plane and another somewhere on the real axis if the z plane, within the unit disc. However, suppose the s domain zero is non-minimum phase, as follows

$$\frac{s-3}{(s+1)(s+2)}$$

In the z domain we obtain

$$\frac{z(z+4e^{-2h}-5e^{-h})}{(z-e^{-h})(z-e^{-2h})}$$

Again there is a zero at the origin of the z plane and another somewhere on the real axis. This latter zero is outside the unit disc for most values of sampling period, but lies inside the unit disc for fairly large sampling periods as is shown in Table 3. This is the other side of the coin. Whether practical advantage can be taken of this occurrence is not clear. In this example a sampling period of two seconds is too great for the discrete model to satisfactorily represent the dynamics of the continuous system

Table 3. The variation in location in the z plane of zeros resulting from minimum phase and non-minimum phase zeros in the s plane

Sampling period h (seconds)	Minimum phase zero	Non-minimum phase zero
3	−0.0448	+0.239
2	−0.0987	+0.603
1	−0.097	+1.298
0.7	−0.0034	+1.497
0.5	+0.129	+1.561
0.25	+0.434	+1.468
0.1	+0.733	+1.249
0	+1	+1

7. OBSERVATIONS ON CONTINUOUS SYSTEMS

It may be argued that we almost always have non-minimum phase zeros in the s domain, although they are often at infinity. For instance, considering the following transfer function

$$\frac{K}{s(s+1)(s+2)}$$

if a loop is closed around this, the denominator of the closed loop transfer function becomes

$$s(s+1)(s+2)+K$$

A root locus plot for K rising from zero to infinity would show the root at -2 approaching $-\infty+\mathrm{j}0$ (the far left-hand end of the real axis). This zero is certainly minimum phase. However, the roots initially at zero and -1 approach one another and on meeting split off the real axis and approach $+\infty\pm\mathrm{j}\infty$ (the top right and lower right corners of the s plane). These are certainly non-minimum phase zeros and they clearly cause instability for some value of K. But, in general finite non-minimum phase zeros cause instability to occur at much lower values of K.

8. AN OBVIOUS METHOD OF OVERCOMING THE PROBLEM

One of the major aims of digital simulation is to predict the onset of instability in dynamic systems, analytical tools for nonlinear systems being unsatisfactory, and clearly the z transform representation of systems, and the resulting difference equations leave much to be desired.

There are, of course, other methods for obtaining a difference equation from a differential equation. One of these is called the pole-zero mapping technique. Here all finite zeros are mapped by the function $z=\mathrm{e}^{sh}$ just as the poles are, whilst zeros at infinity are mapped to $z=-1$. This is however, a totally arbitrary method.

Another method is that of backward differences, where $\mathrm{d}/\mathrm{d}t$ is simply equated to $(1-z^{-1})/h$. This is an acceptable approximation, provided h is small, and it seems to avoid the aforementioned problem of spurious zeros. Another similar method is to assume that there exists a sampler between all first order dynamics so that the z transform of all first order systems in cascade can be simply multiplied together. Again, this method is only accurate if the sampling frequency is high, but it is very easy in that only two transform pairs are usually required. These are

$$\frac{K}{1+sT}-\frac{K(1-\mathrm{e}^{-h/T})}{1-\mathrm{e}^{-h/T}z^{-1}}$$

and

$$\frac{1}{s} - \frac{h}{1-z^{-1}}$$

These result in difference equations as follows

$$x_t = ax_{t-h} + bu_t$$

and

$$x_t = x_{t-h} + hu_t$$

where x is the output of the dynamic element and u is the input, $a = e^{-h/T}$ and $b = k(1-a)$. This is the method which was used for simulation of the large load system previously referred to. The simulation was successful in that it produced results which were applicable to the real system. This method also has the advantage of all states being directly available to the modeller, there being a one–one correspondence between the continuous system equations and the derived difference equations. This makes the method very convenient for the modeller. Each dynamic element is defined by one program statement. If it is required to simulate an underdamped second order system, then this is done by connecting the two first order systems mentioned above in cascade, and then closing a loop around them. Variation of the gain constant k then produces the required degree of damping. This method of producing an underdamped second order system is particularly attractive as very often it is exactly the way such underdamping occurs in real systems. In the application mentioned, inherent internal feedback resulting from the reaction of the load on the motor, through the gearbox, produced the underdamping which gave rise to a particularly awkward resonance.

9. CONCLUSION

A serious difficulty encountered in the simulation of continuous dynamic models by discrete methods has been considered and a method of avoiding the problem has been suggested. It appears that at least for the simulation of dynamic systems it is better to use somewhat naive discretization techniques rather than formal z transform methods.

REFERENCES

Clarke D. W., 1981, Introduction to self-tuning Controllers, in *Self-tuning and Adaptive Control* (Ed. C. J. Harris and S. A. Billings).

Landers P., 1982a, Adaptive Control of Tank Guns, in Applications of Adaptive and Multivariable Control, IEEE Conference at University of Hull.

Landers P., 1982b, Control Systems with Pass Band Resonances, Second International Conference on Systems Engineering at Coventry Polytechnic.

Wellstead P. E., Edmunds J. M., Prager D., and Zanker P., 1979, *Int. J. Control*, **30,** 1.

Exploratory Data Analysis—A New Field of Applied Mathematics

R. Biehler
Universität Bielefeld

1. INTRODUCTION

This chapter deals with Exploratory Data Analysis (EDA) and is based on a detailed theoretical analysis of the latter (cf. Biehler, 1982). On the one hand, EDA is controversially discussed among statisticians because it questions quite a lot of principles that underlie traditional statistical work. On the other hand, EDA begins to play quite a role within discussions on the teaching of probability and statistics because it seems to be both rather elementary in a mathematical respect as well as widely applicable and close to practical work with real data (cf. for example Gnanadesikan *et al.* 1983). These features will probably, or so we hope, make EDA interesting enough for a conference devoted to problems of teaching applications of mathematics. The paper is not concerned with questions of teaching directly. As EDA is as yet not well known or, sometimes, is apparently known but misunderstood, the paper concentrates on clarifying general principles of EDA, on discussing its relation to subject matter fields and to traditional statistics. Subsequently, some propositions are formulated which argue that the results of studying EDA might be more generally important for a theory of mathematics education which is—to formulate this as a challenge—not only concerned with teaching mathematical modelling, but also with teaching applied mathematics.

Exploratory Data Analysis is a recent scientific development attained within the efforts to find new tools and principles for the practical analysis of data. EDA and its rise are closely linked to the textbook bearing the same title published in 1977 by J. W. Tukey, from whose introductory lectures at Princeton University it evolved. The textbook contains techniques for the handling and the representation of data, in which probabilistic concepts have only a subordinate role. The focus of EDA is on the exploration of data, i.e. on the search for peculiarities and structures in data sets, and for simple comprehensive descriptions of the phenomena discovered. Graphical displays are the main tools for this activity. In the next section, an example will be discussed to illustrate EDA's approach. It is primarily a vehicle for general ideas and should not be misunderstood as an example which can directly be transferred into the classroom.

2. SOME PRINCIPLES AND TOOLS OF EXPLORATORY DATA ANALYSIS—DISCUSSED IN THE CONTEXT OF AN EXAMPLE CONCERNING SUICIDE RATES

The example is concerned with a set of data representing suicide rates or, more precisely, death rates due to suicide. It has been taken from Erickson and Nosanchuk (1977, pp. 14) and is also discussed by Biehler (1982, pp. 70). Fig. 1 contains a table of suicide rates related to sex, country, and age.

The aim of EDA is to explore data in order to reveal structure and anomalies, so-called 'indications'. Indications are something to be given to the subject matter expert, who is to think about their interpretation and relevance in connection with his knowledge and with problems of the respective subject matter, i.e. the sociological problem of suicide in this case. It is claimed that the results of EDA can contribute independent new perspectives to the analysis of subject matter issues. This claim is justified because general experience with data analysis in very different fields is incorporated in the general tools and principles of EDA. By applying EDA methods, this experience is implicitly used to analyse new data. It is not claimed, however, that the results have some objective truth or meaning independent of the concrete context. Rather the results are thought to be instruments used to explore the respective context further.

Returning to our concrete example, we can get quite a lot of information from Fig. 1 by visual scanning of the data, e.g. the information that generally male suicide rates are higher than female ones or that suicide rates seem to increase with age etc. The question is how this activity can be made more efficient. The general approach of EDA consists of developing new graphical representations for data to facilitate their exploration. Also, general orientations about what to look for in such displays are given. This approach can be understood if we assume that representations have a double function: they do not only store known information, but they also have an exploratory function, i.e. they are tools to develop knowledge. Different representations usually differ widely with regard to their sensitivity to certain relations and structures in data sets. That is why we find two principles underlying the practice of EDA. Firstly, a *principle of multiplicity of representations*, respectively a *principle of varying representations* is applied, which allows one to explore data from different

		Age				
Country	Sex	25–34	35–44	45–54	55–64	65–74
Canada	M	21.6	27.3	31.1	33.5	23.5
	F	7.8	11.5	14.8	12.3	9.2
Israel	M	9.4	9.8	10.2	14.0	27.3
	F	7.6	4.2	6.7	22.9	19.1
Japan	M	21.5	18.7	21.1	31.1	48.7
	F	14.0	10.3	13.2	21.0	40.1
Austria	M	28.8	40.3	52.3	52.8	68.5
	F	8.4	16.4	22.4	21.5	29.4
France	M	16.4	25.2	36.1	47.3	56.0
	F	6.6	8.9	13.0	16.7	18.5
Germany	M	28.3	34.6	41.3	49.1	51.8
	F	11.3	15.6	24.2	25.6	27.3
Hungary	M	48.2	65.0	84.1	81.3	107.4
	F	12.7	18.4	26.9	34.7	47.9
Italy	M	7.1	8.3	10.8	17.9	26.6
	F	3.5	3.7	5.5	6.7	7.7
Netherlands	M	7.8	10.6	17.9	20.2	28.2
	F	4.7	8.2	10.5	15.8	17.3
Poland	M	26.2	29.1	35.9	32.3	27.5
	F	4.4	4.7	6.6	7.3	7.0
Spain	M	4.1	7.0	9.6	15.7	21.9
	F	1.4	1.6	3.8	5.4	5.7
Sweden	M	27.6	40.5	45.7	51.2	35.1
	F	13.0	17.5	19.6	22.4	17.1
Switzerland	M	21.7	33.6	41.1	50.3	50.8
	F	10.4	15.9	18.2	20.1	20.6
UK	M	9.6	12.7	14.6	17.0	21.7
(England and Wales)	F	5.1	6.5	10.7	13.0	14.1
United States	M	19.6	22.2	27.8	32.8	36.5
	F	8.6	12.1	12.5	11.4	9.3

(Table taken from: Erickson and Nosanchuk, 1977, p. 14; their source of data: *World Health Statistics Annual 1971*, vol. 1; World Health Organization, 1974.)

Fig. 1. Suicide rates 1971; unit: number per 100.000

points of view. Secondly, a *principle of making representations more efficient* underlies the construction of new and the transformation of old schemes of representation. The latter principle takes into account both the human mind's specific but limited capability to recognize patterns and the specific structures for which the representations should be sensitive. The latter reflects general practical experience as to which structures seem to be most relevant in exploring data. Below, several representations of our data set will be constructed to illustrate use and usefulness of these principles.

In the following, we shall concentrate on male suicide rates. First, we will abstract from country-dependence and treat the data in each age category as a one-dimensional data set. For the exploration of one-dimensional data, Tukey has invented a new display, the stem-and-leaf display. Fig. 2 shows such a

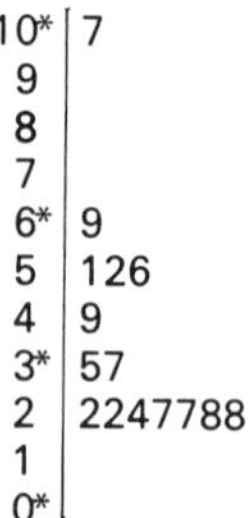

10* | 7
9 |
8 |
7 |
6* | 9
5 | 126
4 | 9
3* | 57
2 | 2247788
1 |
0* |

Fig. 2. Male suicide rates, age 65–74; stem-and-leaf display

display for the data of age group 65–74. The data have been sorted. The first digit has been put in the so-called 'stem', the second digit in the so-called 'leaf'. The third line from the bottom, for example, represents the (rounded) data 22, 22, 24, 27, 27, 28, 28. In fig. 2, several indications can be seen. Among others, it indicates where the values are centered, that the distribution is highly skewed and that there is an extreme outlier (value 107). All these indications can hardly be seen in the original table. Obviously, the stem-and-leaf display is related to the histogram. The following advantages of the new display are notable.

The stem-and-leaf display enables us to perceive the distribution of data values within each interval. Also, it is simpler to proceed from a value in the display to the datum that produced it. For example, it is easy to identify the country belonging to the outlier (Hungary) and to the lowest values. The change to the stem-and-leaf display reflects a change in goals with regard to the exploration of data. Usually, a histogram displays relative frequencies and is used to get a first impression of the underlying probability distribution. The individual data values do not matter and are not 'distinguishable' because it is usually assumed that the data have been generated by independent repetitions of an experiment under similar conditions. EDA is concerned with situations where these conditions are not satisfied and where it might be important to get more information about individual data, e.g. about the circumstances that have produced an outlier.

Stem-and-leaf displays can also be used to compare several sets of data. Fig. 3 shows a so-called 'back-to-back' stem-and-leaf for two age groups. The back-to-back arrangement of the stem-and-leaf displays follows the principle of making representations more efficient, because it would be more difficult to compare two distributions if the two displays were arranged, for instance, one beneath the other.

Several new phenomena can be seen in the new representation:

— the average level of suicide rates is higher in the older age group,
— the spread of the data has increased,

age 25–34		age 65–74
	10*	7
	9	
	8	
	7	
	6*	9
	5	126
8	4	9
	3*	57
98862220	2	2247788
60	1	
9874	0*	

Fig. 3. Male suicide rates, age 25–34 and age 65–74; 'back-to-back' stem-and-leaf display

— both data sets contain one upper outlier, which seems to be more extreme in the older age group.

One might have expected the rise of level after having explored the table of Fig. 1. The increase in spread is a new discovery. One may begin to speculate about the interpretations of the indications revealed. One reason for the greater diversity of suicide rates for older men might be that perhaps the experience of old age and the older people's values related to suicide differ more strongly from country to country than those of younger men. This may be due to the fact that they grew up in earlier periods when countries were less 'homogenized' by industrialization and higher levels of international communication (cf. Erickson and Nosanchuk 1977, pp. 18). We shall not follow these lines of research, which are the responsibility of the subject matter expert, the sociologist, but shall continue to explore the data in order to reveal more indications.

Fig. 4 is another version of the stem-and-leaf display, in which the numbers have been substituted by coded country names. Again, several new indications can be noticed:

— the outlier is Hungary in both age-groups
— some countries have hardly changed their rank and still cluster together
— France, for instance, has dramatically changed its rank.

These indications could be used as a starting point for detailed investigations into differences and similarities among the different groups of countries in order to see which variables might affect suicide rates. We shall now try to compare all the five age groups by their stem-and-leaf displays. The general increase of level can be seen in Fig. 5, but it is difficult to get a clear impression with regard to spread and shape. Besides, the huge variety of numbers impairs visual comparison. EDA has developed some new graphical displays to

age 25–34		age 65–74
	10*	HUN
	9	
	8	
	7	
	6*	AUS
	5	SWI, GER, FRA
HUN	4	JAP
	3	SWE, USA
AUS, SWE, GER, POL, CAN, JAP, SWI, USA	2*	SPA, UK, CAN, ISR, ITA, NET, POL
FRA, UK	1	
ISR, NET, ITA, SPA	0	

Fig. 4. Male suicide rates,
age 25–34 and age 65–74;
'back-to-back' stem-and-leaf display
with additional information (country codes)

	25–34	35–44	45–54	55–64	65–74
10					7
9					
8			4	1	
7					
6		5			9
5			2	310	621
4	8	01	161	79	9
3		54	166	4123	57
2	22986802	9275	18	0	4778822
1	60	0913	01805	4867	
0	9784	87			

Fig. 5. Male suicide rates;
5 stem-and-leaf displays

optimize such comparisons. In Fig. 6 we see five so-called box plots for the suicide data. The line in the middle of each box marks the median of the data, the two lines at the end of each box mark the lower quartile q_1 and the upper quartile q_u. Thus, the length of the box represents the interquartile range, which is used as a measure of spread. The nonsymmetry of the box gives a rough indication of skewness. The meaning of the lines starting from the box can be explained as follows: A rule of thumb for outlier identification is used. First, so-called 'fences' are defined by

$$\text{upper fence } f_u = q_u + 1.5(q_u - q_l)$$
$$\text{lower fence } f_l = q_l - 1.5(q_u - q_l)$$

Values outside the fences are called outliers and have been marked by a small circle in Fig. 6. The most extreme values inside the fences are called 'adjacent

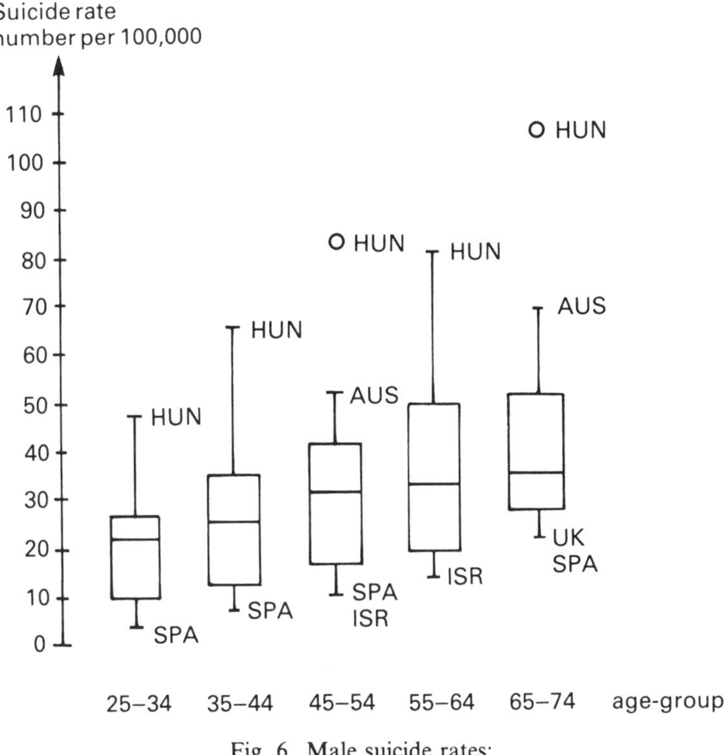

Fig. 6. Male suicide rates;
5 box plots

values'. The lines go from the end of the box to the respective adjacent values. Both outliers and adjacent values have been identified in Fig. 6 by the first three letters of the country they belong to, because general experience suggests that they deserve special attention. At this place, it is not possible to tell more about the genesis and the justification of this rule of thumb; for more details see Tukey (1970, ch. 5, pp. 14) and Biehler (1982, pp. 285).

Box plots are visual displays of numerical summaries of data. They reflect an important change as compared to traditional descriptive statistics. The arithmetical mean as a measure of location and the standard deviation as a measure of spread are no longer applied in EDA to summarize data. Instead, the median and the interquartile range are used because they are simply calculable *resistant* measures of location and spread. Resistance means, roughly speaking, resistance with regard to outliers. The traditional measures are strongly affected if there are outlying observations. To use resistant measures and to define outliers with regard to resistant measures of location and spread is a double strategy which serves to achieve a clear separation of the data in a summarized 'main group' and in a set of outliers. This is particularly important in exploratory work, where so-called 'dirty data', which usually contain outliers, are to be analysed. The special attention paid to outliers is based on the experience that outliers are frequently interesting indications of relevant variables which have not been thought of in advance. It should be noted that the concept of resistance is related to the statistical concept of robustness, but it is conceptually distinct from that concept (cf. Biehler 1982, pp. 57).

Now let us return to Fig. 6. Among other things, we see the following new indications:

— the spread (height of the box) shows an upward trend; the oldest age group, however, does not fit into this pattern;
— the type of skewness changes systematically from one direction to the other;
— the character of Hungary as an 'overall outlier' shows up very clearly;
— Spain is always at the bottom of the distribution.

All these indications raise a lot of questions. We shall explore only one of them because the data can be exploited even more in this respect. The discovery that the oldest age group does not follow the trend of spread pertaining to the other ones can be interpreted as follows. The group 65–74 is an outlier on a "higher level". As the general strategy of EDA in dealing with outliers is to search for discriminating variables, this line of enquiry seems to be promising. A peculiar feature of the group 65–74 is that 65 is the age of retirement in most countries. Hence, might there be a particular retirement effect, which decreases variation? To explore this hypothesis, which has typically been generated by the combined effort of EDA and subject matter considerations, we shall construct two more representations of our data. Pursuing the principle of varying representations one step further, they will contribute still another perspective on our data.

Figs. 7 and 8 show profile displays. To avoid visual confusion, i.e. following the principle of making representations more efficient, separate pictures for countries with and without strictly increasing suicide rates have been constructed. Even in those countries with strictly increasing suicide rates 'something happens' in connection with retirement. In Germany and Switzerland, for instance, the 'rate of change' decreases. In some other countries, it increases. Most interestingly, some countries in Fig. 8 show even a pronounced absolute decline of suicide in the oldest age group. Besides, Fig. 8 shows another anomaly which could not be seen in the other displays. Japan differs from all other countries because we perceive a decrease from the first to the second age group. This particular outlying character of Japan surely deserves special considerations as to its social structure.

After having explored Figs. 7 and 8 we shall be surer that something particular happens in connection with retirement and that the decrease of spread noticed in Fig. 6 is probably no mere chance fluctuation. Above that, the

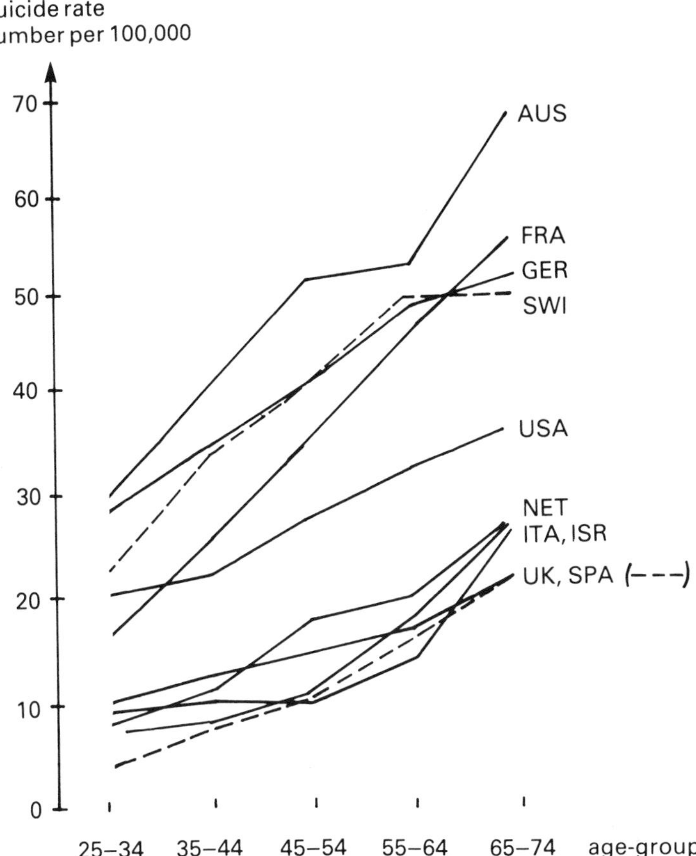

Fig. 7. Male suicide rates;
countries with strictly increasing rate;
profile displays

last two figures have revealed some further structure in the data. They allow us to classify the countries in a new way, namely with regard to the type of profile and not only with regard to the average level of suicide rates. This might open some new lines of further investigation. Such investigations may be carried through on the subject matter side, but one might also think of exploring the female rates or data from other years than 1971 along the lines suggested by the results of the above analysis.

Let me summarize the general aspects of EDA which have been illustrated by our example.

First, some new displays have been introduced, such as the stem-and-leaf and

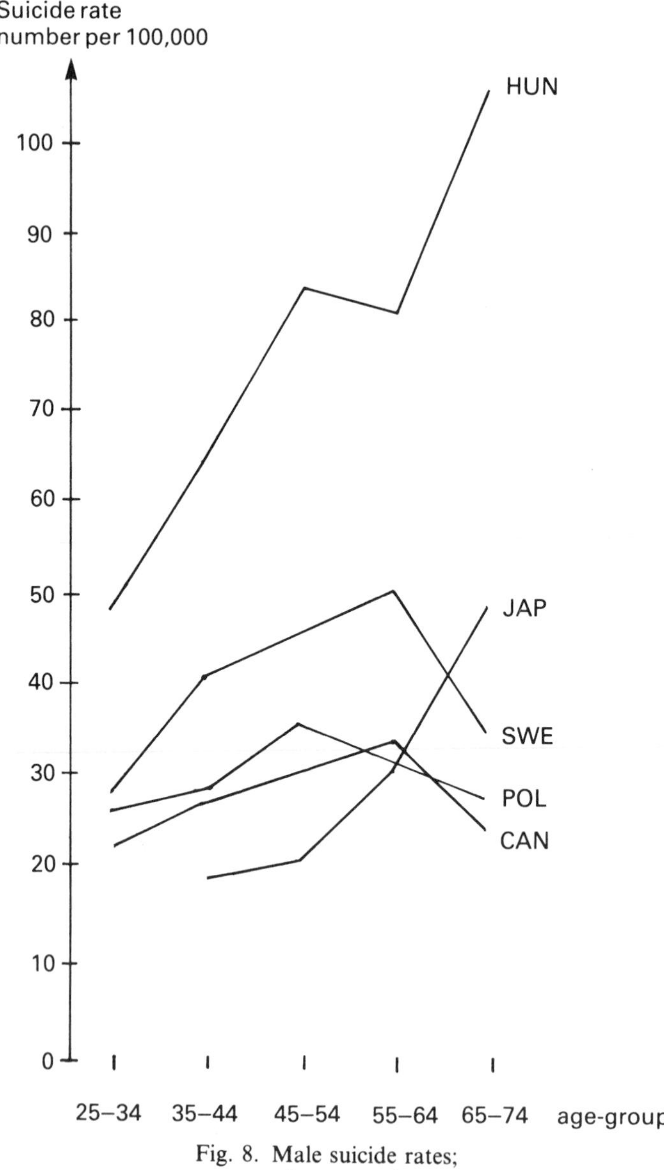

Fig. 8. Male suicide rates;
other countries;
profile displays

the box plot. It has been argued in which respect they are results of the attempt to make representations more efficient.

Second, the usefulness of the principle of multiplicity of representations has been illustrated.

Third, it has been shown what it means to discover indications in data sets, which function they have for providing new perspectives for the investigations of the subject matter expert. It must be added that the textbooks on EDA develop a systematic knowledge about the question, which indications in the different displays should be looked for. This is part of the general competency of the EDA-expert.

Fourth, the fundamental role of graphical tools for exploring data has become clear. This is mainly due to the fact that graphical displays are particularly suited for using the pattern recognition capability for discovering new phenomena (see MacDonald-Ross, 1979). It is only a slight exaggeration if we state that EDA is an important element in a recent historical revolution concerning the importance of graphical displays for the analysis of data (cf. Beniger and Robyn, 1978; Fienberg, 1979). In practical work with large data sets, the computer plays an important part in calculating and in constructing graphical displays on the basis of the latter. This development can be interpreted as organizing a particular man–machine interaction. It is not aimed at a complete automatisation of data analysis, rather, the exploration of graphical displays and the decisions what to do next are genuine human activities, which are not formalized.

3. EXPLORATORY DATA ANALYSIS AND TRADITIONAL STATISTICS

I shall now turn to the question of the relation of EDA and traditional probabilistic statistics. The example of suicide rates will provide an illustrative background for the following general considerations. There are two related, but different approaches to this problem:

- —the type of data which can be analysed;
- —the type of problems with regard to data which can be solved and the role of data in the analysis.

From the first perspective, it has to be stated that, roughly speaking, classical objectivist statistics has to presuppose that data are a random sample from a wider population. As our example illustrates, EDA has a wider range of applicability. Its objects are data sets which may be whole populations or samples for which the broader population and the way of sampling is not known. One motive for developing EDA was to construct general methods for the analysis of such so-called 'dirty data'. EDA represents a constructive answer to practical demands which can be distinguished from two other complementary ways of responding. There are statisticians who accept the limited applicability of probabilistic statistics and deny the possibility of general methods for the analysis of dirty data. Similarly, there are subject matter experts who favour subject specific intuition where statistical methods cannot be applied.

Second, the general questions answered by statistical methods are problems of parameter estimation, confidence intervals and testing of hypotheses. Even if the data are a random sample, certain conditions as to the state of knowledge have to be fulfilled before statistical methods can be applied. For example, a set of hypotheses which should be tested has to be known or an estimator must have been chosen whose reliability should be determined. Such 'pre-data-decisions' usually rely heavily on subject matter knowledge and on considerations about which aspects of the data might be relevant for the subject matter problem. Now, EDA has developed methods which are useful in situations where such definite assumptions and questions do not exist but rather should be generated by means of exploring the data. Hence, even if the suicide rates had been a random sample, it would not have been reasonable, for instance, to apply analysis of variance techniques to estimate and test for effects. For it could not be assumed that the data sets only differ with respect to means, or that such a difference would be the most relevant one for the problem. Thus, the second difference between EDA and classical statistics is not an ontological one, i.e. not related to different types of data, but an epistemological one, i.e. related to the level of pre-data knowledge. It is central for EDA that the pre-data knowledge or hypotheses do not determine the data analysis completely as in traditional statistics; rather, the data themselves influence their analysis with the effect that the pre-data knowledge may be transformed and enriched by unforeseen elements. This strong claim becomes clearer when we recall the principle of classical statistics which forbids that hypotheses be formulated after data inspection because that would render the significance levels invalid. So, EDA breaks with the statistical tradition that error probabilities should be controlled, re-establishing the positive principle that revealed indications should be interpreted by the subject matter expert and evaluated on the basis of other data and in other experiments. EDA has been compared to the activity of a detective who collects indications that later are to be carefully considered by the court of statistical inference (cf. Tukey 1977, p. 1). This metaphor is a bit misleading as it seems more promising in many applications of EDA, just as in our example, to relate the indications to the context of the data or to use them as guidelines for analysing similar data. The direct test of observed differences with regard to statistical significance is only one option among several and cannot have the conclusive power present in the case of pre-data hypotheses.

4. EXPLORATORY DATA ANALYSIS AND TEACHING APPLIED MATHEMATICS

In this concluding section I shall take a rather abstract view of the problem how to make school mathematics more 'applied'. I interpret the new emphasis given to mathematical modelling partly as a response to the exaggerated alignment of curricula to pure mathematics. But two problems seem to me to remain unsolved by the new approach. First, this approach does not touch the problem what type of mathematics should be taught in the rest of the curriculum which is not devoted to solving modelling problems. Second, the emphasis on solving realistic problems, i.e. on model formulation, interpretation and evaluation, often faces the problem, at least in contexts of general education, of limited knowledge of subject matter which seems to be highly relevant in these processes. Although I have no firm conclusions to offer, the analysis of EDA leads to some 'indications' which suggest that it might be promising to rethink some issues concerning the desirable type of applied mathematics in the classroom.

EDA has two peculiar features, one concerns its internal structure, its particular type of mathematical activity, the other concerns its relations to subject matter fields. As to the first, some mathematicians have said that EDA does not belong to mathematics. Indeed, EDA is an experimental activity with systems of numbers which is organized by consciously optimized tools and general heuristic principles. Vague mathematical concepts play an important part in exploring displays. No theorems are proved, the ultimate justification of the tools and procedures lies in their success in practice, and it is even more important to have rich experience with applying EDA techniques than to know mathematical properties of the techniques used. Last but not least, graphical methods play a substantial role and they do not fit into the picture of mathematics as a formal science. I agree with this diagnosis but I maintain that EDA is part of mathematics. It is just these particular features of informal and experimental mathematics which, in my opinion, deserve more attention within the efforts of changing curricula towards a more practical or applied interpretation of mathematics.

As to the relation of EDA to subject matter problems, it might be objected that in the textbooks on EDA, just as in our example, we find no genuine modelling problems discussed. Again, I agree with the diagnosis but not with the presumably implied therapy. EDA accepts that there exists a certain level of division of labour in practice. It starts not 'from scratch' but from a certain state of a problem, then, by means of exploring data, it makes a contribution to an overall problem solution, i.e. the genuine subject matter problem is not *solved*, rather it is *transformed* according to the general experience incorporated in the mathematical techniques of EDA. This attitude, namely not to restrict oneself to a mathematical microworld but to point out new perspectives for the enquiry of subject matter problems—while seeing that an overall problem solution is not attainable in the context of developing a general data-analytical or mathematical competency—would seem to me worth considering when the question how to integrate applications of mathematics in the mathematics classroom is being discussed.

REFERENCES

Beniger J. R., Robyn D. L., 1978, *Amer. Statist.*, **32**, 1.

Biehler R., 1982, Explorative Datenanalyse—Eine Untersuchung aus der Perspektive einer deskriptiv-empirischen Wissenschaftstheorie. Universität Bielefeld, IDM-Materialien und Studien Bd. 24.

Erickson B. W., Nosanchuk T. A., 1977, *Understanding Data*, McGraw-Hill.

Fienberg S. E., 1979, *Amer. Statist.*, **33**, 165

Gnanadesikan R., Kettenring J. R., Siegel A. F., Tukey P. A., Symposium on Exploratory Data Analysis, In Proc. Fourth Int. Congress on Mathematical Education (M. Zureng et al., edr), 1983, Birkhäuser, 344.

MacDonald-Ross M., 1979, *Instructional Science*, **8**, 223.

Tukey J. W., 1970, *Exploratory Data Analysis* (limited preliminary edition vols. 1, 2, 3), Addison-Wesley.

Tukey J. W., 1977, *Exploratory Data Analysis*, Addison-Wesley.

Experiences of Mathematical Modelling at Sheffield University

D. M. Burley
Sheffield University
and
E. A. Trowbridge
The Royal Hallamshire Hospital Sheffield

SUMMARY

We have had experience of teaching mathematical modelling at Sheffield University over the past ten years. This has been primarily for advanced undergraduates whose mathematical and mechanics training is reasonable. Against this background we discuss three questions. (1) How do we teach modelling? Here we discuss the successes and failures of the system that we have used. (2) How do we examine modelling? We make a strong plea for project work to complement traditional examination methods. (3) Finally we ask, what do we teach? We describe our conclusions on what we should teach and give a list of examples of project titles that we have used and a long list of annotated references of our seminar sessions.

1. INTRODUCTION

We have taught a course in mathematical modelling in Sheffield over the past ten years and we would like to describe the methods we have used, some successful, some abortive. First, however, we would like to make some general comments about the philosophy of teaching applied mathematics.

One important question to be answered is: 'Do we lecture and write too well?' Here we are not referring to our technical skill as presenters of material but of the material presented. In traditional mathematics teaching we present our lectures and books in such a way that every problem is carefully stated and an answer is always obtained by a beautifully logical and concise argument. At no time do we ever suggest that we might have made a mistake in our thought processes, certainly not a fundamental one; the only mistakes we own up to are trivial algebraic or arithmetical ones. This whole approach gives students a great inferiority complex and inhibits a spirit of adventure and inquisitiveness in their work. In contrast to *our* presentation, students normally cannot understand the problem or the reason it was set. They usually make innumerable errors both trivial and fundamental and the possible significance of their final result is normally a complete mystery. To a large extent we, as teachers, are to blame for this situation since we set questions which test skill in manipulation only and usually neglect the development and interpretation of the problem. This is particularly true when it comes to examinations, a point to which we will return later.

Let us contrast our approach to lecturing and the writing of books with our approach to research or even our lecture preparation. Usually we start with an idea about a real problem and try to develop it. Nine out of ten of these ideas are unsuccessful and are discarded. We work at what we think is a successful one, we find flaws in our arguments, we remove the flaws, we give up the idea, we come back to the idea, we restrict the hypothesis, we relax the hypothesis, we try again, etc., etc. Provided we do not finish up in a 'give it up' mode we come to a model of reality or a suite of theorems that are related to, but possibly somewhat different from, our original idea.

How do we present the material to students? Do we show the mistakes, the blind alleys, the muddled thinking? Almost certainly not! We present our beautiful logical formulation that gives the impression that our thoughts at all times are in perfect working order. This is how students think *we* think!

The production of a precise set of lecture notes which are regurgitated at examination time provides the student with little experience of the real world of mathematical application. The traditional methods of teaching does little to develop the skills of self investigation, critical appraisal or literature searching. It does not suggest to the student a quantitative language which enables a model of a real problem to be described. He has no comprehension, for example, that continuum physics theories are approximations, or that inviscid and viscous fluids are idealizations of a fluid which can be used to describe the same real fluid in different circumstances. Traditional lectures and the added 'corset' of examinations tend to extinguish any creative flair that a student might possess. He is not encouraged to use his own subjective judgement to develop, solve and then interpret real problems which are amenable to mathematical exploitation.

Having made the case essentially for teaching mathematical modelling, it is recognized that students must acquire the technical skill to be able to solve the mathematical problems that arise in the construction of mathematical models. The formal method, which is criticized above, is an efficient and essential method of teaching mathematical techniques. Our only plea is that alongside our beautiful logical expositions, we give students a chance to see the tortuous processes of our minds in our lecturing and writing and that we illustrate the way we work and think. Our major interest is to produce flexible, adventurous mathematicians. Although our remarks arise from our interests in specific areas of applied mathematics, most of our comments are applicable to other sections of mathematics after suitable amendment.

We will now try to give the answers to three questions that arise when adopting an alternative approach to traditional teaching through lectures about mathematical modelling

How do we teach it?
How do we examine it?
What do we teach?

Before launching into our answers to these questions we should describe the group of students with whom we have developed our experience. They are all studying in their final year of a mathematics degree and all have a uniform background in applied mathematics. In particular they have a reasonable background in techniques and they have all studied an elementary course on mechanics.

2. HOW DO WE TEACH MATHEMATICAL MODELLING?

In this section we will discuss our general approach to teaching modelling and later we will try to fill in details of the material that we have used. The two sections will naturally overlap a little.

Over the years we have developed three main precepts

(i) modelling should not dominate other courses,
(ii) modelling should be limited in the first year,
(iii) modelling needs maturity and technical skill and should be reserved mainly for second- and third-year students.

Students find the alteration in emphasis introduced by modelling ideas a bit of a shock so they need to be weaned slowly from traditional teaching methods. This first observation gives rise to our comment (ii). In the first year it is sufficient to introduce applications that highlight modelling ideas into existing courses. Students should be encouraged to criticize and compare different models of the same phenomena and to interpret any solutions obtained. Our experience is that three or four of these exercises are quite adequate to stimulate interest without interfering with the main work of the course.

The choice of topic for introductory models needs to be made carefully. In physically orientated applied mathematics courses the most natural place for modelling is within mechanics. Unfortunately in mechanics and traditional physics, modelling aspects have been lost in antiquity and it is necessary to find situations that emphasize the idea of a model. Indeed at this introductory level problems in kinematics (e.g. traffic light queueing problems) are much more suitable than detailed mechanical problems or work in biology or medicine. Any problem that is not instantly understandable to the student should be avoided at this stage. There are, however, some realistic ideas in mechanics that can be used. For instance one that we have used successfully for a *general* discussion concerns models of the Earth. A whole range of models can be considered from a point particle when we want the Earth's orbit, to a sphere if we want to study eclipses, to a sphere with a heated fluid layer if the weather is of interest. Even the very traditional type of mechanics question, which usually has no immediate relevance to a real problem, can be used to show modelling concepts. The following example illustrates this.

Example: *A weightless hoop has a heavy particle embedded in its rim. It is placed on a perfectly rough floor and released from rest in position A. When it reaches position B the velocity v is given*

(*a*) *by energy*, $v^2 = 2gh$
(*b*) *since B is the instantaneous centre of rotation*, $v = 0$.
Explain!

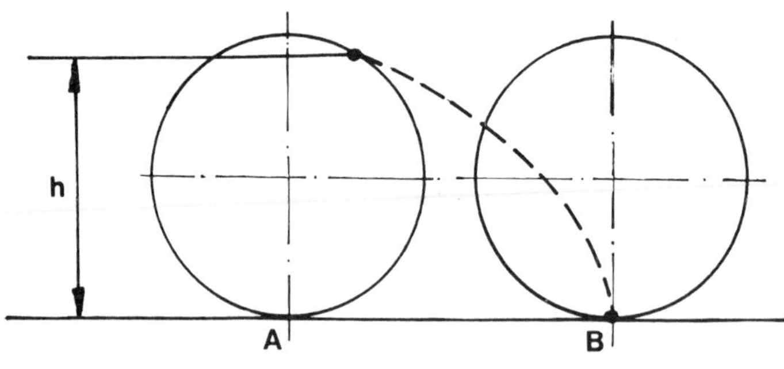

Fig. 1.

The obviously anomalous results need explanation and can be used to ask questions about modelling via point particles, weightless hoops, rough floors etc.

Once modelling is attacked fully the only way that students learn the necessary skills is by attempting to develop simple models themselves and this almost certainly requires a project approach. Each student has to go through the trauma of setting up his own model, trying to obtain results and finding results for verification. To do this from 'cold' is difficult and we have found it necessary to introduce the ideas into preceding years, as noted in earlier paragraphs.

The major drawback of projects is that they are very time consuming for both students and staff. Students put a lot of effort into the work, usually because they find it interesting, but great care must be taken that this work is not to the exclusion of all else. We have found it useful to specify an amount of time that we consider reasonable for their project, although we suspect that we give a gross underestimate in many cases. It will be certain that several staff will need to be involved since students will require a half to one hour per week for consultation with someone more experienced than themselves. For these reasons we re-emphasize our earlier comment that modelling and project work must not dominate courses.

One of our other experiences is that projects, although allowing the development of creative skills, introduce very little *new* mathematics. Indeed our general philosophy is that this year's models should be done with last year's mathematics since students must feel in control of at least one aspect of the modelling process otherwise they flounder completely. We have found, however, that students have learned a great deal about mathematics through projects. For instance, in one particular project, students came to grips with the significance of the Fourier series and with the even more mysterious δ-function and why it is useful. Detailed manipulative skills are well taught in traditional courses and our aim is to add to this knowledge by showing how the skills can be used in applications to problems of real interest.

Besides project work, which we consider to be essential for a proper modelling course, we have found it necessary to run a series of seminar sessions. Here material is taken from the literature and is given in a predigested form. It is a slow, laborious process to extract this material and convert it into a useful form, particularly in non-conventional areas like biology and medicine; some comment will be made on this in the last section. The difficulty of seminar sessions is that you give your model and it does not come from the student. However, the student needs to see an 'expert' at work but it is very important to point out the failures and inadequacies as well as the successes. We have tried to involve the students as much as possible, in particular in considering extensions, modifications, restrictions to the model to see what is lost and gained by them. The main problem is that good material for this type of session is often hard to find.

In either project or seminar work it is important that material is well documented and easily accessible. Good library facilities are essential; all key papers or sections of books should be copied and put into the library reserved section. If one person removes the original reference from the library it is lost to anyone else for the rest of the term.

The programme mapped out above is difficult to follow and only rarely have we been able to satisfy all our criteria and fine words.

In practice there will inevitably be that group of students who show little concern for modelling aspects of the course. They see their job as obtaining a correct answer to the mathematical problem and it is someone else's job to tell them how the mathématical problem arose and what the results mean. This is a comforting philosophy which is greatly encouraged in some circles. We believe that this unadventurous and restrictive approach is encouraged by our examining system.

3. HOW DO WE EXAMINE MODELLING?

Our last comment leads directly to the problem of examining modelling courses. In our educational system it is certain that a course will need to be examined and a mark for the course obtained in some way.

In formal examinations students are tested against a restricted syllabus and they are set specific technical tasks that can be completed in thirty minutes. Such a test can be performed correctly (100 %) or totally incorrectly (0 %) and can rarely account for originality or creativity. Can you imagine ever giving 100 % for an essay on the causes of the French Revolution? Would you ever think of setting an example which is not stated with absolute precision or where the answer is not clear-cut? In real life, imprecision and lack of conciseness are exactly the difficulties that occur in both setting up a problem and in interpreting the solution. For an extreme view of the formal examination system in this country we would like to refer to Professor Scorer's (1971) article.

Projects, on the other hand, are much more realistic in their attempt to test a student's ability to explore real life problems. They also have the great merit of testing *communication* skills—recall McLone's (1973) survey of students and employees. Where else do we give students an opportunity to write up an extensive piece of work? No matter whether a student continues in mathematics or not, reading and writing reports is one of the important jobs that many will have to do. In fact, after a little gentle criticism of their efforts, we have found that students cope well with the writing of reports.

When it comes to examining projects we have not found this particularly difficult; it is *very* time consuming and is one reason that we believe that modelling should not totally dominate undergraduate courses. We have two markers on each project who assess the work independently and after consultation come to some agreed mark. In our situation several students investigate the same problem and we have found it essential to give a *viva voce* examination on their work. We are interested in whether a student understands what he has written, whether or not he has cooperated with others in the construction of the model. Also we encourage the use of material from books or papers, as long as the student understands what he has written and makes due reference to his sources. Indeed talking to others, working in groups, reading the literature form an essential part of this type of work. Examining by project also tests communication skills in both written and spoken word as well as standard mathematical and modelling expertise and we believe it complements traditional examining procedures.

The major difficulty with project work is *large numbers* of students. We have found it impossible to set individual projects for our groups of 15 to 50 students. Thus we have allocated the same project(s) for the whole group and hence our consequent *viva voce* examination. Most authors (e.g. Symposium on Mathematical Modelling (1979)) who write about modelling seem to deal with small numbers, usually less than 10, and are playing a different game from our own.

Our current system is to provide 20 lectures for 10 weeks giving the seminar sessions described in the previous section. The students are also given:

One or two projects to be completed during the lecture course. These projects are closely allied to the lectures. They count for about 30% of the total marks for the course and there is *no viva voce* examination. Extensive comments on these projects are returned to the students as constructive criticism and enables modelling maturity to develop as the course progresses.

A final project to be completed in about four months after the lecture course has finished. The project does not follow directly from the course but is in the general area covered by the lectures. A set of starting references are supplied. The project counts for 70% of the marks for the course and a *viva voce* is given but this rarely changes the mark by more than $\pm 5\%$.

In assessing the projects we use as a rough guideline the following mark scheme: Presentation 10, Discussion of the real problem 15, Formulation of the model 30, Mathematical computation 15, Interpretation 30. This scheme is known to the students in advance. It is often difficult to be completely precise in marking each category but the scheme has been found to be satisfactory.

Typical examples of projects that have been set are:

Pulsatile blood flow
Diffusion through kidney walls
Blood–alcohol in the body
Animal locomotion
Spin dryer mechanics
Tumble dryer mechanics
Wheel balancing
Models of jumping in track and field events
Mechanics of valves

In previous years more formal methods of examining have been tried, largely to reduce the amount of work in marking. There were two parts to the examination. The first part consisted of giving the students a general topic on modelling around Christmas time. They were then asked to prepare an essay on the topic which was to be written during the formal examination in the following summer. The essay could be based on the seminar sessions but at least one substantial example was expected from outside the lectures. Little technical detail was expected in the essay, it was largely a philosophical discourse on modelling aspects and interpretation of results. Typical topics used were:

Modelling with springs and dashpots
The use of balance equations
Problems in modelling in biology and medicine
Linearization of equations in mathematical modelling
The concept of a lumped parameter model

This idea, on the whole, was very successful, one or two of the essays were good enough for publication.

Together with the essay, unseen models of a type similar to the seminar sessions were set. The student is given a model and he is expected to work it through, consider extensions and interpret any results obtained. In a formal examination unless an initial model is given, a crop of zero or very low marks resulted; poor students, in particular, never started. Because of the open-endedness of the questions set, full marks were given for about half of an 'ideal' solution produced by the lecturer.

This part of the examination proved to be a very poor test, firstly because the model was the lecturer's and not the student's. Even worse was the fact that most students tended to do the mathematical bits and leave the modelling and interpretation aspects entirely alone. For this reason the decision to move to project work was made, despite the vast increase in staff time needed.

4. WHAT TO TEACH?

We would like to discuss and give examples of the seminar sessions that we have used over the past ten years, many of which have formed the basis for one of the projects. Our main conclusions are:

(i) Suitable material is hard to find.
(ii) The material must be well documented.
(iii) The material must be capable of extension and variation without significantly increasing the difficulty.
(iv) The work must be capable of analytical solution. Only limited computed solutions or quoted solutions are acceptable.
(v) Our own interest is in deterministic work so we include no stochastic problems.

Fortunately over the past five years or so an increasing amount of material has been published which provides a starting point for models. Sources that we have used extensively are Noble (1967), various articles in Symposium on Mathematical Modelling (1979), Andrews & McLone (1976) and James & McDonald (1981) and we welcome new initiatives that bring together appropriate material in a convenient form. As pointed out earlier it takes considerable effort to find the material and adapt it for classroom use. We have found it particularly difficult to find models which have comparatively simple analytic solution. We feel this to be important since by varying parameters a great deal of understanding is achieved and provides the bulk of the exercises that we set after our seminar sessions. It is beneficial to set extensions and modest variations to the work of the lectures but such adaptations are very difficult to find without producing mathematics an order of magnitude more difficult or requiring a computer solution.

A final difficulty in finding material is to obtain experimental results that can be used for verification purposes. Even in some of our better and most successful models it is this aspect that has troubled us most. Often we have been thrown back on comments such as 'these results are plausible', 'these results show the correct trends', 'these results are intuitively satisfactory'. Certainly this aspect has been the weakest part of our courses. On the other hand it has produced some of our best discussions since students have been prepared to argue what is 'plausible' or what are 'correct trends'.

At the more advanced level of final-year courses, where most of our experience has been gained, we have often introduced material that involves substantial background reading. We have found models in the biological and medical area popular with the students and these, of course, need a lengthy introduction. We have found it not particularly fruitful to delve deeply into the biological or medical aspects from the teaching viewpoint but to give the broad picture of a phenomenon and then to concentrate on one particular aspect. If you try to go beyond this rather simplistic approach the technical detail of the system dominates and almost becomes a lecture course in itself. This is interesting *per se* but it is not what we are aiming at. Indeed we have found that any model given in our seminar sessions should be restricted to not more than two lectures. After this length of time so much technical detail has been produced that the topic has almost become a research effort. Students soon lose interest in this detail, largely because they have not had sufficient time to absorb the background material. When actively engaged in their own modelling there is no problem in maintaining interest but in the more passive seminar sessions this is not so, especially for the less able or poorly motivated student.

We conclude our discussion by quoting a series of models that we have used at various times in our lecture courses, some of which we have used as project material. The ones we quote are the more successful models but we have many more.

Biological models

Mathematics of a kidney machine, Burley (1975)

Simple equation $\mathrm{d}u/\mathrm{d}x = cu$, good comparison with experiment.

Valve in the use of hydrocephalus, Burley (to appear), Hakim & Burton (1974)

Three-stage model. Mainly mechanics involving springs and dashpots.

Reservoir model of the heart, Rashevsky (1964), Cope (1965)

An old model, easy to motivate. Simple mathematics. Plenty of experimental data in Cope.

Hormone–glucose interaction, Ackerman *et al.* (1966), Davies (1976)

Typical lumped parameter model, essentially concerned with stability. Davies article a difficult extension.

Blood flow models. Open University (1974), Singer (1969)

Rigid wall model. Flexible wall model giving transmission line equations. Both essentially one-dimensional.

Vibrational models

Seismograph, Volterra and Zachmanoglu (1965), Thomson (1966).

Simple forced vibrations. Need to amplify vibrations.

Vehicle suspension, Volterra and Zachmanoglu (1965).

Spring and dashpot model. Several variants possible. Interesting to ask, what do we want from our solution?

Foundation vibration/vibration transmission, den Hartog (1956), Thomson (1966).

Effect of vibration of foundations and how vibrations are transmitted. Many examples of earthquakes on buildings, engine vibrations transmitted to engine housing.

Effect of nonlinearities in springs and dashpots; dry friction, den Hartog (1956), Davis (1962).

More demanding parts of vibration theory. Of considerable practical interest.

Vibrations caused by moving loads, Fryba (1973).

Again a more demanding part of vibration theory. Fascinating study of cable cars on overhead cables, trains on railway tracks, etc.

Crankshaft drives, den Hartog (1956), Thomson (1966).

How a crankshaft drive causes vibrations. Some simple mathematics can be used here. Reasons for 4, 6 or 8 cylinder engines can be deduced.

Heat flow

Theories of cooling fins, Bird, Stewart & Lightfoot (1960), Kern & Kraus (1972).

Simple, essentially one-dimensional heat conduction. Plenty of examples can be constructed.

Ablation of heat shield, Andrews & Athey (1976).

Ablation of a semi-infinite solid (we have no reference for this) is simple. Can get orders of magnitude of thickness of heat shield for re-entry capsules. Next step difficult.

Electrical, electromechanical interaction

Ignition system of an internal combustion engine, Noble (1967).

First- and second-order differential equations. Complicated situation, analysed piecewise. Interpretation interesting.

Cone loudspeaker, McLachlan (1951), Beranek (1954).

Simple model giving electromechanical interaction. Nice to set up but interpretation not easy, experimental results sparse.

Capacitor microphone, White & Woodson (1959).

Familiar in tape recorders. Good example of linearization and stability. Need to understand frequency response to compare results.

Mechanics

Mathematics in sport, Trowbridge & Paish (1981).

Models of running, jumping, throwing can be used to illustrate many mechanical principles.

Satellites, Burghes & Downs (1975).

Traditional applications of mechanics but nonetheless useful.

Spin/tumble dryers.

No references for this topic. Industrial firms supply data willingly. Simple models easy to construct. See value of industrial contact.

Miscellaneous

Model of grazing, Noy-Meir (1975).

Extends scope of predator/prey models. Several variants possible. Interesting to set up and interpret. Good experimental data difficult to obtain.

Model of the ear, Zwislocki (1965), Viergever & Kalker (1975).

Facinating model of how the inner ear works. Difficult and difficult to extend but worth the effort for advanced students.

REFERENCES

Ackerman E., Gatewood L. C., Rosevear J. W., Molnar G. D., 1966, in *Concepts and Models in Biomathematics.*, Ed. F. Heinmets, Marcel Dekker.

Andrews J. G., Atthey D. R., 1976, in *Mathematical Modelling*, Ed. J. G. Andrews and R. R. McLone, Butterworths.

Andrews J. G., McLone R. R. (Editors), 1976, *Mathematical Modelling*, Butterworths.

Beranek L. L., 1954, *Acoustics*, McGraw-Hill.

Bird R. B., Stewart W. E., Lightfoot E. N., 1960, *Transport Phenomena*, Wiley.

Burghes D. N., Downs A. M., 1975, *Modern Introduction to Classical Mechanics and Control*, Ellis Horwood.

Burley D. M., 1975, *Math. Spectrum*, **8**, 69.

Cope F. W., 1965, *Adv. in Biol. & Med. Phys.*, **10**, 278.

Davis H. T., 1962, *Introduction to Nonlinear Differential and Integral Equations*, Dover.

Davies M. J., 1976, in *Mathematical Modelling*, Ed. J. G. Andrews and R. R. McLone, Butterworths.

den Hartog J. P., 1956, *Mechanical Vibrations*, McGraw-Hill.
Fryba L., 1973, *Vibrations of Solids and Structures under Moving Loads*, Noordhoff International.
Hakim D., Burton J. D., 1974, *Eng. in Med.*, **3**, 3.
James D. J. G., McDonald J. J. (Editors), 1981, *Case Studies in Mathematical Modelling*, Stanley Thornes.
Kern D. Q., Kraus A. D., 1972, *Extended Surface Heat Transfer*, McGraw-Hill.
McLachlan N. W., 1951, *Theory of Vibrations*, Dover.
McLone R. R., 1973, *The Training of Mathematicians*, Social Science Research Council, London.
Noble B., 1967, *Applications of Undergraduate Mathematics in Engineering*, Macmillan.
Noy-Meir I., 1975, *J. Ecol.*, **63**, 459.
Open University, 1974, M321 Units 15 and 16, Parabolic Equations, Blood Flow.
Rashevsky N., 1964, *Some Mathematical Aspects of Mathematical Biology*, Thomas.
Scorer R. S., 1971, *Bull. I.M.A.*, **7**, 50.
Symposium on Mathematical Modelling, 1979, *Bull. I.M.A.*, **15**, 238.
Thomson W. T., 1966, *Vibration Theory and Applications*, George Allen & Unwin.
Trowbridge E. A., Paish W., 1981, *Bull. I.M.A.*, **17**, 144.
Viergever M. A., Kalker J. J., 1975, *J. Eng. Math.*, **9**, 11 and 353; 1977, *J. Eng. Math.*, **11**, 11.
Volterra E. G., Zachmanoglu E. C., 1965, *Dynamics of Vibrations*, Merrill.
White D. C., Woodson H. H., 1959, *Electromechanical Energy Conversion*, Wiley.
Zwislocki J., 1965, in *Handbook of Mathematical Psychology*, Vol. III, Ed. R. D. Luce, R. R. Bush and E. Galanter, Wiley.

The Assessment of Modelling Projects

G. G. Hall
Nottingham University

SUMMARY

The skills involved in modelling are summarized under the headings of content, presentation and drive. These are different aspects of the same project rather than separate entities and consequently it is proposed that their assessments should be combined geometrically rather than arithmetically. The importance of securing acceptable and objective marks is stressed. In practice, double-blind marking followed by a debate to resolve differences has been found a satisfactory procedure of establishing marks.

1. INTRODUCTION

In the final analysis, the justification for including modelling in a mathematics course is the pleasure which the student has in doing it. It is this which will increase the motivation (Hall, 1982) and boost the self-confidence, so important in later careers. Nevertheless, in a warped world, where values are examination-based, some reward in terms of marks is important. Students must be assured, particularly in advance, that their activity is relevant to their course and an integral part of their training. Colleagues may also need to be reminded that this activity has a right to compete for a student's time and attention. Without some reward of this kind modelling will not get started.

The objective of assessment is to evaluate a student's skill and knowledge. In the traditional mathematics examination the candidate is required to reproduce standard arguments, to adapt them to novel ends and to follow standard procedures often in new instances. Since no great original thinking is expected, or found, the time limit for the paper is an integral part of the testing. On the other hand, the only direct method of assessing modelling skills is through a project done over a long period. In this the ability to think originally about a problem, to pursue it to a conclusion and to present the findings cogently and elegantly can be displayed and assessed.

Carrying out the assessment of a project, however, does pose problems for mathematicians. In the traditional examination the examiner exercises considerable judgement about what will be evaluated, and how, through the choice of questions. In assessing a project the examiner must judge these matters separately in relation to each student since projects may differ considerably in the opportunities of displaying particular skills which they offer. The dangers of bias on the part of the examiner are obviously much greater in this situation.

In this paper the problems of assessment are discussed both theoretically and practically. Some suggestions are made on how assessments can be made more reliably and on how they can be defended against various criticisms.

2. WHAT IS TO BE ASSESSED

A recent conference report (UMTC, 1979; see report of Panel U1) has listed and discussed the varied skills involved in modelling. It is relatively easy to decide, for a given project, how it should be rated under these headings. Unfortunately it is difficult to decide what weighting should be given to each skill since the scope for displaying a skill varies greatly from topic to topic. In one project the initial steps may be the difficult ones, deserving of potentially high marks, while in another it may be that the clarity and precision in presenting an argument is the decisive feature. To overcome this difficulty and to simplify the job of marking I have found it helpful to group the skills under three headings and establish the weight of the group rather than of the individual skills. Since the group includes several skills its combined weight varies much less than that of single skills and so can be taken as constant. These groups and their component skills are:

1. Content
   - Ability to handle and make sense of natural or experimental data.
   - Determination of variables and parameters with which to describe observation.
   - Recognition of patterns in data and in processes.
   - Generation of mathematical expressions to summarize observations.
   - Ability to set up a model representing the system and relating its significant variables.
   - Technical ability to manipulate the mathematical expressions of the model to achieve desired objectives.

2. Presentation
   - Representation and interpretation of data.
   - Translation of information into and out of pictorial form.
   - Ability to communicate clearly, especially in writing.

3. Drive Ability to identify situations and to formulate problems.
Ability to consult books etc. for additional techniques or information.
Understanding of when to change a model, method or objective in discussing a problem.
Recognition of what constitutes a solution—evaluation of success of models.
Ability to work effectively in a group.

The first group collects together the more technical aspects of modelling, the second is concerned with the written project itself while the third allows for originality and management.

The evaluation of the content skills is relatively straightforward since these skills are similar in kind to those evaluated in conventional examinations. Realistic measures of what a student is capable of achieving are available. Evaluating the presentation introduces new problems. Some students are barely literate while others produce lively, entertaining English. Some have an untidy scrawl and others pay to have their work typed professionally. Since an object of project work is to give experience of the kind of report writing expected in industry it is important to set high standards of presentation preferably by indicating in advance what is expected. The evaluation of drive is even more difficult since some elements will not be shown in the project and can be judged only by the supervisor. Only the supervisor can say whether the student obtained considerable help or very little, whether or not alternatives were considered before deciding on the topic and the model and whether it was tackled vigorously or reluctantly. Some mathematicians feel that these considerations are not mathematical and so should not enter into a mathematics assessment but, if the project is to prepare students for the demands of research and industry, they must be included. Another difficulty for some markers, is that, because they are familiar with closely supervised projects where the drive is supplied by the supervisor, they may underestimate the achievement of a student who originates and manages his own project.

The relative weights of these components is a matter of judgement. In my work I have given content and presentation equal weight and twice the weight of drive. This is influenced by the difficulty of assessing this last. An equal weighting of all three is another possibility and this may be a fairer representation of the number of skills contributing to each assessment.

3. THE CALCULUS OF MARKS

Most marking of mathematics is based on a simple model of how marks are obtained. There is a marking scheme which lists items such as equations, results, arguments, diagrams etc. and each earns a fixed mark if present and correct. The total mark is the sum of these item marks with, usually, a small bonus for a complete solution or a pleasing presentation. If, at an early stage, a calculation goes wrong most schemes allow 'method marks' or marks for a 'correct continuation' so that an incorrect solution may earn substantial marks. I will refer to this as the *addition model.*

The addition model of marking has an obvious justification when the examination is primarily intended to assess the candidate's memory. When it is used more generally it rests on the assumption that a candidate of greater ability can produce more answers, without much sacrifice of accuracy, inside the severe time limits of the test. This assumption is much more difficult to justify philosophically and could be tested empirically only if another measure of ability could be found.

The marking of a project cannot conform to the addition model. Because of the diversity of projects, no detailed marking scheme can be drawn up in advance and applied to each project uniformly. Furthermore the assumption that quantity and accuracy measure quality is here unacceptable. In a project there is no excuse for incorrect manipulation, illogical argument or illegible scribble even if these are tolerated in the heat of a timed examination. The penalty for these faults must be made more extreme than the addition model allows. The fact that the addition model permits (in the weighting 2, 2, 1 suggested above) a mark of 60% for a project whose presentation and drive are given top marks but whose content is totally wrong is a major defect and a source of genuine criticism of project assessment. Similar comments can be made about a project which is mathematically sound but whose English is impenetrable or unreadable or one where the supervisor contributed every novel feature.

Two criteria can be required of a project-marking model. From the discussion above one desirable condition is that the mark should be zero when any component has a zero mark. The addition model does not satisfy this. The other criterion is that the model should be *homogeneous.* This means that, if each component is given the same mark, as a fraction of the maximum for that component, then the total should be the same fraction of 100%. If the marks for the components are written as a vector (x, y, z), where x is the percentage for content, y for presentation and z for drive, then a homogeneous total for (h, h, h) will be h. The addition model is homogeneous in this sense for, with the weighting 2, 2, 1, the mark for (x, y, z) is

$$A = \tfrac{1}{5}(2z + 2y + z)$$

and, on substitution, this is seen to be homogeneous.

A model which is also homogeneous but satisfies the zero condition is the *product* model. In this model the mark for (x, y, z) is

$$P = \sqrt[5]{(x^2y^2z)}$$

or $\log P = \frac{1}{5}(2\log x + 2\log y + \log z)$

and for $x = y = z = h$ the total $P = h$. If a larger number of components or different weights were required the log form shows, more clearly, how the correct generalization can be written down. Although this model may appear quite different from the addition model there are connections. When the components x, y, z are approximately equal P is almost equal to A. From the familiar relation between arithmetic and geometric means P can never exceed A. When the variance of the component marks becomes larger, P becomes much smaller than A till it vanishes as each component becomes zero. Table 1 shows some comparisons. This table can be readily extended using the homogenous property and also a *scaling* relation, satisfied by both A and P, whereby if all components are multiplied by the same constant then the combined mark is also multiplied by it. The table shows numerically the extent of the penalty which P imposes on an unequal performance.

It is important that this penalizing effect of P should not be exaggerated. It is readily shown that it increases with the number of components. The choice of three components penalizes a large variance sufficiently enough to direct a student to rectify it by improving his weak skills but, provided the least mark is more than 1, without becoming overwhelming.

4. PRACTICAL CONSIDERATIONS

The marking of projects depends on making judgements. To avoid the dangers of bias attendant on 'impression' marking it is important to ensure that these judgements are based on a critical appraisal of specific aspects of the project. The procedure suggested here is designed to achieve this.

An advantage of first assessing the project under three headings is that it forces the marker to measure the work against some standard in very specific aspects. It is highly desirable that the mark should be supported by a brief comment drawing attention to good or bad features and so explaining the mark. The numerical scale of marks also requires attention. In many degree schemes project marks are combined with marks from written papers and so ought to conform to the same range of marks for the same standard of performance. Thus, if an upper second standard on a paper has marks in the range 60–70, then a project whose content is considered to be upper second should be marked in the same range. Judgements in terms of degree classes are often found easier to make and a translation into marks can be performed later.

Projects should be double-blind marked. This is obviously so where markers are inexperienced or where sceptical colleagues have to be convinced of the reproducibility of the marks but it is also necessary as a guarantee to students that their work will be objectively assessed. The two marks should be compared and discussed so that a single mark can be returned. Often the ranking orders of the marks, given by two markers, for a set of projects are found to be the same but slight adjustments of scale are needed to align the marks. Occasionally the project supervisor will have comments to add on aspects, relevant to drive, not visible in the project which the other marker may have ignored or misunderstood. The comments, mentioned above, provided by each marker are specially useful at this age. If necessary to resolve differences of marks the project should be read by a third marker.

5. CONCLUSION

As our colleagues in the humanities know, marking that relies on judgements is difficult and time-consuming. If project work is to survive in mathematics courses it must be shown that it can be done objectively. The procedure outlined here involving a three-fold assessment of content, presentation and drive followed by a product model of combining them tries to make the marking procedure more uniform and more self-critical. The discipline of double-blind marking provides a feedback mechanism so that the system cannot drift too far away from good standards.

REFERENCES

Hall, G. G., 1982, *Intern. J. Math. Ed. Sc. Tech.*, **13,** 599.

UMTC (University Mathematics Teaching Conference), 1979, *Using and Learning*, Shell Centre, Nottingham.

The Hewet Project

J. de Lange and M. Kindt
Utrecht University

SUMMARY

The Hewet project is a project that will lead to a new curriculum for pre-university students. Applications and modelling play a vital role in this curriculum.
In this chapter we will sketch the following aspects of this project.

1. Historical background
2. The new programs
3. The experiments—the organization
4. The development of student materials
5. Testing
6. The program in action
7. An example: Leslie matrices
8. Conclusions

1. HISTORICAL BACKGROUND

In 1968 a new curriculum was introduced in the Netherlands for mathematics in secondary high school (pre-university students). Students got the choice for the last two years: math I and math II. The possibilities were: none of them; math I; math I and math II. Math I consisted of calculus and statistics. Math II was mainly linear algebra. Math I aimed at students heading for exact sciences, agricultural sciences and medical schools. Math II aimed at students interested in a wider scope of mathematics. Students heading for economical and social sciences didn't need any math-education. But the times were changing and there was identified a need for mathematics in these disciplines too. So these disciplines now request math I as well. This was a bad situation because math I was never meant to be a preparation for these students, and many were unable to complete the math I course successfully.

During the nineteen-seventies several reports were written on the problem, but nothing actually happened. But, in 1979 a commission was installed that published a report—the so called Hewet-rapport—in February 1981.

The conclusions were: there is a need for two new courses in mathematics: Math B: for students who are heading for exact and technical sciences; Math A: for students who are heading for economic and social sciences. Experiments should start right away (summer 1981) and the experiment should be ended within six years.

The minister of Education accepted the conclusions and so the start of the Hewet project took place in 1981.

2. THE NEW MATH A AND MATH B PROGRAMS

Math A: aimed at students who are heading for economical and social sciences. They need mathematics as a tool. The four areas that are part of the program:

(1) elementary and applied calculus;
(2) matrices with applications;
(3) probability and statistics;
(4) automatic data processing.

(1) Graphs of functions, and the conclusion you can draw.
The derivative; the rules for differentiation.
Optimalization problems.
Periodic and trigonometric functions.
Exponential and logarithmic functions, especially as they occur in biology and economy.
(2) Matrices, but not as mappings.
Sum and product of matrices.
Incidence matrices; connectivity matrices; migration matrices; probability matrices; data matrices; population projection matrices.
Systems of linear equations.
Linear programming.
(3) How to look critically at statistical material.
Frequencies; histograms; cumulative frequencies.
Permutations; combinations.
Stochasts, binomial and hypergeometrical distribution.
Testing hypotheses.
(4) Use of standard programs within (1), (2) and (3).
Some elementary programming.

Math B:

(1) Calculus, as it is now within Math I.
(2) Three-dimensional geometry.

This is back in the curriculum after some 15 years. It is, more or less, a mixture of vector geometry and solid geometry.

3. THE EXPERIMENTS—THE ORGANIZATION

The need for careful experiments was clear as the proposed program for math A was rather revolutionary—but not as far as the mathematical contents are concerned. But the fact that applications, mathematization and modelling should play a vital role in this new curriculum was a matter of major concern. This became clear when the teachers and schools reacted to the proposals (in 1980/81): although a big majority was in favour for the new curriculum, concern was shown for this applications aspect. The importance of in-service training for teachers was stressed as well.

Eventually the following model was chosen:

In Aug '81 the first *two* schools start with the new experimental program. (first central examination: May '83).
In Aug '82 the first two schools continue.
In Aug '83 the next *ten* schools start. (first central examination: May '85).
In Aug '84 the next *forty* schools start. (first central examination: May '86).
In Aug '85 the remaining ($\simeq 430$) schools start. (first nationwide final central examination: May '87)

It was clear from the start that the in-service training should be started *before* the schools entered the experiment. This, however, was impossible for the two schools, as there was no time between the outcome of the report and the start of the experiment.

But during the year 1982/83 the teachers of the ten schools were prepared for the things to come during sixteen $3\frac{1}{2}$-hour sessions. This experimental training course will be the model for the next round of these in-service training courses: during 1983/84 some 180 teachers—mainly from the 40 schools—will do the course, and from 1984/85 there are courses planned for some 600 teachers a year.

The $2+10+40$ schools will work with student material that has been developed by the so-called Hewet team. This Hewet team consists of Martin Kindt and Jan de Lange who were advisers to the committee that made the report, developed the student materials, organized and taught the first in-service course, followed all the lessons involved in the two experimental schools, made reports to the officials and the field and other things. Since 1982 they are being assisted (part-time) by Heleen Verhage. Guus Vonk assists for the automatic data processing part and Ellen Hanepen does a lot of organization—altogether more or less a team consisting of four people.

Furthermore there is a group consisting of people from universities, high schools, the ministry of Education and inspection. They see the Hewet team every two months, give support, criticize where necessary and check that the experiments keep within the boundaries of the Hewet report.

4. THE DEVELOPMENT OF STUDENT MATERIALS

To name the subjects for a new curriculum is one thing, to make student material for it is quite another, especially since there was hardly any time: the message that experiments in school could start in August 1981 came not earlier than April 1981. But ever since the instalment of the committee (1979) the Hewet team had been looking for material that could be used in the books. Since the booklet 'Matrices' was available at the conference in English we concentrate on that subject.

Since applications had to play a vital role the usual mathematics textbooks didn't offer much inspiration. One had to look at books used in other disciplines. Browsing through many books on economy, social sciences, biology, medicine, geography and others led to a list of possible subjects. The problem was, of course, that many were at too high a level. But maybe those can be used in the future. (Markov-chains, Leontief input–output matrices) Applications that were—almost—fit to use were:

—distance tables
—stock matrices
—connectivity matrices
—frequency matrices
—direct-routes matrices
—relation matrices
—probability matrices
—migration matrices
—population projection matrices

This seems to many people a (too) wide scope. But the mathematical aspect is very limited. The problem was, as stated in the Hewet report: 'Attention should be paid to the meaning of the matrix-operations within different contexts'.

The 'matrix operations' involved are just:

—scalar multiplication
—adding matrices
—multiplying matrices

Is this all there is? In a way: yes. But the pure mathematical content is just a minor part of the subject.

The first concept of a textbook was ready by August 1981, just one week before the experiments started. This is the way it has been ever since. It was tested in the two schools, and the developers of the text were present all the time. They took notes, interviewed the students and teacher, helped the teacher to prepare the lessons. This led to a new text, the second version which was ready one week before the second round started August 1982. The same procedure led finally to a third version (May 1983). From this version an English translation is available.

In this way booklets with the following titles were produced for Math A.

—Probability (three versions)
—Functions on two variables (three versions)
—Differentiation (three versions)
—Graphical representation (three versions)
—Probability distributions (two versions) (including: testing hypotheses)
—Automatic data processing (three versions)
—Linear programming (two versions)
—Periodic functions (one version)
—Exponential and logarithmic functions (one version).

For Math B:

—Three-dimensional geometry (two versions).

5. TESTING

Right from the start of the experiments it was clear that one of the problems should be: How to test mathematics A?

Many teachers are afraid of math A just because of this fact. It seems impossible by just changing some numbers in an exercise that you get one for a test. During the first year this turned out to be a real problem. The teachers involved made tests together with the Hewet team. This worked out satisfactorily but is not a solution for the future. But all tests were made into a booklet and this is available for all teachers. This book will continue to grow in the next years and form a source of inspiration for future teachers. As a matter of fact, the teachers of the second round of the two schools are making the tests all by themselves. Besides ordinary written tests—certainly not multiple choice—there were some experiments on other forms of testing. At both schools children made a booklet on a problem—at one of them, on the possible distinction of the Hooded Seal, where the questions in the book 'Matrices' lead the way. In the other school the students were asked to rewrite an article on 'Transmigration in Indonesia' and to use graphical representation whenever this seemed to fit. The original article contained lots of numbers, but no graph(ic)s. Both reports turned out very satisfactorily. In May 1983 the first central examination took place.

6. THE PROGRAM IN ACTION

After two years of experiments—at only two schools—it seems possible to give a first impression.

The students seem to enjoy math A: in the first weeks they have problems. They are not so sure as they used to be. More than one answer seems possible. And all the texts you have to read! And the thinking you have to do. And there are no tricks to master. You have to work through the book by yourself. There is no use in copying from your neighbour.

But the advantage is: you use your mathematics, and you understand what you're doing.

The results in terms of tests were—at least—satisfactory. The program as it worked out in action is more or less as follows:

MATH A; 4 hrs per wk; 2 years	
Line I (algebra, calculus) ≏ 3 hrs per week 2 years	Line II (prob. + statist.) ≏ 1 hr per week 2 years
automatic data processing	
matrices periodical functions differentiation 1 functions of 2 variables exponential and log. functions linear programming differentiation 2	graphical representation permutation + combination prob. distributions (binomial + hypergeometr.) testing hypothesis (normal distribution)

time ↓ (both sides)

7. AN EXAMPLE FROM MATH A: LESLIE MATRICES

In the two final chapters of 'Matrices' migration and population projection matrices are treated. A real-life problem—the possible extinction of the Hooded Seal—forms the final chapter of the booklet. Before giving a report on the classroom activities with this chapter we will describe the Leslie matrix.

The first step is to arrange the information about age composition of the population in a single column of figures. Next, a square grid of rows and columns—the matrix proper—is constructed, with one row and one column for each age-group in the adjacent population column. Across the top row of the matrix are entered the number of children born to mothers in each age-group during the chosen time interval. Then the number of individuals in each age-group that are alive at the end of the chosen time interval, and have thus advanced from one age-group to the next, are entered diagonally on the grid, from top left to bottom right, in the spaces just below the main diagonal.

An example (from the book 'Matrices').

American women: (numbers in thousands)

Age	Females	Surviving daughters (1940–55)	Surviving females (1955)
0–14	14459*	4651	16428
15–29	15264	10403	14258
30–44	11346	1374	14837

* 14·5 millions.

This gives the following Leslie matrix:

$$L = \begin{pmatrix} \frac{4651}{14459} & \frac{10403}{15264} & \frac{1374}{11346} \\ \frac{14258}{14459} & 0 & 0 \\ 0 & \frac{14837}{15264} & 0 \end{pmatrix}$$

Or, more in general:

$$L = \begin{pmatrix} f_0 & f_1 & f_2 & \cdots & f_{n-1} & f_n \\ p_0 & 0 & 0 & & 0 & 0 \\ \vdots & p_1 & \ddots & & & \vdots \\ \vdots & \vdots & & \ddots & & \vdots \\ 0 & 0 & & & p_{n-1} & 0 \end{pmatrix}$$

f_0: reproduction

p_0: surviving probability (promotion probability).

Multiplying the initial composition (as a vector) with this matrix gives future population compositions.

From the textbook:

From earlier research surviving probabilities were known for the situation without hunting Hooded Seals.

For newborn puppies there is a 85 % chance to survive the first year.

For the other animals the chances are slightly better: 89.4 %. This is also the chance for animals in the class of 11 years and older to stay in that class.

The reproduction of the Hooded Seals:

class	0	1	2	3	4	5	6	7	8	9	10	11^+
‰	0	0	53	240	249	352	369	418	422	422	430	438

Make the Leslie matrix from this information.

The students had no problems in constructing this matrix. The result, as computer output is:

```
0.000 0.000 0.053 0.240 0.249 0.352 0.369 0.418 0.422 0.422 0.430 0.438
0.850 0.000 0.000 0.000 0.000 0.000 0.000 0.000 0.000 0.000 0.000 0.000
0.000 0.894 0.000 0.000 0.000 0.000 0.000 0.000 0.000 0.000 0.000 0.000
0.000 0.000 0.894 0.000 0.000 0.000 0.000 0.000 0.000 0.000 0.000 0.000
0.000 0.000 0.000 0.894 0.000 0.000 0.000 0.000 0.000 0.000 0.000 0.000
0.000 0.000 0.000 0.000 0.894 0.000 0.000 0.000 0.000 0.000 0.000 0.000
0.000 0.000 0.000 0.000 0.000 0.894 0.000 0.000 0.000 0.000 0.000 0.000
0.000 0.000 0.000 0.000 0.000 0.000 0.894 0.000 0.000 0.000 0.000 0.000
0.000 0.000 0.000 0.000 0.000 0.000 0.000 0.894 0.000 0.000 0.000 0.000
0.000 0.000 0.000 0.000 0.000 0.000 0.000 0.000 0.894 0.000 0.000 0.000
0.000 0.000 0.000 0.000 0.000 0.000 0.000 0.000 0.000 0.894 0.000 0.000
0.000 0.000 0.000 0.000 0.000 0.000 0.000 0.000 0.000 0.000 0.894 0.894
```

The next problem is: what is the present composition of the population?

The textbook:

The most simple solution, counting the animals for a longer period of time, is not possible. The animals swim in deep water, and aerial photography during the short time that the puppies are born is impossible because they are unreachable by light planes. So the composition is estimated from counting marked animals, the section on dead female animals, and finally the composition of caught animals.

The catch of one ship during 1975 is:

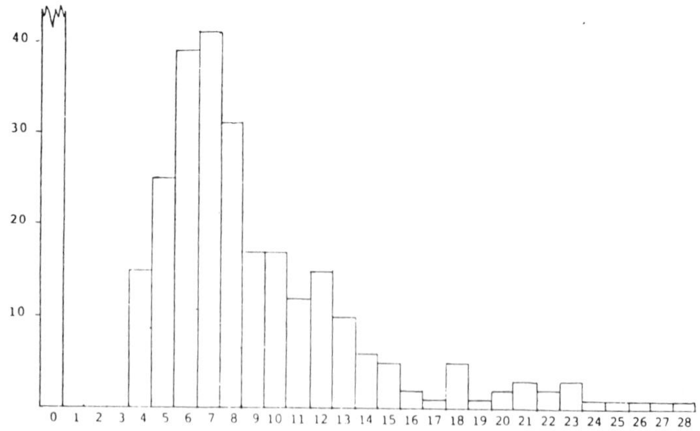

Give an explanation for the hole in the histogram on the left side.

This hole finds its origin in the fact that it is impossible to hunt 1- and 2-year old animals: they are unable to produce puppies at that age and stay in the water all the time. This fact has to be found by the students from the general zoological description, that is given in the textbook.

The fact that the Leslie matrix is 12 × 12 makes it impossible to work with it without a computer.

The start-composition, as guessed by the students, was:

This gives the following output:

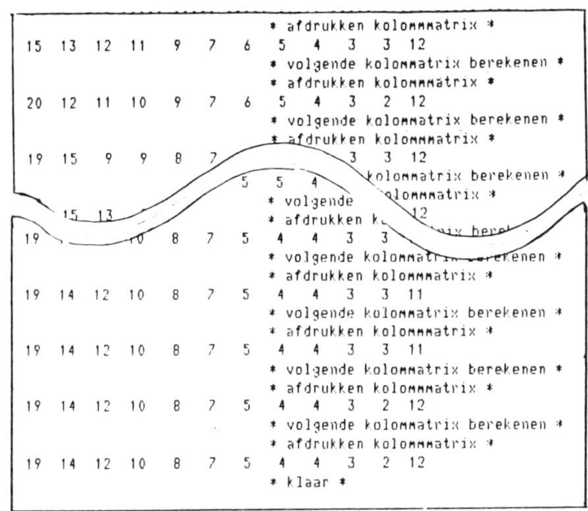

```
                          * afdrukken kolommatrix *
15  13  12  11   9   7   6   5   4   3   3  12
                          * volgende kolommatrix berekenen *
                          * afdrukken kolommatrix *
20  12  11  10   9   7   6   5   4   3   2  12
                          * volgende kolommatrix berekenen *
                          * afdrukken kolommatrix *
19  15   9   9   8   7               3   3  12
                                       kolommatrix berekenen *
                         5   5   4
                          * volgende      olommatrix *
    15  13                * afdrukken k      12
19              10   8   7   5   4   4   3   3
                          * volgende kolommat          enen *
                          * afdrukken kolommatrix *
19  14  12  10   8   7   5   4   4   3   3  11
                          * volgende kolommatrix berekenen *
                          * afdrukken kolommatrix *
19  14  12  10   8   7   5   4   4   3   3  11
                          * volgende kolommatrix berekenen *
                          * afdrukken kolommatrix *
19  14  12  10   8   7   5   4   4   3   2  12
                          * volgende kolommatrix berekenen *
                          * afdrukken kolommatrix *
19  14  12  10   8   7   5   4   4   3   2  12
                          * klaar *
```

After some 15 years the situation (in percentage) became stable, and in absolute numbers:

```
                R E S U L T A A T

                            * waargenomen gegevens *

12
0      0      0.053 0.240 0.249 0.352 0.369 0.418.0.422 0.422 0.430 0.438
0.850
       0.894
              0.894
                    0.894
                          0.894
                                0.894
                                      0.894
                                            0.894
                                                  0.894
                                                        0.894
0      0      0     0     0     0     0     0     0     0     0.894 0.894

1900   1400   1200  1000  800   700   500   400   400   300   200   1200
```

```
                           * afdrukken kolommatrix *
1900  1400  1200  1000   800   700   500   400   400   300   200  1200
      totaal:  10000
                           * volgende kolommatrix berekenen *
                           * afdrukken kolommatrix *
2008  1615  1252  1073   894   715   626   447   358   358   268  1252
      totaal:  10864
                           * volgende kolommatrix berekenen *
                           * afdrukken kolommatrix *
2181  1707  1444  1119   959   799   639   559   400   320   320  1359
      totaal:  11806
                           * volgende kolommatrix berekenen *
                           * afdrukken kolommatrix *
2371  1854  1526  1291  1000   857   715   572   500   357   286  1500
      totaal:  12829
                           * volgende kolommatrix berekenen *
                           * afdrukken kolommatrix *
2586  2015  1658  1364  1154   894   767   639   511   447   319  1597
      totaal:  13951
                           * volgende kolommatrix berekenen *
                           * afdr            mmatrix *
2808  2198  1802  1482  1219                     571   457   400  1713
      totaal:  15167
                                                    nen *
```

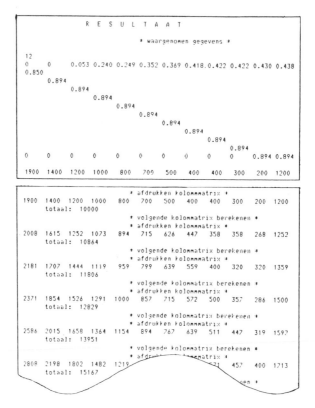

The growth is a little more than 8.7 % per year. From this situation—without hunting—we proceed to the situation with hunting.

The textbook:

> Estimated survivability chances are 81.1 % for all animals of the age of 4 years and older.
> For puppies some 28.6 % are caught. This means that reproduction in all age-classes has to be lowered with this percentage.

In the classroom this led to some surprising results:

It was clear to most students (from the histogram) that on the diagonal the entries 0.894 had to be changed into 0.811 from the 4-year class on. This led to protest from some students. 'One can conclude from the histogram that hunting takes a toll of animals of age 4 and older, but I think some 3-year-olds are caught as well'—hesitation among teacher and the Hewet team-member.

But the student made his point clear: 'From our matrix-without-hunting' it is evident—see first row—that 3-year-old animals reproduce as well. So they have to go into the ice. Hence, they can be hunted'.

The teacher changes the 0.894 into 0.811 and is proud of his students.

Later the correction proved to be correct. But there were more surprises. The text 'prescribes' to lower the reproduction rates by 28.6%. This means: the caught puppies 'are never born'.

But some students preferred a different solution: they want to lower the survivability–probability by 28.6%. This is a correct solution as well.

From the reaction of the students to the presented material one certainly can conclude that they really understand what a Leslie matrix is. The 'matrix-with-hunting' becomes:

0.000	0.000	0.038	0.171	0.178	0.251	0.263	0.298	0.301	0.301	0.307	0.313
0.850	0.000	0.000	0.000	0.000	0.000	0.000	0.000	0.000	0.000	0.000	0.000
0.000	0.894	0.000	0.000	0.000	0.000	0.000	0.000	0.000	0.000	0.000	0.000
0.000	0.000	0.894	0.000	0.000	0.000	0.000	0.000	0.000	0.000	0.000	0.000
0.000	0.000	0.000	0.811	0.000	0.000	0.000	0.000	0.000	0.000	0.000	0.000
0.000	0.000	0.000	0.000	0.811	0.000	0.000	0.000	0.000	0.000	0.000	0.000
0.000	0.000	0.000	0.000	0.000	0.811	0.000	0.000	0.000	0.000	0.000	0.000
0.000	0.000	0.000	0.000	0.000	0.000	0.811	0.000	0.000	0.000	0.000	0.000
0.000	0.000	0.000	0.000	0.000	0.000	0.000	0.811	0.000	0.000	0.000	0.000
0.000	0.000	0.000	0.000	0.000	0.000	0.000	0.000	0.811	0.000	0.000	0.000
0.000	0.000	0.000	0.000	0.000	0.000	0.000	0.000	0.000	0.811	0.000	0.000
0.000	0.000	0.000	0.000	0.000	0.000	0.000	0.000	0.000	0.000	0.811	0.811

The output:

```
                                  * afdrukken kolommatrix *
1500  1300  1200  1100   900   700   600   500   400   300   300  1200
      totaal:  10000
                                  * volgende kolommatrix berekenen *
                                  * afdrukken kolommatrix *
1555  1275  1162  1073   892   730   568   487   406   324   243  1217
      totaal:   9931
                                  * volgende kolommatrix berekenen *
                                  * afdrukken kolommatrix *
1539  1322  1140  1039   870   723   592   460   395   329   263  1184
      totaal:   9856
                                  * volgende kolommatrix berekenen *
                                  * afdrukken kolommatrix *
1519  1308  1181  1019   843   706   587   480   373   320   267  1174
      totaal:   9777
                                  * volgende kolommatrix berekenen *
                                  * afdrukken kolommatrix *
1502  1292  1170  1056   826   683   572   476   389   303   260  1168
      totaal:   9696
                                  * volgende kolommatrix berekenen *
                                  * afd            matrix *
1490  1276  1155  1046   857                     386   316   246  1158
      totaal:   9616
                                                          en *
```

It becomes clear that extinction threatens.

Finally some alternatives were studied (by the students).

The textbook:

We have two Leslie matrices now:

—the 'situation-without-hunting';

—the 'situation-with-hunting'.
The first is not acceptable, especially not for the local population.
The second one leads to extinction.
Some acceptable solutions might be:
A: —no hunting of puppies;
hunting on 3-year-olds and older.
B: —the hunting on puppies will be cut in half.
C: —the hunting on puppies will be cut in half;
no hunting on other seals.

In a test the students were asked to construct the Leslie matrix involved. They succeeded very well in solving this problem.

8. CONCLUSION

The experiment is still in its early stages. The student material seems more or less balanced out.

The first final examination took place. The first round of in-service training was a success. But it is too early to draw any conclusions, although it *seems* that children and teachers are, in general, happy with the new curriculum and the way it was filled out with the student material. But we have to wait for more conclusions. The next ten schools join in August 1983 and this will give us more insight. But one can only be satisfied with the results so far. If the experiment succeeds it will certainly have some important consequences for the rest of secondary education (in the Netherlands). In the lower years of secondary education in particular one will try to prepare the students better for the new program, so that applications (Math A) and three-dimensional geometry (Math B) will play a more important role.

Discrete Modelling, Difference Equations and the Use of Computers in Mathematical Education

J. Ziegenbalg
Paedagogische Hochschule, Reutlingen

SUMMARY

It is the author's opinion that the teaching of mathematical modelling should not be deferred until the students have command over tools from the areas of calculus or differential equations. Difference equations lead to very elementary descriptions of many real world problems. The computer is an ideal tool for evaluating difference equations by straightforward iterative computation. Some examples from various areas are given. An extremely elementary method for obtaining the solutions of a second order linear difference equation using nothing but secondary school mathematics is given.

It is argued that difference equations can be seen as a thread running through all stages of mathematical education (from elementary school to university). Furthermore it is stressed that the use of difference equations can contribute to a more unified way of teaching mathematics and its applications. They can help integrate some seemingly 'disparate' branches of knowledge and methodology in mathematical education.

1. GENERAL REMARKS

With the arrival of microcomputers, mathematical modelling has considerably gained momentum in mathematical education. Genuine real world applications, which have hitherto been beyond the scope of mathematical work in the schools, have become accessible through the enormous computational power of the microcomputer.

Owing to the characteristics of the digital computer, methods pertaining to the fields of finite and discrete mathematics have become increasingly important as tools for model building by way of using computers. Methods from the field of difference equations lend themselves particularly well to mathematical modelling, because:

—they offer very elementary but nevertheless adequate tools for the mathematical description of real world problems;
—they can be dealt with on several levels of sophistication; on the one hand, simple iterative computation and, on the other hand, analytical treatment;
—they offer motivations for and links to such areas as calculus, differential equations and linear algebra;
—by refinement processes, there is a continuous transition from discrete to continuous methods;
—descriptions by means of difference equations are ideally suited for translation into computer programs (very often, continuous models have to be discretized anyhow, when they are going to be evaluated on a computer).

Difference equations, as the theory of iterative processes, can be viewed as a thread running through all stages of mathematical education:

—elementary school level: compound arithmetic operations as iteration of primitive operations (e.g. multiplication as iterated addition, etc.); Euclidean algorithm; iterative and recursive games and strategies (e.g. tower-of-Hanoi game);
—lower secondary level ('middle school'): progressions, sequences, series; basic economic, ecological and other real world problems; stochastic simulations (in particular, random walks);
—upper secondary level: difference equations in combination with tools from calculus, linear algebra and stochastics, as a universal method of dealing with refined models in the fields of natural science, technology, economics, social sciences and linguistics.

Difference equations also can contribute to a more unified way of teaching mathematics and its applications.

—Inner-mathematical integration: methods and subject matter from the areas of calculus, linear algebra and stochastics interweave in the description and solution of dynamical processes by difference equations.
—Integration of mathematics and the natural sciences: frequently in the teaching of physics, chemistry and also biology there are applied means of mathematical description (e.g. differentiation, integration, differential equations) which have not yet been developed in the parallel mathematics courses. Thus, special courses on 'mathematics for biologists, chemists and physicists' are occasionally offered. Because of their low formal prerequisites difference equations can be used as a universal method in many of these areas. It can be hoped, that the process of departmentalization in the teaching of the sciences can be slowed down.

—Integration of problems transcending the area of mathematics and the natural sciences: difference equations in conjunction with computer simulations can be used as a general method for describing and evaluating dynamical processes in these fields.

In Fig. 1 an attempt is made to show the 'interweaving' of difference equations with related fields of knowledge.

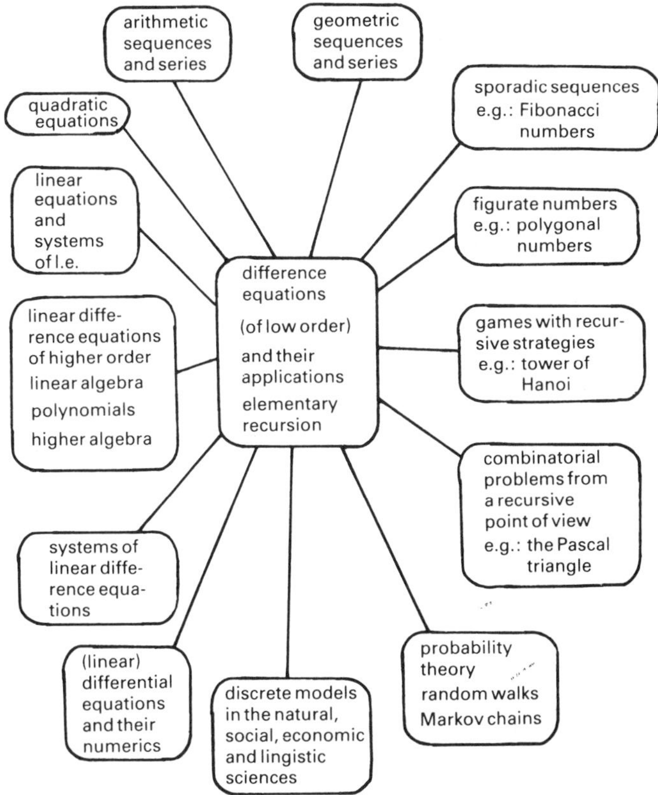

Fig. 1. The 'interweaving' of difference equations with related fields of knowledge

2. THE CONCEPT OF A DIFFERENCE EQUATION

In general, any equation setting up a relation between every $n+1$ consecutive elements $y_k, y_{k+1}, \ldots, y_{k+n}$ of a sequence $(y_k)_{k\in N}$ is called a difference equation of order n.

From the point of view of teaching it is natural to start with difference equations in some of their simplest forms:

$$y_{k+1} = y_k + B \quad \text{(arithmetic sequence)}$$
$$y_{k+1} = A * y_k \quad \text{(geometric sequence).}$$

Typical applications of these equations are well-known in lower secondary education.

3. LINEAR DIFFERENCE EQUATIONS OF THE FIRST ORDER

By combining the difference equations belonging to the arithmetic and geometric sequence, we get the equation

$$y_{k+1} = A * y_k + B. \tag{1}$$

(In technical terms it is called a first order linear difference equation with constant coefficients and constant inhomogeneity; the coefficients being 1 and A; the inhomogeneity being B.)

3.1 Some of the numerous applications of this type of equation

(a) In economic contexts: annuity-type problems. Whereas the equation belonging to the geometric sequence describes the growth process of an account under compound interest, equation (1) describes the growth of an account under compound interest to which additional periodic payments in constant height (B) are made.
(b) Funds subject to a constant interest rate and constant periodic deductions (here B is negative).
(c) Amortization.
(d) Harvesting: suppose an animal or plant 'population' is under geometric (exponential) growth when undisturbed (with growth factor A). For the purpose of harvesting, a fixed part (B) of the population is removed.
(e) Games. For instance: the number of moves in the tower-of-Hanoi game (when applying the standard recursive strategy) is:

$$U_{k+1} = 2 * U_k + 1$$

(with k being the number of disks in the game).
(f) Economic dynamics (see Goldberg (1958) or Dürr & Ziegenbalg (1984)).

3.2 Explicit form

Equation (1) lends itself easily to straightforward computation. For mathematical purposes it is desirable to have a non-recursive ('explicit') representation. We get it by direct computation with the general parameters A and B and the initial value y_0:

$$y_k = A^k * y_0 + (A^{k-1} + A^{k-2} + \ldots + A^2 + A + 1) * B;$$

yielding the following theorem:

The difference equation $y_{k+1} = A*y_k + B$ *has the following solution (depending on the initial value* y_0*):*

$$y_k = A^k * y_0 + B*(1-A^k)/(1-A) \quad (\text{if not } A = 1)$$

$$y_k = y_0 + k*B \quad (\text{if } A = 1).$$

3.3 Graphic representation and convergence

In addition to the standard graphic representation in (k/y_k)—cartesian diagrams, the representation in (y_k/y_{k+1})—diagrams ('cobweb-diagrams') gives much insight. In Fig. 2 this type of representation is given for different values of A.

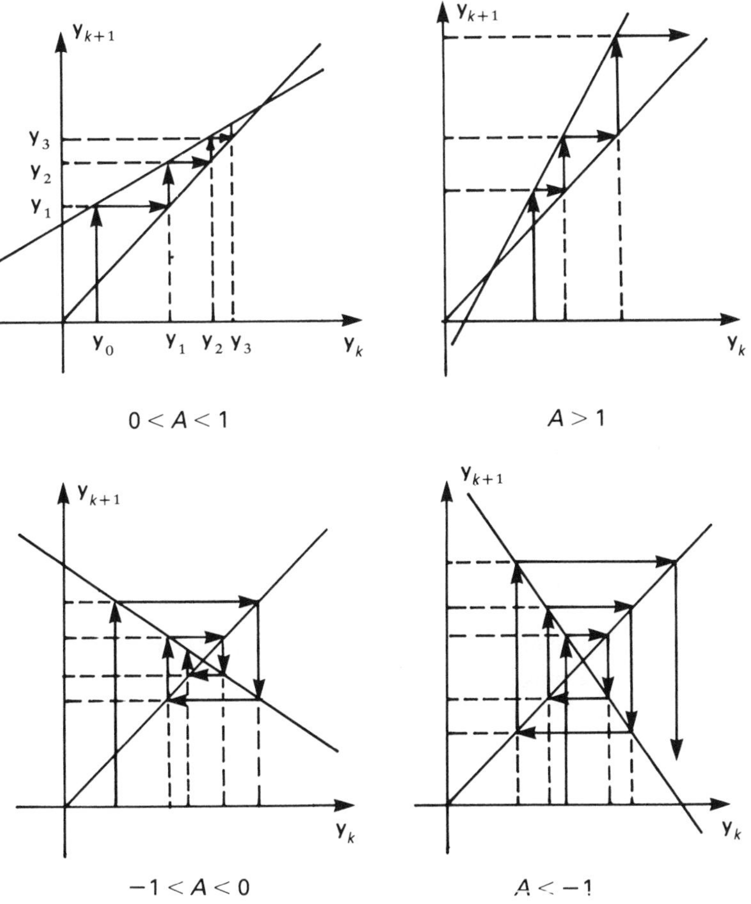

Fig. 2. 'Cobweb'-diagrams for first order linear difference equations

From this representation we get the idea of moving the origin to the point P with coordinates $(B/(1-A); B/(1-A))$, i.e. to the intersection of the lines given by the equations $y = A*x+B$ and $y = x$.

The coordinate-transformation

$$v_k = y_k - B/(1-A)$$

gives rise to the following form of the original equation (1):

$$v_{k+1} = A*v_k$$

showing that each difference equation of type (1) is equivalent to a geometric series. In particular, the criteria for convergence of the linear difference equation with constant coefficients are identical to the criteria for the corresponding geometric sequence; i.e. convergence holds for $\text{abs}(A) < 1$.

4. LINEAR DIFFERENCE EQUATIONS OF ORDER TWO

In this case, too, we will only consider the simplest type here:

$$y_{k+2} + a_1*y_{k+1} + a_0*y_k = b. \tag{2}$$

4.1 Examples

(a) The Fibonacci numbers: $F_{k+2} = F_{k+1} + F_k$.

(b) Random walks on a straight line (a particular Markov chain):
We consider a stochastic process on $r+1$ states as follows.

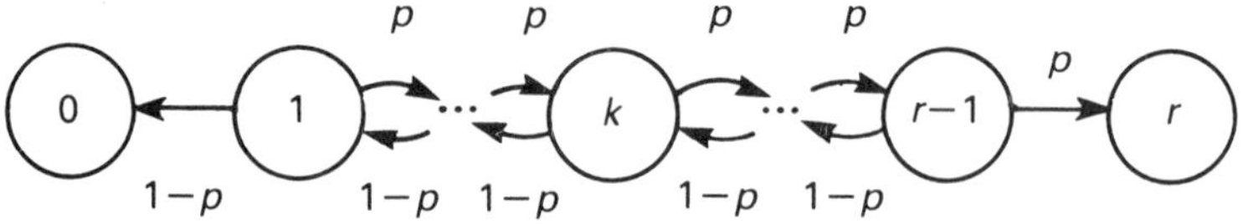

Transition probabilities: p: from state k to state $k+1$
$1-p$: from state k to state $k-1$.

Let p_k be the probability of reaching state r when the process is in state k;
d_k be the average number of (transition) steps until termination of a 'walk' (either in state 0 or state r).

Using the backward equations for Markov chains (see, for example, Feller (1957) or Engel (1976) we obtain the relations

$$p_k = p*p_{k+1} + (1-p)*p_{k-1}$$
$$d_k = p*d_{k+1} + (1-p)*d_{k-1} + 1.$$

(c) Distribution of temperature in a metallic rod.
By slicing the rod into flat disks and assuming that the temperature of a

disk is homogeneous and equal to the arithmetic mean of the adjoined disks we get

$$t_k = t_{k+1} * 1/2 + t_{k-1} * 1/2.$$

The temperatures at the ends of the rod are supposed to be predetermined.

(d) Figurate numbers; polygonal numbers.
The recursive construction of polygonal numbers of type E:
step 1: one dot
step 2: a regular E-gon
step $k+1$ ($k > 0$): given the dot pattern of stage k, the pattern of stage $k+1$ is constructed by adjoining an open frame of as few as possible dots to the old pattern such that the vertices of the new pattern make up an E-gon with $k+1$ dots on every side of the new E-gon.
This construction is best explained by an example (with $E = 5$):

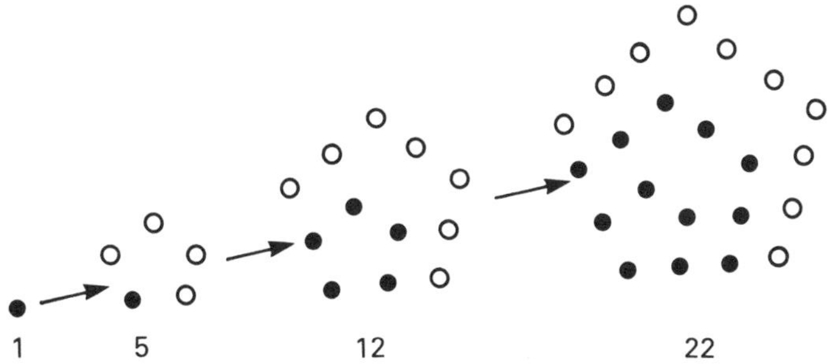

Let $G_k(E)$ be the total number of dots in the stage k pattern. From the construction it follows that

$$G_{k+1}(E) = G_k(E) + (E-2)*(k+1) - (E-3)$$

By applying the method of homogenization (see Dürr & Ziegenbalg, 1984) we get

$$G_{k+2}(E) = 2*G_{k+1}(E) - G_k(E) + (E-2)$$

a second order equation which can either be solved directly or further homogenized.

(e) Models for national income (see, for example, Dürr, & Ziegenbalg, 1984).

(f) Models for inventory analysis (see, for example, Goldberg, 1958).

4.2 Explicit form

Direct computation with the symbolic parameters (a_1, a_0 and b) will lead to very complicated terms in this case. Therefore in many publications the solutions are given more or less without comment (sometimes, however, with an exercise to verify that they are indeed solutions).

I will, therefore, in somewhat greater detail describe a method using a standard technique which seems to be little known. This will be done for the homogeneous equation (i.e. for $b = 0$) to begin with. By adding the zero term $m*y_{k+1} - m*y_{k+1}$ we get the following equation:

$$\underbrace{y_{k+2} + (a_1 + m)*y_{k+1}}_{\text{(I)}} \quad \underbrace{-m*y_{k+1} + a_0*y_k}_{\text{(II)}} = 0.$$

By setting both of the terms (I) and (II) to zero we get:

$$\left.\begin{aligned} y_{k+2} + (a_1 + m)*y_{k+1} &= 0 \\ -m*y_{k+1} + a_0*y_k &= 0 \end{aligned}\right\}. \tag{3}$$

The solutions of these first order equations are well-known by now. If we succeed in finding a simultaneous solution for the two equations in (3) this will clearly also solve (2) for $b = 0$.

Equations (3) will have nothing but simultaneous solutions if they are identical. We can use the parameter m to make them identical. Comparing the coefficients of the equations in (3) leads to the relation

$$-(a_1 + m) = a_0/m.$$

It follows that m will have to satisfy the so-called 'characteristic equation'

$$m^2 + a_1*m + a_0 = 0.$$

From (3) with roots m_1 and m_2 we get the formal solutions

$$(-a_1 - m_1)^k,\ (-a_1 - m_2)^k,\ (a_0/m_1)^k,\ (a_0/m_2)^k.$$

By use of Vieta's theorem we realize that these solutions contain identical pairs and that furthermore they can be written as

$$m_1^k \quad \text{and} \quad m_2^k.$$

As it is the case with all (linear or differential) inhomogeneous equations we get all the solutions of the inhomogeneous equation by adding one fixed solution of the inhomogeneous equation to the set of solutions of the homogeneous equation. As for the technique of homogenizing a linear difference equation by transforming it into a linear homogeneous equation of higher order which contains the original solutions we refer to Dürr & Ziegenbalg (1984).

4.3 The characteristic polynomial in the context of linear algebra

At first glance it might be suspected that the above terminology was not well-chosen, because the concept of a characteristic polynomial is used already in linear algebra. However, by applying matrix notation, the homogeneous equation

$$y_{k+2}+a_1*y_{k+1}+a_0*y_k=0$$

can be written in the form

$$\begin{pmatrix} y_{k+1} \\ y_{k+2} \end{pmatrix} = \underbrace{\begin{pmatrix} 0 & 1 \\ -a_0 & -a_1 \end{pmatrix}}_{\text{'transition matrix': } M} \begin{pmatrix} y_k \\ y_{k+1} \end{pmatrix}.$$

The characteristic polynomial of the transition matrix M (obtained by computing the eigenvalues of M) is

$$\lambda^2+a_1\lambda+a_0=0.$$

So, in retrospect, the choice of the terminology can be explained, although from the perspective of the student it might be somewhat confusing.

5. SOME FINAL REMARKS

By what is known about the world today we may assume that it is a discrete system consisting of a finite number of elementary objects. Therefore, continuous functions and methods can be considered useful for an elegant, but only approximative description of the world. When, however, continuous functions are approximated by discrete methods in turn, one may wonder why discrete mathematics should not be used as a means of description in the first place, particularly since most of the discrete methods can be dealt with in extremely elementary ways. One has to take into account, however, the specific heuristic situation when applying finite methods. In particular the widespread practice of just giving the solutions without any clues as to how they might be found, ought to be avoided.

The book Dürr & Ziegenbalg (1984) was written with the intention of avoiding arcane tricks (arcane at least for the beginning student) and with the particular heuristic situation of mathematical modelling by the use of difference equations in mind.

REFERENCES

Dürr R., Ziegenbalg J., 1984, *Dynamische Prozesse und ihre Mathematisierung durch Differenzengleichungen*, Ferd. Schoeningh-Verlag, Paderborn, Germany.

Engel A., 1976, *Wahrscheinlichkeitsrechnung und Statistik* (Band 2), Ernst Klett Verlag, Stuttgart, Germany.

Feller W., 1957, *An Introduction to Probability Theory and its Applications*, John Wiley, New York.

Goldberg S., 1958, *Introduction to Difference Equations*, John Wiley, New York.

The Modelling Game: The Uses of Broadcast Television in a Modelling Course

J. Jaworski
BBC/The Open University

SUMMARY

One Friday morning in early January 1982, three mathematicians came together and solved some problems. What distinguishes this act of collective problem solving from the thousands of similar events that occur in the working life of any professional mathematician, is that it occurred in a television studio. It was an exercise in eavesdropping, requiring some courage on the part of the participants. Their agreement to take part implied a contract to allow the resulting material to be used no matter how embarrassing or unprofessional it might appear to be. In the lively environment of a university department or an industrial setting, it is easy to be relaxed about the mis-remembered formula, the corner-cutting technique that usually works or the untidy and imprecise jotting. There was every reason to believe that the eventual audience for the four television programmes that were eventually made in this way would be far less forgiving.

It is to the credit of all who took part that having given their initial agreement, their commitment never wavered.

This chapter describes how these television programmes came to be made, what they were designed to achieve, and what the reaction to them has been. In some sense, it is less a presentation of conclusions than a subjective account of how we came to make these programmes. That seems appropriate, as the act of creating television programmes is very much a subjective act: almost as much so as the doing of mathematics.

1. BACKGROUND

In 1982, the Open University offered to students for the first time, the course

MST204 *Mathematical Models and Methods*.† This course is the principal introduction to applied mathematics for OU students and follows the generally well-known pattern of OU course presentation in which students receive a number of printed course units, are assigned to a part-time tutor who grades assignments and conducts tutorial sessions, attend a week-long Summer School and take a final course examination. In addition, and importantly for this account, students view a weekly television programme, 25 minutes long, broadcast nationally throughout the United Kingdom.

MST204, as the course title suggests, contains a significant proportion of mathematical modelling. It had always been felt that the modelling activity would appear sterile unless students participated in it themselves, and from the earliest stages of planning, the course had contained a substantial student modelling project, occupying some 40 hours and accounting for one-eighth of the student assessment. The process of designing and implementing an assessment procedure for this component of the course has been described elsewhere (*Berry & O'Shea*, 1982). This chapter examines the debate and eventual decisions taken on how to support the modelling project via television. It looks at the production style adopted, and at student reactions.

2. WHY USE TELEVISION AT ALL?

Television is a scarce and expensive resource. Students are intolerant (and vocal) about television that is perceived by them as peripheral or simply motivating. They are often required to view at unsocial hours in the early morning or late evening, are often themselves under serious time pressure, and can react badly against teaching carried by a programme that does not appear to be well-integrated with their study patterns.

These problems are not unique to the MST204 course. They are well-known and taken to heart throughout the BBC–Open University partnership. Within the Mathematics Faculty in general, and the MST204 course team in particular, the belief is that tightly paced study patterns are most productive. Accordingly, each regular text unit of MST204 contains a clearly identified TV section: this involves some preparatory work, a viewing of the programme and some follow-up work. The position of this section in the unit may vary, but the student is clearly advised about interdependencies between the sections, and about a suggested study order.

Such close management of student behaviour ceases to be possible with the four weeks of study allocated to the student project. The only supporting text is the *Project Guide*, a handbook setting out the methodology adopted by the course team, and containing preparatory exercises, sample scripts and general advice. The student does not 'study' this in the accepted sense.

† For a fuller discussion of the design of the MST204 project scheme see the chapter by Berry and Le Masurier in this volume.

There were three possible choices of action:

(a) to omit the four TV programmes scheduled for the weeks of project activity;
(b) to use the otherwise vacant slots for programmes of more general applicability;
(c) to produce programmes to support the students' modelling.

The first option had its supporters, but the view eventually prevailed that a regular transmission pattern (already interrupted by Bank Holidays, Royal Events and the like) was of great importance in keeping students active on the course. Option (b) was felt to be likely to result in students perceiving 'irrelevant' broadcasts. In fact, this option was partly adopted, in that the last project slot was filled with a broadcast of advice on exam revision, which has proved most popular and has regularly been given extra transmissions as a result of student demand.

However, it was option (c)—the production of three programmes to support the student's own modelling—that was adopted. A subset of the course team, the project group, who were also managing the development of the more conventional components of the project, took television devising under their collective wing and began to explore the options.

3. THE DEBATE

The MST204 course was produced over a period in excess of two years; for the majority of that time, the project group were considering television, which had been scheduled at the end of the production period, both to capitalize on early feedback and also to allow for full consideration of what might be unusual programmes. In the event, the later programmes were made as regular students were already studying the early units of the course. Many avenues were explored. It was anticipated that students would ask for examples of what would be expected from them, or would demand assistance with the specific projects that they had chosen (from a list, changed annually). Offering sample modelling—by observing students at work, for example—was not felt to be logistically possible. The genuine student modelling activity was believed to be continuous over the major part of the life of the course, regardless of the allocated weeks on the course calendar, was often purely cerebral, and carried with it no guarantees that simply watching a student throughout the year would provide anything of use to anyone, even if the money and time could have been found.

Offering specific advice on the projects available was not popular. It seemed to deny the essential nature of modelling in that the student would now be given predigested facts. Further, it would not allow the course team to readily change the list of available topics, and would set a premium on one or two of the projects, which would be different from the others through being on television.

One option that was actively and thoroughly considered was that the programmes should abandon any attempt to assist the students directly, but should exhibit modelling as an activity of some intellectual substance. It was proposed that a number of exemplary (and highly individual) modellers should be invited to show how it might be done. These exemplars should not be interpreted as advice to the students as to how they should themselves organize their modelling, but rather should be seen as substantiating the claim that modelling was important. No approaches were made, but the names of Seymour Pappert and Edward de Bono were high on our lists.

The reasons for abandoning this approach were complex ones. It was felt that such a message would be diffused, and possibly misunderstood by students. The value of exhibiting behaviour that the student might expect to replicate, when this was never the intention, was doubted. It was strongly felt that students (largely new to any modelling activity at all) would be apprehensive, would misjudge what was expected of them, and would rely heavily on the straightforward methodology itemized in the *Project Guide*. Under such conditions, exposing them to skilled and eccentric behaviour would confuse rather than clarify. For many students, such programmes would be the last straw, and increase the student drop-out.

Other alternatives were considered. A study of applications of modelling enjoyed a majority of support for some while. It was proposed that a study such as traffic-flow through the Humber Tunnel should be carried out. It was even (quite seriously) proposed that predictions should be tested by hiring a fleet of cars that would anonymously drive at restricted speeds through the tunnel to demonstrate platooning effects. It was felt that the cooperation of the Police and Tunnel authorities could not be guaranteed!

A suggestion that proportional representation voting systems should be studied as models of electoral preferences resulted in a lively debate as to the precise meaning (within the context of the course) of the word ‘model’. As a result, this avenue was left alone.

The mechanical modelling of athletic events was an active candidate for inclusion, and in the event appeared in the course in an even more novel form to be described below.

Such debate unavoidably took time, which was beginning to run out. The needs of the students were being more and more frequently re-iterated, and while no-one wished to simply exhibit good project reports, it was felt that students would over-estimate the precision involved in modelling, would not appreciate how messy and apparently casual the process can often be. Open University materials are generally of a very high standard, and it has often been found that a simple mistake or printing error can cause severe student distress. While full-time mathematicians in industrial or academic circles will often be cynical about the quality of printed texts, and will not be seriously disturbed by errors, students learning at a distance are fairly vulnerable and place great confidence in the 'rightness' of material in front of them. It was felt that their likely 'failure' to provide the elegant solution first time through a modelling project would be a major stumbling-block.

Accordingly, the project group proposed a format that would expose students to 'real' modelling. They would be permitted to see just how untidy the notes of a working mathematician are in practice, to eavesdrop on the corner-cutting techniques that are learned only by experience, and to see that a working mathematician learns as much by being wrong as by polishing a paper for publication.

4. THE MODELLING GAME

At the time of discussion there was a popular early-evening BBC television series, *The Great Egg Race*. This placed teams of contestants in a studio with the sort of 'junk' that might be found in a well-equipped workshop owned by an enthusiastic do-it-yourselfer. Against the clock, these teams would attempt to build from scrap devices such as a separator to divide a mixture of peas, rice and sand into separate piles within a given period and without manual intervention.

The skill and invention shown by contestants is great. Understandably, the programme concentrates on the final product rather than on the process of producing it. While our focus would be necessarily different, it was felt that here was a format that could be adapted to suit our ends.

The debate was far from over: trial runs of the proposed format were fraught with bloody argument. The proposal to include a scoring element was diplomatically dropped. Such radical steps might have been acceptable with live students undergoing the soul-baring experience, but the project group had courageously decided that they would ask the course team members to volunteer, as professional mathematicians and modellers, to expose their methods on the screen. It is greatly to the credit of those who supported the venture that, having made the initial and difficult choice of participating, they never thereafter wavered in their commitment—even when they were exposed as inefficient or mistaken: characteristics that are viewed with relaxation in a lively face-to-face environment, but could easily be misinterpreted at a distance.

Eventually, one Friday morning in early January 1982, the modelling game became a reality. The first team consisting of three course-team members sat apprehensively in a TV studio strewn with paper, calculators, rulers and the like, and boasting a 'hot-line' to the researcher who would search out any required data from the university library. Staff of the BBC visual effects department were on hand to build any simple devices or experiments that might be requested. No-one had any real idea of what to expect, least of all the camera crew, the vision-mixer and the director. Excepting meal breaks, we had determined to record everything that happened. An overlapping scheme of video-tape changes had been worked out to avoid a significant exchange occurring while we were not recording. Team-members would lunch separately for the same reason. The project (chosen to avoid lengthy data-collection, or visits to locations outside the studio) was specified in the traditional sealed envelope, the clock started, and we began.

5. THE PROJECTS

We were careful to point out to students that the Modelling Game paralleled their own project, and that the parallel was not always close. Teams had two hours for each project; they worked in groups of three and collective problem solving is an easier activity than being a distance student. The teams exercised no choice over their projects which were assigned to them. The problems were chosen as being capable of 'solution' in a short period of time. In many cases, this meant that the problem actually had a solution. The very first project was indeed of this sort: given a dozen ropes of varying thickness, the teams had to predict the shortening effect of tying an overhand knot in a specified rope. In the last minutes of their time allowance, they could test this prediction. It was a problem with a 'right answer'. This choice was deliberate: we wanted there to be a resolution of the problem by the end of the programme, and the more open-ended (and realistic) problems could have resulted in apparent disarray.

In all, we recorded around 24 hours of video-tape, using two teams on four problems, over a period of some three months. The second problem invited the teams to explore the positioning of a cycle lamp on the front of a cycle and the third asked them to predict and explain the angle at the top of a conical heap of poured sand. These three problems were edited into the three broadcast television programmes, while a fourth—modelling the action of a high-jumper—was used as a Summer School activity, with film being used together with stop-frame facilities for students to take measurements from. At the Summer School, students used the first part of this tape as a resource and viewed the course-team performance as 'light relief' at the end of the week.

6. THE PROGRAMMES

Editing the 24 hours of tape took nearly three months. Although the teams had tackled only one or two projects a day, and different teams had visited the studio on different days, we cut the resulting material as if both teams were racing against each other as well as the clock. We bathed the studio in yellow light for one team, and blue for the other, as a simple visual clue as to who was who.

The first programme, concerned with predicting the effect of a knot in a rope, was edited as a straightforward narrative account of the modelling process. We watched the teams compiling feature lists (the course methodology) and pruning or otherwise refining these. We saw the measurements taken and the subsequent processing of these measurements. We overheard the teams reasoning their way to an abstract model, and also plotting data points to gain an empirical value.

There were some beautiful insights that we felt gave students a good picture of the reality of modelling. It will come as no surprise to professional mathematicians, but astounded our students, that PhDs actually *do* compute the slope of a graph by counting the 'squares up divided by the squares along'; that line-fitting is done by eye, rather than least-squares; that an academic can look at a scattered array of points and murmur pleadingly: 'well, it *could* be a straight line . . .'. We showed how mistakes can be made in the midst of humour. We showed that correcting mistakes is not always done with total scientific detachment. In summary, we showed modelling as it really is.

The full course team generally approved of the first programme. They enjoyed it as a piece of television, and felt that it aided the students with their own modelling. There was still dispute over whether what had been seen was modelling, which resulted in the acid comment from one viewer that he was personally pleased that Newton had bothered to give us some Laws of Motion as well as some numbers . . . .

The second programme was to be viewed by students when their own project was under way. We felt that the particular week in question was one where students would be vulnerable to loss of enthusiasm, procrastination, and misuse of time. Accordingly, we edited the programme on placing a lamp on a bicycle to show where the teams had made major errors. Again, there were beautiful insights. The team that propounded wild theory and didn't check it with the studio bicycle were to be found with their time three-quarters gone discovering that 'this is a badly-designed bike . . . '. As the narration observes a little tersely: 'If it *is* badly-designed, it was badly-designed ninety minutes ago . . .'. The face of the contestant who realizes that only five minutes remain and that the theory requires major revisions, speaks volumes to students who may overestimate how much remains of their own 40 hours.

The last broadcast was scheduled for the week in which students would begin writing up their final report. The problem, of predicting the angle at the top of a sand-heap, had added to it the requirement that the teams undergo an inquisition as a form of proxy report. We felt strongly that it was a misuse of TV to simply watch someone writing.

7. THE RESULTS

All newly presented courses are well-surveyed by the University's Survey Research Department in their first year. While prompt feedback is useful, there is a feeling that in the first year of presentation of a course, tutors as well as students are at a disadvantage, and that students do not receive the more considered advice that tutors can offer after some experience of the course. In this case, such an observation may help to explain the poor overall feedback. Few students responded to the opportunity to comment on the programmes, but those that did generally disapproved. It is tempting to think of a vast satisfied silent majority, but anecdotal evidence suggests otherwise.

In part, the reaction against the programmes is a reaction against the project as a whole. While administration of the project was smooth, and student performance good when compared to similar project schemes elsewhere in the Open University, many students were confused and apprehensive and, receiving no reassurance from tutors (who were unable to guarantee that a particular level of participation would be adequate in assessment terms), transferred their anguish to the TV programmes, which were perceived as being unhelpful.

From the anecdotes, and talking to individual students, it emerges that the first programme was generally well-received. The most negative reaction was that the programme showed how enjoyable it was to solve problems with three of your friends when there was nothing else to do, but offered no assistance to students alone in the early morning, with the family expecting breakfast after the programme was over!

The second programme, deliberately edited to show confusion and lack of progress, was perceived by the students as confusing and without direction. In part we feel that a stronger narration is called for; we had expected the students to make their own connections, but clearly more direction is required. Similar reactions are recorded to the third programme.

No viewing of an OU TV programme is an isolated act. Each is an experience involving preparation and pre-work. The accompanying TV notes must share some of the blame for the low student opinion of the product. In 1982 it was only possible to provide full notes for the first programme, with simple stop-press announcements of subjects sufficing for the remainder. In 1983, we have corrected this. Students viewing the bicycle programme will have tackled the problem themselves (the most positive feedback to this programme came from a tutor who had found it worthwhile working through the problem with his students before they had viewed it). They will further be asked to identify the problem areas as they view, and also to suggest how they would extricate themselves from similar situations. In this way, we hope to provide students with a more personal and hence more relevant experience.

With adequate supporting text, we shall look again at the performance of the programmes in 1983. Their existence is an act of will and courage on the part of the course team, and we wish to be sure that student reaction is accurately assessed before anything as drastic as cancellation is envisaged.

THE MODELLING GAME

A first look at the problem: the first programme

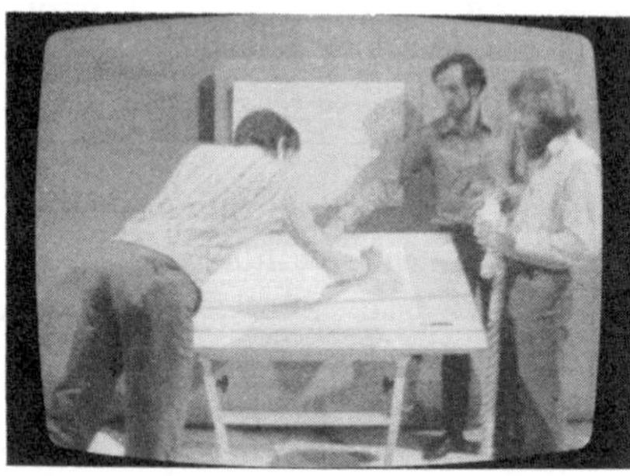

Modelling knots: plotting data points

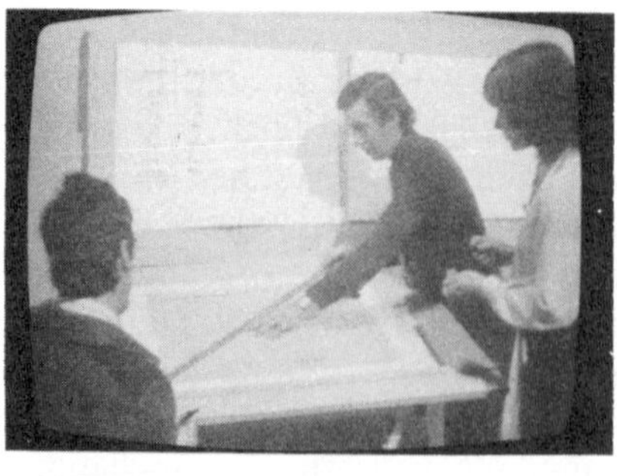

'It could be a straight line!'

These photographs illustrate the four full-length modelling programmes produced by the BBC for the Open University. Programmes 1 to 3 are transmitted twice each year as part of the MST204 "Mathematical Models and Methods" course. The fourth programme, incorporating a commercial athletics training film on high-jumping, is used at the one-week residential Summer School with students taking measurements and timings from the screen, and then viewing the tutors tackling the same problem.

Institutions with a licence to record Open University broadcasts may wish to make off-air copies of the three broadcast programmes. Alternatively, viewing copies of all four programmes may be borrowed from the BBC. Contact John Jaworski, Producer, Mathematics, BBC Open University Production Centre, Walton Hall, Milton Keynes MK7 6BH (0908–655442).

To see or be seen? The second programme

Before the envelope is opened: the third programme

Light relief: measurement of stopping distance

The Summer School Film—modelling high-jumping

Experiment is cheap: thinking in the saddle

'Action Man' helps the high-jump modellers

8. ACKNOWLEDGMENTS AND REFERENCE

Many people contributed to the programmes. Ted Smith, director and Moira Leatham, vision mixer, bore the brunt of twenty-four hours of studio work of greater complexity than is usual. The camera crews responded to the challenge with enthusiasm beyond their professional calling, as did all the technical staff, in particular Pete Matheson who edited the full programmes. But we must acknowledge the work done by the teams: John Berry, John Bolton, Mike Thorpe, Mick Bromilow, Ann Walton and Roger Duke. Tim O'Shea presented, and John Trapp stepped in as reserve when Roger Duke moved to Australia. And especially, for their philosophy as well as their more practical help, we would wish to thank John Mason and Professor Oliver Penrose.

The assessment scheme for MST204 modelling is described in Berry J. and O'Shea T., 1982, Assessing mathematical modelling, *Int. J. Math. Educ. Sci. Technol.*, **13**, 6, 715–724.

Experiences with Modelling Workshops

I. D. Huntley,†
Sheffield City Polytechnic,
and
D. J. G. James
Coventry (Lanchester) Polytechnic

1. INTRODUCTION

In the late seventies a number of institutions in the U.K. were attempting to introduce courses on mathematical modelling within their undergraduate degree programme. This was seen as an attempt to remedy deficiencies highlighted by various reports (McLone, 1973; Gaskell and Klamkin, 1974) regarding the ability of mathematics graduates to formulate problems and communicate results to the non-specialist.

Two such courses were those offered at Coventry (Lanchester) Polytechnic (James and Steele, 1980) and at Paisley College of Technology (McDonald, 1977). In both these courses a conscious effort was being made to teach modelling skills and staff were finding the task far from easy. The difficulties arose from many directions, for example, the uncertainty as to what and how to teach in this field and, perhaps most of all, at the time a lack of good readily available source material. This led to collaboration between the staff at the two institutions and in order to instigate collaboration on a wider scale the two institutions initiated the idea of a National Workshop on the Teaching of Mathematical Modelling.

Over the last four years a series of such workshops have been organised and have been partly financed by a grant from The Nuffield Foundation. The purpose of this paper is to outline briefly the activities of the various workshops and to discuss some of their findings.

2. THE INITIAL WORKSHOP

The initial workshop was directed towards producing documented case studies which could be used as support material for the teaching of introductory

† Formerly at Pariley College of Technology.

courses in mathematical modelling. In their planning the organisers' intention, as expressed in the background information sent to potential participants, was to place the emphasis on understanding real systems, formulating problems and interpreting the results of the analysis. The amount of mathematical analysis required would be appropriate to the stage at which the material would be presented and the mathematical techniques would be already known to the students.

The workshop was organised in the form of two three day sessions. The aim of the first session was to establish objectives and guidelines for the preparation of material to be presented at the second session. Consequently the first session was arranged so that participants could present their views on the nature of mathematical modelling and the teaching of the subject. These presentations were then followed by group discussions on the nature of the material to be prepared for the second session.

At the outset of the workshop some of the participants were of the opinion that the process of model building was sufficiently capable of definition to be described at some length and that such a description could be called a methodology. By the close of the first session, however, it had become apparent that different people used different approaches to modelling and the approach sometimes depended on the context. Participants were not generally in favour of being too prescriptive on method of approach, as experience with students had shown that there is a fine dividing line between a framework and a straight jacket. For example, it was believed that the framework approach adopted by the Open University Mathematics Foundation Course (Open University, 1978) was seen by the students as a 'modelling algorithm' prohibiting them from developing the initiative and flair which is necessary in modelling. There was, however, general agreement amongst the participants on the nature of the material to be produced; in particular it was agreed that in most situations communication was extremely important, first in developing a 'verbal model' of the situation under study and then in interpreting the results of the study and forming a report.

At the end of the first session it was agreed that each participant should prepare for the second session, a case study illustrating a problem situation and how mathematical modelling would help in the study of that situation. It was agreed that the case study should not consist solely of a model or models for that situation but should be written to demonstrate the processes used by the mathematician in developing the model(s). The proposed method of presentation of the material to students should also be detailed as well as extensions in the form of exercises for the students. Many participants were strongly of the opinion that only situations in which mathematics could make a positive contribution should be considered. However, it was agreed that modelling an apparently trivial everyday situation was acceptable provided it was made clear to the reader how this represented a simplification of a more complex realistic situation.

Guidelines for the contents of a typical documented case study issued at the end of the first session, may be summarised as follows.

(i) A statement of the situation.
(ii) A way of modelling the situation.
(iii) An outline of *other* ways of dealing with the situation (wherever applicable.)
(iv) A description of how the problem situation and analysis might be presented to a group of students.
(v) Indications of extensions to be followed up by the teacher or students.
(vi) Ideas on assessment (where applicable).
(vii) All necessary reference.

As an aid to the development of the material participants were organised into groups of 4 or 5 individuals. The problem statements were circulated to all the participants a month prior to the second session. As well as developing their own proposal each participant was asked to give serious consideration to the way in which he/she would approach the proposals of fellow group members. If time allowed they could then turn their attention to any other proposal of interest. As a general principle it was agreed that participants should avoid revealing to the other participants, complete details of their approach to their own proposal prior to the second session. It was hoped that these arrangements would create a climate for the second session in which all participants would be able to make constructive suggestions for the improvement or development of case studies prepared by others. Equally each participant should benefit from what should, effectively, be constructive refereeing of their own work.

The intention at the second session was to first discuss the case studies in detail by the groups and then in plenary sessions. During the group sessions the examination of each case study was extensive and constructive and it quickly became apparent that the small group discussions were likely to be more fruitful than the presentation of papers in plenary sessions. The planned plenary sessions were therefore curtailed in favour of the former and used instead to coordinate the activities of the individual groups and to define policy on the structuring of the case studies.

By the end of the second session all the authors had received advice from their fellow group members on the detailed content and presentation of their case studies. In addition agreement had been reached on a policy concerning the final form of the case studies which can be summarized as follows.

(a) The case studies should be aimed mainly at lecturers or teachers embarking on courses in modelling.
(b) The format of the written case study should be left free to reflect the nature of the problem and the authors' own style of modelling and writing. The following guidelines should however, be followed.

(i) Each case study should start with a statement of the problem to be considered. This statement should be accompanied by other material providing any necessary background and motivation. This section should form a quite distinct part of the case study.

(ii) Authors should provide, wherever appropriate at points throughout the study, suggestions to the prospective lecturer or teacher on how to present the case study and handle the class. The workshop participants were conscious of the possibility of being dogmatic but it was felt that, on balance, there were instances where the author's advice—based on experience—could be valuable.

(iii) It is likely that all case studies will be of a form where, following presentation of the statement of the problem (see (i)) students would be encouraged to think carefully about the situation and come back with reactions and suggestions on how to proceed. Case studies should be written in such a way that they take account of this method of approach. Clearly, any lecturer making use of the case study will be faced with a variety of student reactions and will proceed on the basis of some discussion with the class. The author of the case study will have to adopt a particular stance from those available but should make it quite clear that this may only be one approach to the problem.

(iv) Based on a clear statement of the author's stance (see (iii)), there should be a formulation of the problems leading to an appropriate model(s).

(v) The analysis of the model should include an interpretation of the 'solution' and wherever possible some validation of the model and 'solution'.

(vi) Authors should provide, as examples for further work, ideas for extending the model(s) and/or problem situation.

(vii) A list of references should be provided.

(viii) If the authors have ideas on further case studies in related areas these may be summarized at the end of the case study.

The revised case studies were sent to three referees, one of whom was selected from the group involved in its development. Referees were asked to complete a detailed questionnaire for each case study reviewed. Comments received were incorporated in the final versions which have been published (James and McDonald, 1981).

3. FURTHER WORKSHOPS

It was generally agreed that the format adopted for the second sessions led to fruitful discussions and was a good method of approach. Participants were strongly of the opinion that active small group sessions was an important ingredient of a successful workshop. However, completion of detailed questionnaires by referees outside the group associated with the case study added little and could be dispensed with. Instead thorough vetting of each case study was best left to members of the associated group with an 'outsider' being used to give a final overview.

The activities of the workshop have been ongoing and a number of successful workshops, adopting the revised format, have been organized. Over 60 individuals have participated in various workshop sessions. A second collection of documented case studies is currently with a publisher and a third collection is at a development stage.

As mentioned earlier the above workshops were primarily directed towards producing documented case studies which could be used as support material for the teaching of introductory courses in mathematical modelling. However, the lack of source material is just one of the problems facing the teacher of a modelling course and a recent workshop was devoted to considering two particular problem areas.

The first of these was to investigate the existence of a common methodology of approach to a modelling exercise. In particular to attempt to identify the thinking processes of experienced modellers when they are exposed to a new problem situation—with particular emphasis being placed on initial reaction. If a methodology of approach could be identified then valuable guidance could be given on how much courses should be taught.

The second issue was to investigate the existence of alternative 'solutions' to the same problem statement. In the case studies documented to date a particular stance towards the problem statement has been adopted by the author. However, workshop participants have been conscious of the danger of appearing dogmatic and have always emphasized that there is no one 'answer' to any particular problem statement; others may take quite different, but equally acceptable, approaches to that of the authors. In order to reinforce this important facet of modelling teaching, it was felt that there would be merit in attempting to document how experienced modellers, working independently, responded to the same problem. It is believed that such documentation would serve as confidence builders for both students and less experienced staff; all too often they seek 'the correct solution' in the literature rather than being prepared to develop their own thinking towards the problem situation. A secondary advantage of such documented case studies is that they would provide excellent material for someone wishing to use the problem in a group teaching situation; in this context it would be particularly valuable if the documentation gave an indication to the user of various possible roles for individuals working within a group.

In an attempt to meet these objectives the workshop was restricted to twelve experienced modellers who were also familiar with the activities of the workshops. Participants were divided into groups of four with each group having a nominated chairman. The groups worked independently of one another and each was presented with the same five problem statements. Each chairman was asked to prepare an agreed group report covering the two aspects:

(a) methodology of approach adopted by the group,
(b) attitude towards the problem statements indicating their understanding of it and possible directions they would take towards developing a suitable mathematical model.

During the plenary session concluding the workshop it was unanimously agreed that it had been a most stimulating exercise; even the most experienced modellers expressed the view that they had benefited by participating in such a group activity. A follow up workshop is being planned following which a detailed report will be published. However, already some interesting points have already arisen from the chairman's reports. Particularly noteworthy was the fact that none of the participants seemed inclined, in the first instance, to start actually building models. Initial reaction usually concerned enquiry about the physical mechanism behind the operation of the system under study and elucidation of the data and experiences that was available to help in its solution. It was generally agreed that this initial stage was the most enjoyable part and certainly convinced everyone that students, when exposed to a new situation, should be given an opportunity to talk about it within a group situation thus reinforcing the belief held by many, that mathematical modelling is a group activity.

4. CONCLUDING REMARKS

The activities of the Workshops have received recognition overseas as well as throughout the U.K. To date over 60 participants have attended the various sessions.

Courses in Mathematical Modelling are appearing more and more within the curriculum of degree and sub-degree courses in mathematics and the workshops are providing an excellent forum for the exchange and ideas on these developments. It is believed that the existence of the workshops has already played an important part in the development of these courses, as they have provided the necessary stimulus and confidence for individuals associated with them. A survey to investigate the growth of modelling courses in British higher education is currently being undertaken as one of the Workshops' activities.

To date the Workshops have been primarily concerned with the preparation of documented case studies, which can be used as background material for mathematical modelling courses. The lack of such material is but on problem facing the teachers. Much work remains to be done on teaching methodology and the work commenced at the recent workshop is only a first step. Another problem area for the teacher is that of assessment; further consideration needs to be given to this problem if courses in mathematical modelling are to have the same academic standing as other courses within the curriculum. A further major problem restricting the growth in modelling courses is the lack of confidence and experience in teaching such courses amongst teachers of mathematics and there is a need to provide in-service courses to remedy this problem.

It is intended that future workshops will be directed towards these problem areas.

5. ACKNOWLEDGEMENTS

The authors would like to thank all those individuals who have participated in the activities of the workshops and helped to make them successful. A particular thank you is due to John McDonald, formerly attached to the Department of Mathematics and Computing at Paisley College of Technology, who played a major part in establishing the Workshops. On behalf of all those who have participated in the activities of the Workshops the authors thank the Trustees of the Nuffield Foundation for their financial support.

REFERENCES

Gaskell R. E., Klamkin M. S., 1974, *Amer Maths. Monthly*, **81**, 699.

James D. J. G., McDonald J. J., 1981, *Case Studies in Mathematical Modelling*, Stanley Thornes Ltd.

James D. J. G., Steele N. C., 1980, Mathematical Modelling in Undergraduate Courses, Proc AAMT 8th Biennial. Conf. Canberra.

McDonald, J. J., 1977, *Int. J. Maths, Edu. Sci. Technol.*, **8** (4), 453.

McLone R. R., 1973, The Training of Mathematicians, SSRC Report.

Open University Course M101: Block, 1978, *Mathematical Modelling*, Open University Press.

Wallpaper Group Design Package

P. Strachan
Robert Gordon's Institute, Aberdeen

SUMMARY

The GROUP structure is well known to mathematicians. The following is an attempt to introduce it to a wider audience, composed of artists, designers and architects.

1. DEFINITIONS OF A GROUP

(i) $\forall\, x, y \in G,\ x * y \in G.$
(ii) $\forall\, x, y, z, \in G\ x * (y * z) = (x * y) * z.$
(iii) $\exists\ e \in G$ such that, $x \in G,\ x * e = e * x = x.$
(iv) $\forall\ x \in G,\ \exists\, y \in G$ such that $x * y = y * x = \mathrm{e}.$

where G is a set of objects and a binary operation.

This interesting structure can be discovered in many different disguises from areas as far apart as geometry and campanology (Budden 0000). One interesting application is in two-dimensional geometric design. The groups of particular interest are

(i) The cyclic and dihedral groups C_n, D_n, $n = 1, 2, 3, \ldots\ldots$
(ii) The frieze groups, $f_1, f_2, f_3, \ldots, f_7$.
(iii) The wallpaper groups, $w_1, w_2, w_3, \ldots, w_{17}$.

1.1 The cyclic and dihedral groups

For the cyclic groups, the elements of G are the set of rotations about a common centre which leave the pattern invariant. The dihedral group D_3 has

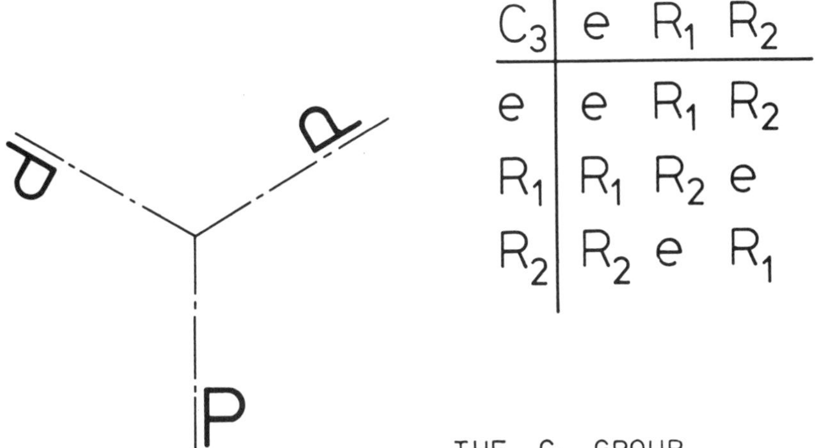

C_3	e	R_1	R_2
e	e	R_1	R_2
R_1	R_1	R_2	e
R_2	R_2	e	R_1

Fig. 1.

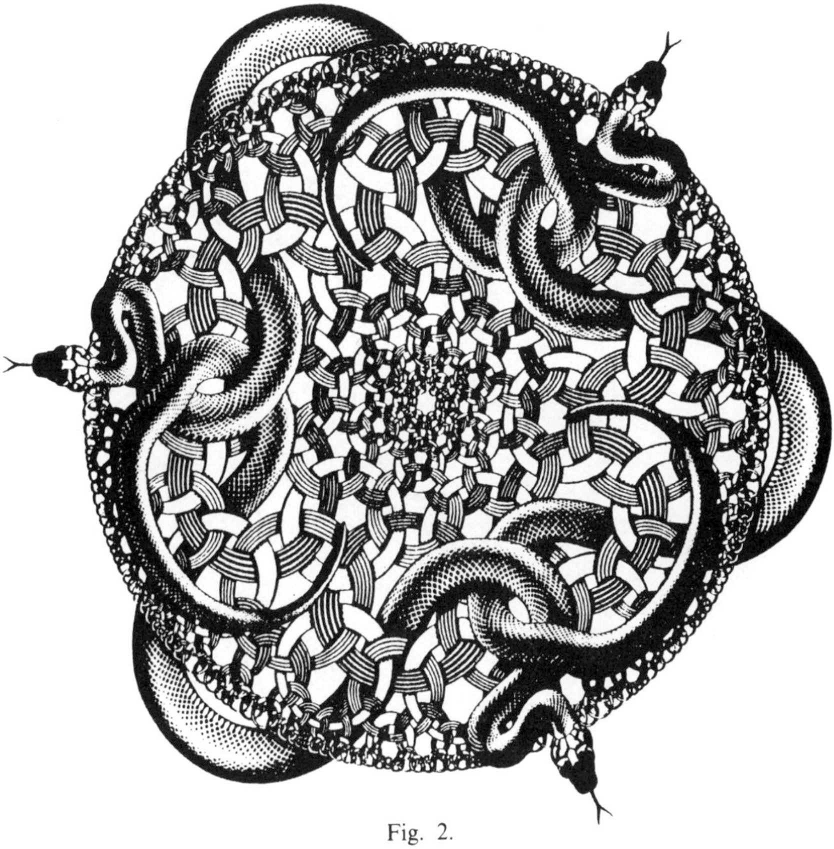

Fig. 2.

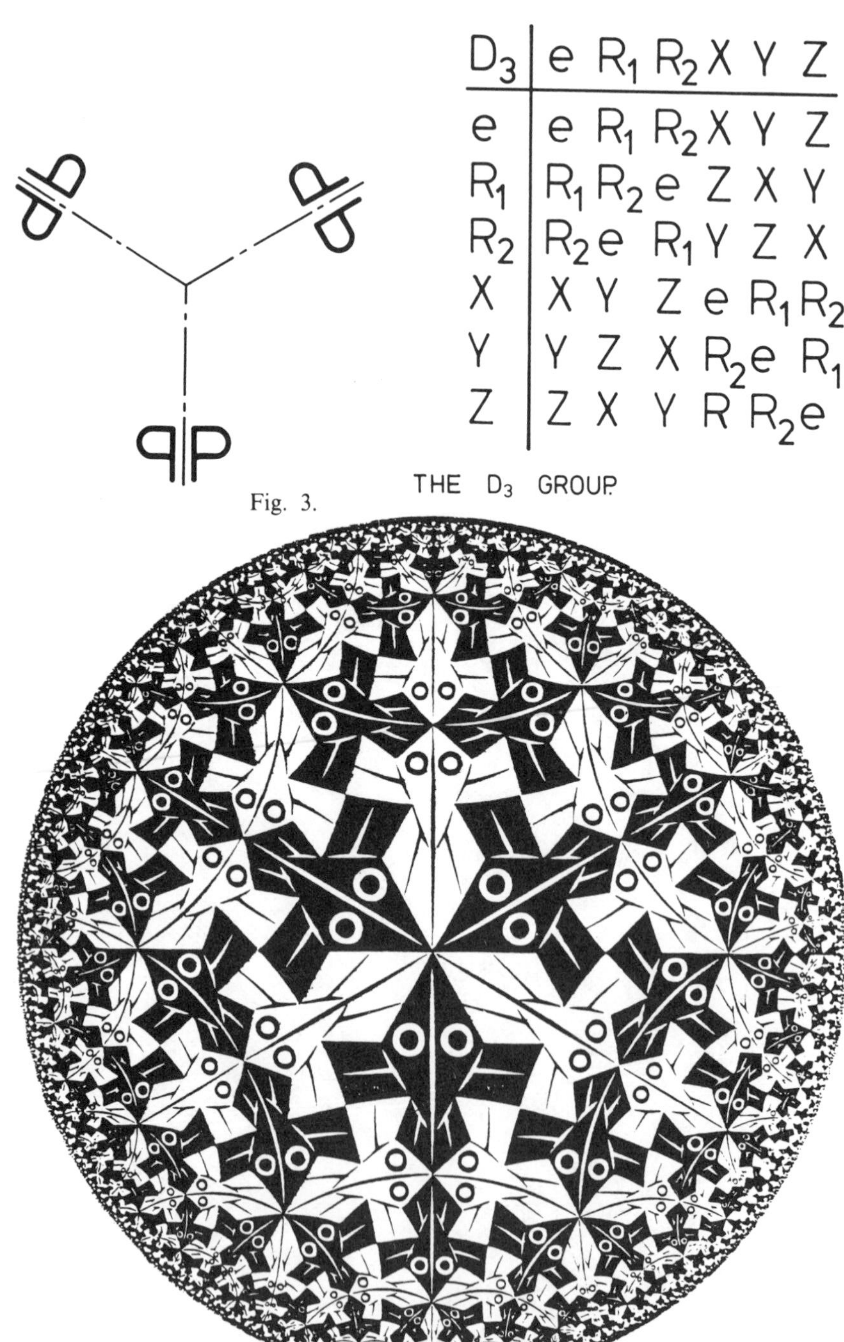

D_3	e	R_1	R_2	X	Y	Z
e	e	R_1	R_2	X	Y	Z
R_1	R_1	R_2	e	Z	X	Y
R_2	R_2	e	R_1	Y	Z	X
X	X	Y	Z	e	R_1	R_2
Y	Y	Z	X	R_2	e	R_1
Z	Z	X	Y	R	R_2	e

THE D_3 GROUP

Fig. 3.

Fig. 4.

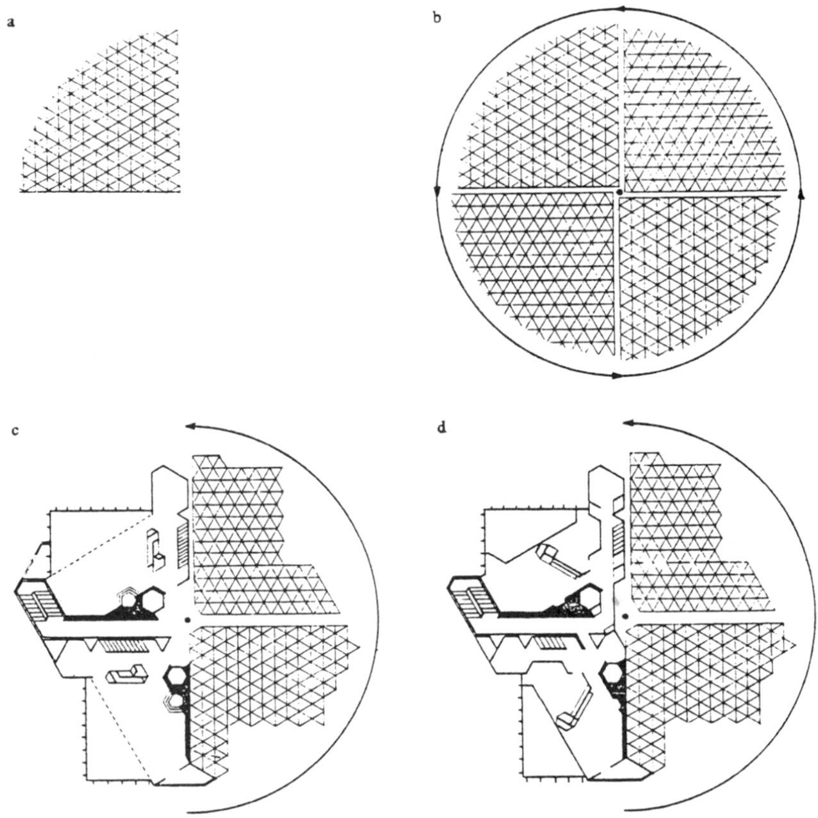

Fig. 5.

other symmetry operations, X, Y, Z, mirror reflections in the fixed axes x–x, y–y, z–z, respectively

1.2 The friezes (Budden, 0000 515–518)

A frieze is a repeat pattern which extends indefinitely in 1-dimension and is invariant under the symmetry operations of translation, rotation, reflection and glide.

There are only 7 distinct friezes and each one displays the group structure. All 7 can be best remembered by associating each one with the letters of the word SHAPE and the other two with FOOTSTEPS and SINE.

1.3 The wallpaper groups (Budden, 0000 518–538)

If two independent translations are allowed, the pattern can be extended to 2-dimensional space. Such a pattern is called wallpaper, for obvious reasons. The

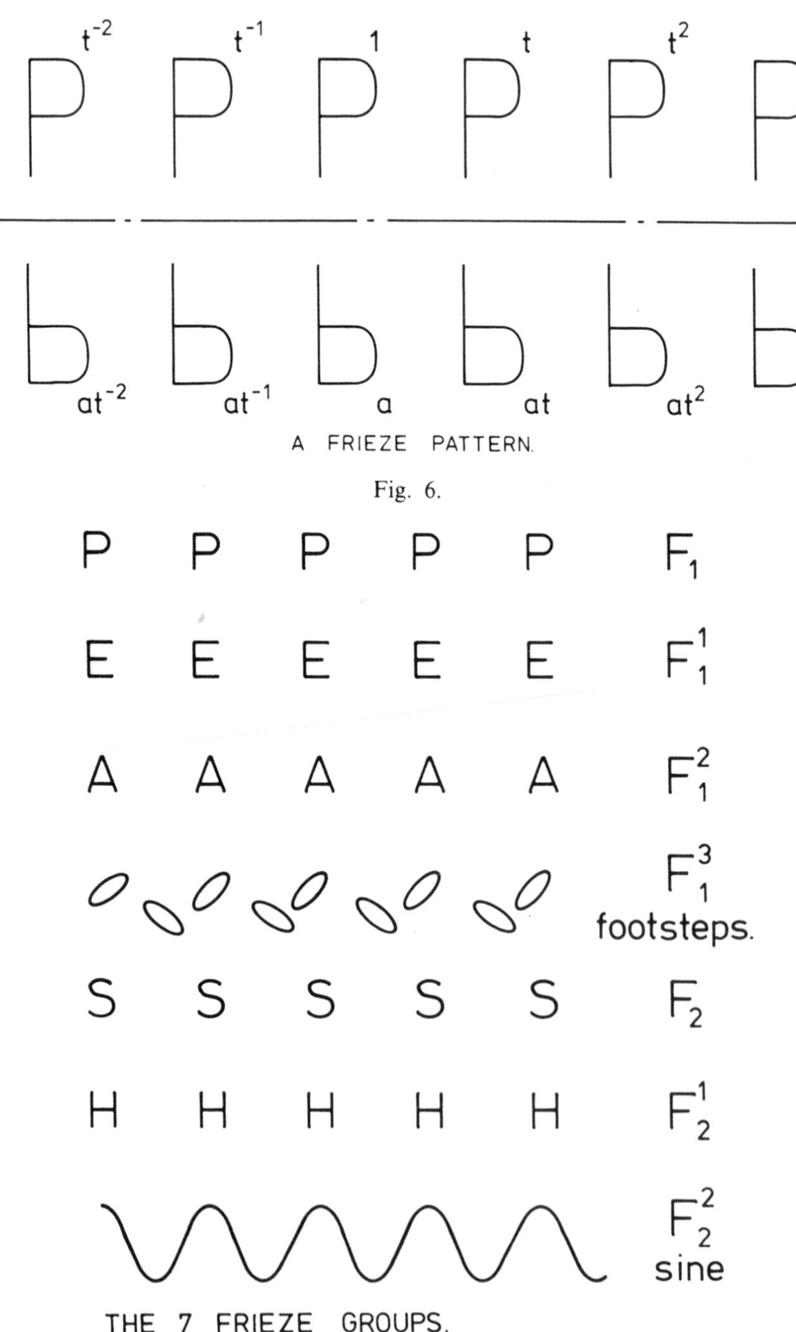

t^{-2}
t^{-1}
1
t
t^2
at^{-2}
at^{-1}
a
at
at^2
A FRIEZE PATTERN.

Fig. 6.

F_1
F_1^1
F_1^2
F_1^3
footsteps.
F_2
F_2^1
F_2^2
sine
THE 7 FRIEZE GROUPS.

Fig. 7.

Fig. 8.

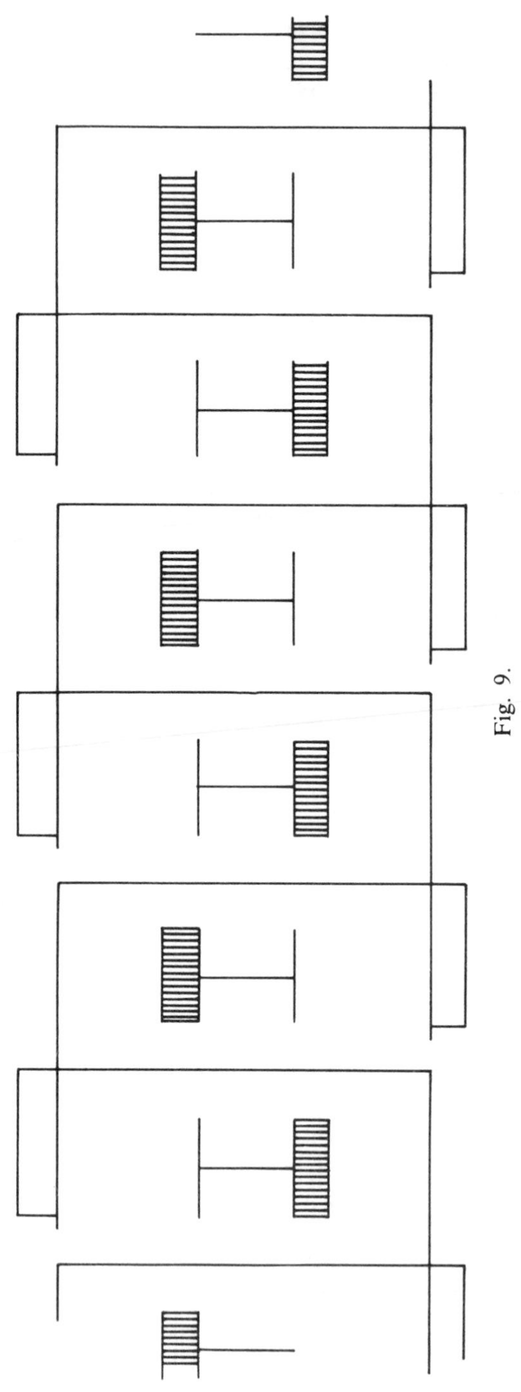

Fig. 9.

symmetry operations allowed are translation, reflection, rotation and glide (the footsteps again).

Artists, architects and designers have used all three types of patterns for centuries. M. C. Esher (see Locher, 0000) noted and sketched the tile patterns in

Fig. 10a.

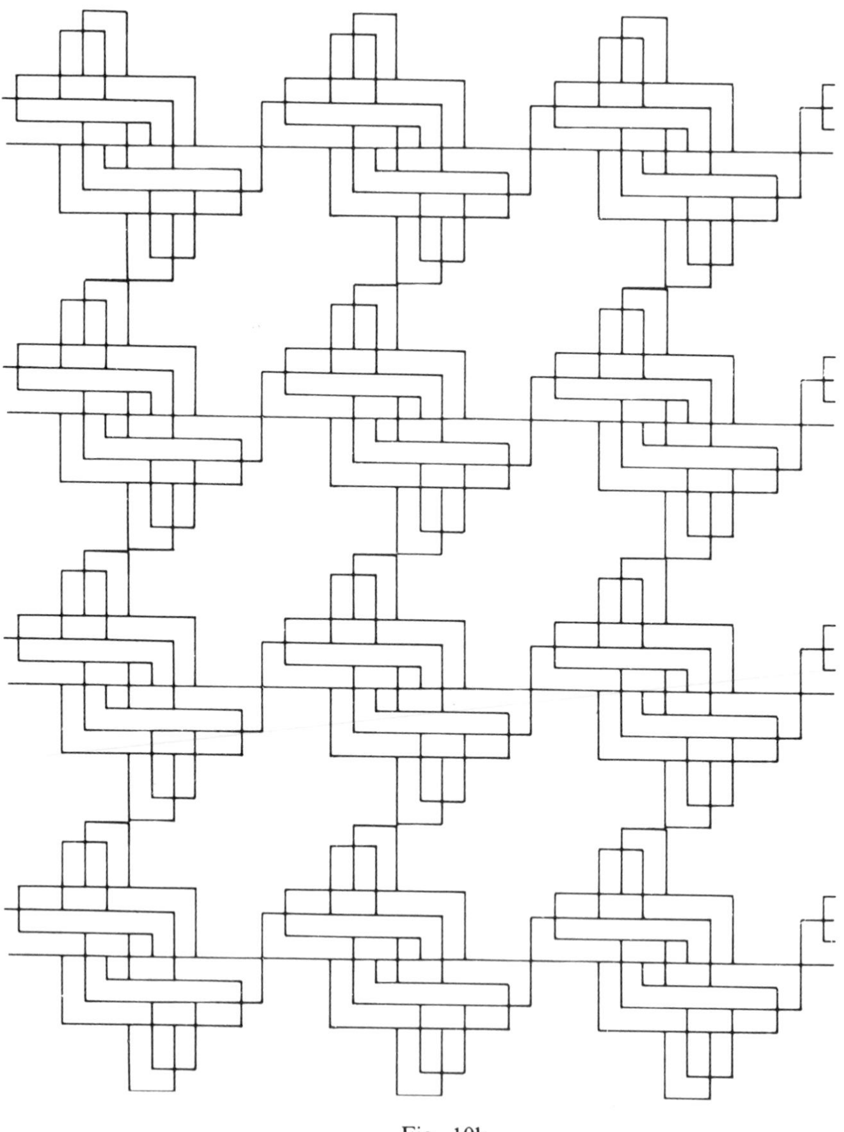

Fig. 10b.

the Alhambra, probably the finest example of Arab architecture in Europe. He showed that these early artists used many of the wallpaper patterns, but not all. Le Corbusier and Frank Lloyd Wright (see March and Steadman, 0000) showed in their designs that they were aware of the patterns, but only used a subset.

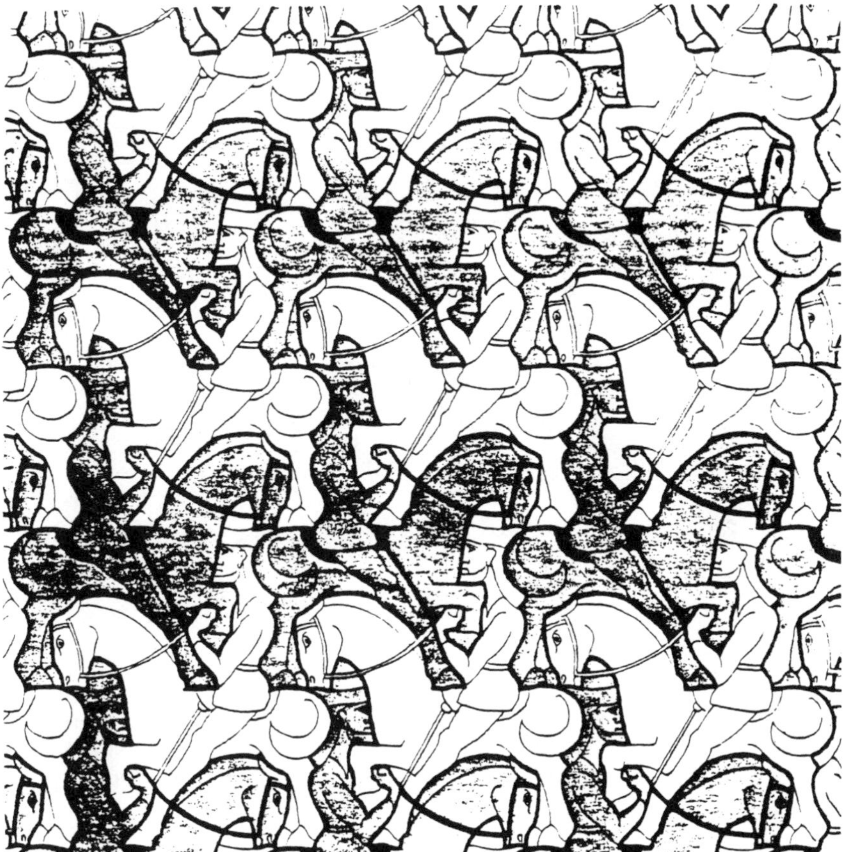

Fig. 11.

Modern designers find these patterns very interesting, but the enormous amount of work required to reproduce even a simple wallpaper, in enough detail to make an aesthetic decision, tends to reduce the variety of patterns investigated. The result is that the optimum solution may not be found.

What is required, is an interactive graphics system which will allow and encourage the artist/designer to reproduce, easily, the patterns chosen. A prototype exists, which works well and has allowed the investigation of the user interface. Extremely valuable advice has been received from an art lecturer and an architect, which has enabled different user interfaces and facilities to be tested.

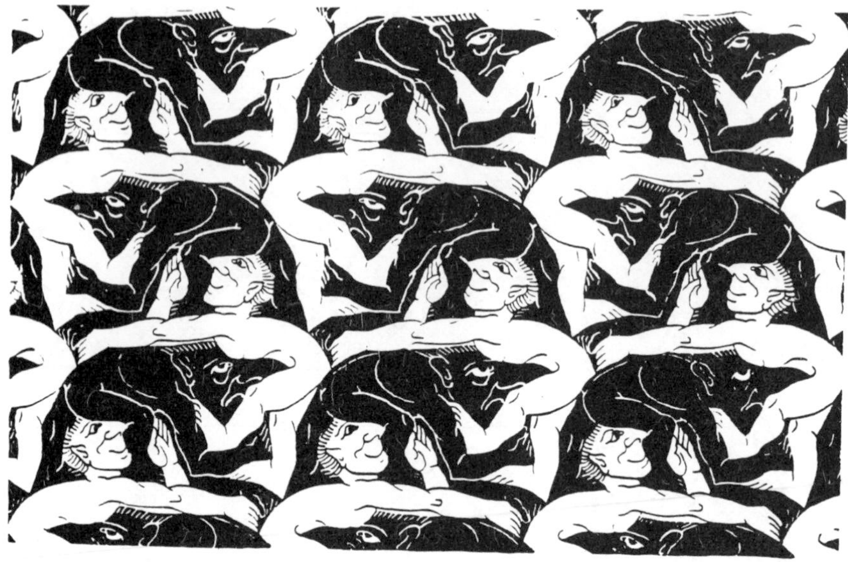

Fig. 12.

2 The W.F.S. SYSTEM (Wallpaper–Frieze–Snowflake)

The main program is written in FORTRAN and uses the Computer-Aided-Design Centre's GINO-F package. This choice enables a wide variety of graphic devices to be used, a very important facility, especially within education. Each device has to have a special purpose-built subroutine interface, but with this achieved, the program is then independent of the device.

At RGIT, GINO-F can drive all Tektronix-compatible devices, Hewlett-Packard HP 2647 terminals, HP 9822 flat-bed plotter, Benson drum-type plotter, Gigi terminals, and Sumnagraphics digitizer. Thus, output can be seen on the screen for temporary scrutiny, or to paper or acetate for permanent hard-copy. Input can be via the screen, keyboard or more usefully by a digitizer. An interactive input module has been developed by a student of the School of Mathematics enabling the user to 'draw' the basic pattern unit (bpu) on the screen and then translate, rotate and dilate the diagram before accepting or rejecting. Previously defined patterns can be altered, combined or split into components.

Future developments include a hierarchical pattern structure whereby a bpu can itself be a generated pattern, e.g. a cyclic group could be the bpu for a frieze, a segment of which can, in turn, form the bpu for a wallpaper and so on. The essential feature of any useful system must be that it must be completely transparent to the user, he must see clearly what he is doing without his view being obscured by programming conventions and keyboard gymnastics, and

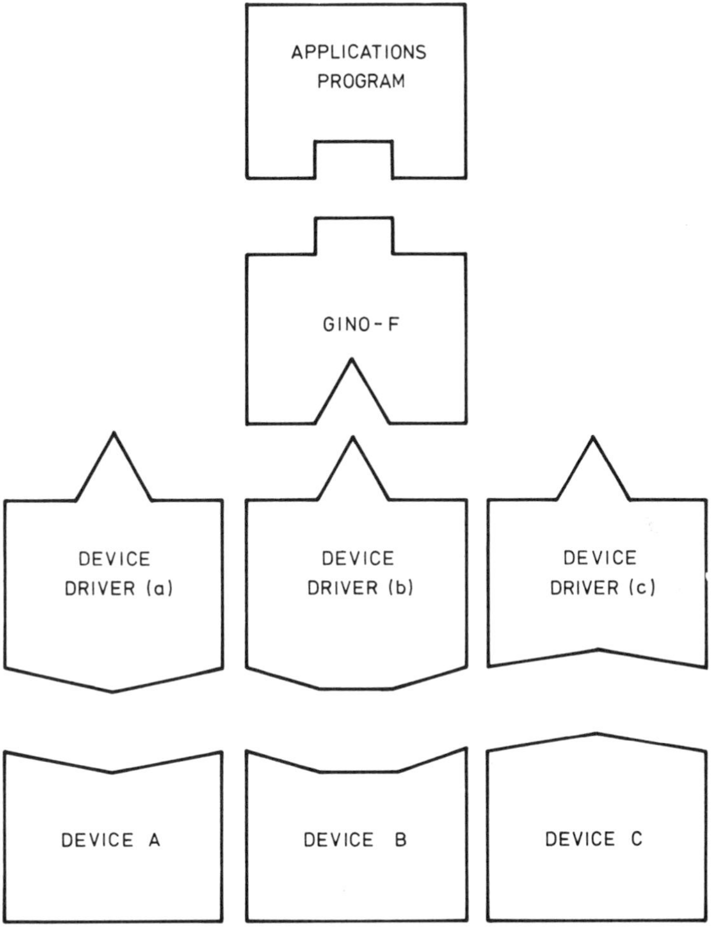

Fig. 13.

the final resulting output must be aesthetically pleasing. At the moment the fine control of colour is difficult with present hardware, but the quality of the coloured line-drawing is certainly acceptable.

It is hoped that the final system will be useful to a wide range of artists, graphic designers, architects and all those interested in 2-dimensional patterns.

ACKNOWLEDGEMENTS

I wish to express my thanks to: Mr. John Smith, Gray's School of Art, and Mr. Mark Brady, Student, School of Mathematics, RGIT, Aberdeen.

REFERENCES

Budden, F. J. *The Fascination of Groups*, Cambridge University Press.
Locher J. L. *et al.*, *The World of M. C. Escher*, Harry N. Abrams.
March L. & Steadman, P. *The Geometry of the Environment*, R.I.B.A. Publications.

Analysis of Applications and of Conceptions for an Application-Oriented Mathematics Instruction

Werner Blum and Gabriele Kaiser
University of Kassel (F.R.G.)

SUMMARY

In connection with the (revived) demand for considering applications in the teaching of mathematics, various schemata or lists of criteria have been developed since the end of the sixties, which set up requirements about closeness to the real world or about the type of mathematics being used, and which have made it possible to analyze the available applications in their light.

After having stated the problem (in section 1), we present (in section 2) a sketch of some of the best known of these and of some earlier schemata, although we are not aiming for a complete picture. Then (in section 3) we distinguish among different dimensions in the analysis of applications. With this as a basis, we develop (in section 4) our own suggestion for categorizing types of applications and conceptions for an application-oriented mathematics instruction. Then (in section 5) we illustrate our schemata by some examples of performed evaluations. Finally (in section 6), we present some preliminary first results of the analysis of teaching conceptions.

1. INTRODUCTION

Since 1975, we (that is, a research team at the University of Kassel) have been involved with questions about an application-oriented mathematics instruction. Early on, one goal of our work became the creation of a *bibliography* and a clear *categorization* of important publications with examples, analyses and general considerations about application-oriented mathematics instruction. In the course of this work it became obvious that *schemas* for *analysing examples of applications*, as well as relevant *instructional conceptions*, were not only helpful but indeed necessary for enabling us,

—to better evaluate publications at a glance with regard to their usefulness
—to index the literature in question clearly and in a unified way, keeping the essential aspects in mind
—to better compare publications with each other and thus to describe and evaluate trends and changes as well as shortcomings
—to point up the discrepancies between the claims of general theory and concrete suggestions
—to better direct the course of our own concrete developmental projects.

Our analyses comprise in the sense just mentioned general considerations and examples on the conference-topic '*teaching of mathematical modelling*', too. For example, our schemas also help us

—to better comprehend different aspects of the topic 'teaching of mathematical modelling', such as the process of modelbuilding, goals, functions of applications or their connection to the real world, and to keep these aspects separate from one another.

With our schemas, we have tried to consider, among other things, the following *aspects*:

—discrimination between the analysis of teaching conceptions and the analysis of examples of applications and therefore (see section 3) discrimination between a 'normative' and a 'descriptive' use of different dimensions.

Here we want to proceed in a *pragmatic* manner; that is, we want to propose *practical* schemas. We do *not* intend to develop either something like a '*theory* of the analysis of applications' or a very *detailed* pattern of categories but

—attainment of a middle way with regard to the degree of precision: the schemas should be both as precise and as practical as possible, both in its dimensions and with reference to their number.

2. SOME CURRENT AND EARLIER SCHEMATA FOR ANALYSING APPLICATIONS

In literature, one firstly can find

(1) *Schemata with the (only) dimension 'nature of connection to the real world'* Such schemas analyse *single* applications. Here, the schema of Pollak (1969) can be considered as fundamental. Pollak distinguishes between the following types of applications:

—Immediate use of mathematics in everyday life
—Problems that use words from everyday life, from other scholarly or engineering disciplines, to make the problem sound good, but the connection to reality is often false in one or more ways

—Problems that are made to look like applications, which can be described as pure whimsy
—Genuine applications to real life or to other disciplines whose solution requires mathematization
—Puzzles and games which are totally non-practical

(Pollak, 1969, p. 393–404).

It is not very well known, even in Germany, that some schemas for classifying arithmetic problems had already been developed in the early twentieth century, in the framework of discussions about the 'methodology of arithmetic' ('Sachrechnen'). These schemas are partly more precise and more consolidated than proposals today. The fundamental schema, to which all further classifications refer, was published by Kühnel (1916). He distinguishes between the following types of (arithmetic) problems:

—Applied problems, which can be subdivided into problems whose formulation the students are supposed to work out themselves and problems with prescribed formulations
—Word problems, which can be subdivided into problems which are true to life and problems which are remote from real life
—Puzzles and games
—Exercises using weights and measures and others not using weights and measures

(Kühnel, 1916, p. 16–32, 44–90).

A similar schema was developed by Kruckenberg (1935), wherein the connection to the real world is more differentiated. The criterion of differentiation 'life-relevance of the problems for the students' is an *additional* dimension of this schema. The same holds for another schema from the English-speaking countries which was published by Burkhardt (1981). It emphasizes the relevance of the situations to the present and future life of the students and the role of the applications in the process of learning. Thus, these schemas lead up to

(2) *Schemata with several dimensions*
Such schemas analyse applications as well as teaching conceptions—without always clearly distinguishing this—with regard to *different dimensions.* These dimensions are mostly arranged into graduated *categories.*

The most elaborate schema of this type, which is well-known in the German language area, was published by Becker *et al.* (1979). It is an expansion of a schema of Giles (1966). Becker *et al.* propose the following dimensions relevant for evaluating applications:

—nature of the problem's connection to the real world
—source of the data

—number of possible solutions or type of solutions
—difficulty of translating the problem into mathematical language
—elegance of the solution
—scope of the problem
—difficulty of finding appropriate techniques to solve the problem
—complexity of the mathematical techniques
—relevance to other areas
—transferability of the problem-solution or central ideas
—function of the applications within the course
—relation of the applications to the mathematical part of the course
—connections between the various applications
—relation of the applications to other subjects or to other sciences
—connection to the range of experience of the students
—method of presentation of the problem or of the instructional material

(Becker *et al.*, 1979, p. 11–22).

In 1974 Lörcher developed a schema which tries to evaluate applications with regard to their relevance for students. He proposes the following dimensions:

—realm from which the application is drawn and its connection to the real world
—problem-orientation of the treatment of the topics
—transferability of the problem-solution or of central ideas to other situations
—relevance of the results for the students
—relevance of the data for the students

(Lörcher, 1974, p. 76–82).

3. DISCRIMINATION BETWEEN DIFFERENT TYPES OF DIMENSIONS

A closer look at the schemas shows that these proposed dimensions differ from each other in the following ways:

(a) *Conceptional dimensions*: Some dimensions are not related to applications, but only to *teaching conceptions*. Examples are 'significance of applications in instruction' or 'participation by the students'. Here we are only interested in the '*normative*' use of these dimensions in that we are setting up *requirements for teaching*. We call these dimensions '*conceptional dimensions*'.

(b) *Curricular dimensions*: Some dimensions deal with applications (single applications or sets of applications), but they are only practicable in *connection with teaching conceptions*, in which the applications are included. Examples are 'function of applications', 'connections between the various applications' or 'relations of the applications to the mathematical or extra-mathematical part of the course'. These dimensions can be used '*normatively*' as *requirements for applications* within teaching conceptions; if given applications are included in instructional considerations these dimensions can also be used '*descriptively*' for the *classification of applications*. We call these dimensions '*curricular dimensions*'.

(c) *Situational dimensions*: Some dimensions only concern *single applications* and can be used without reference to teaching conceptions. Examples are 'type of connection to reality' or 'kind of use of mathematics'. Such dimensions can on the one hand be used in a '*normative*', on the other hand in a '*descriptive*' manner. We call these dimensions '*situational dimensions*'.

4. PRESENTATION OF A SUGGESTION FOR ANALYSIS-SCHEMATA

4.1. For analysing *conceptions* for an application-oriented mathematics instruction we use *all* types of dimensions described in section 3, in a *normative* manner:

(*A*1) *Dimensions for analysing conceptions for an application-oriented mathematics instruction*

(1a) Which *goals* should mathematics teaching pursue?

(1b) What *relevance* should applications have?

(1c) How should the *students participate* in this?

(2) Which *function* should applications have?

- —Should mathematics contribute to the solution of real problems or to the comprehension of real situations?
- —Should applications of mathematics contribute to the development of abilities to translate between the real world and mathematics ('translation skills'), to the development of knowledge about applying mathematics ('metaknowledge') resp. to a general view of mathematics?
- —Should the applications illustrate or motivate mathematical concepts? Should they help to organize mathematical areas? Should the use of situations promote general goals?

(3a) What *role* should *mathematical theory* play?

(3b) What *role* should the *extra-mathematical context* play?

(4) How should the *connections* between the various *applications* be?

(5a) How should *mathematics* be *used*?

- —routine or recipe-like use of mathematical concepts
- —intelligent use of mathematical concepts modified according to the situation
- —mathematization of situations and development of appropriate concepts, as also application and interpretation of the results to the original situations?

(5b) Should a *distinction* between *model and the real world* be made? If so, how precise should it be?
(5c) If need be, in what way should the *process of modelbuilding* be considered and which steps does it contain? (for example: idealization, mathematization, mathematical reflections, application resp. interpretation).
(6) Which type of *connection to the real world* should the applications have?
—closeness to real world: real assumptions and real problem-setting.
—remote from real world: real assumptions and unreal problem-setting or unreal assumptions and real problem-setting.
—unreal: unreal assumptions and unreal problem-setting.
(7a) From which *extra-mathematical fields* should the applications be taken?
—situations of everyday life.
—situations from the wider environment such as vocation or business.
—situations from the sciences resp. other subjects.
(7b) From which *mathematical fields* should the applications be taken?
—foundations of mathematics/arithmetic/weights and measures.
—algebra/number theory.
—geometry.
—functions/calculus.
—'applied' mathematics (finite mathematics, numerical methods, stochastics).

To illustrate parts of this schema we use three conceptions: (1) The methodics from *Lietzmann* (1916–1952); Lietzmann is a mathematics teacher and educator, who definitely influenced the German discussion from the beginning till the middle of our century. (2) *Glatfeld* (1983), for the lower secondary level. (3) *Jäger* and *Schupp* (1983), who refer to the stochastic instruction in the German 'Hauptschule'.

(*Ad la*): Lietzmann requires not only formal goals as to train the scientific way of thinking, but also practical goals as to train the students to apply mathematics to real life and to recognize mathematical problems in real life. Glatfeld claims as main goal for applied mathematics to qualify the students to mathematize. Furthermore he demands mathematically oriented goals. Schupp formulates general goals, especially the qualification to mathematize.

(*Ad 5a*): Lietzmann emphasizes intelligent use of mathematical concepts, sometimes situations should be mathematized formulated as 'functional thinking'. Glatfeld requires nearly exclusively mathematizations, also Schupp.

(*Ad 5b*): With Lietzmann the discrimination remains implicit and is not detailed. Glatfeld requires a very precise, theoretically ambitious discrimination of model and real world. Schupp claims a precise distinction between model and real world.

(*Ad* 5*c*): Lietzmann states only the approximate steps of the modelbuilding-process: Picking out the mathematical problem from the environment, problem-solving, translating the results into the real world. Glatfeld differentiates by considering the step from the real situation to the idealized situation. Schupp requires additional steps, which embed the problem-solution in mathematical theories.

4.2 For the analysis of teaching conceptions, it is necessary to evaluate the *accompanying applications*, too. Here we use the *curricular* and *situational* dimensions, in a *descriptive* manner.

(*A*2) *Dimensions for analysing applications included in teaching conceptions*
(1) Which *function* do the applications have? (categories as in A1).
(2a) What *role* does the *mathematical theory* play?
(2b) What *role* does the *extra-mathematical context* play?
(3) What are the *connections* between the various *applications*?
(4a) How is *mathematics* actually *used*? (categories as in A1).
(4b) Is a *distinction* between *model and the real world* made? If so, how precise is it?
(4c) If need be, in what way is the *process of modelbuilding* taken into account and which steps does it contain?
(5) Which type of *connection to the real world* do the applications have? (categories as in A1).
(6a) To which *extra-mathematical fields* do the applications belong? (categories as in A1).
(6b) To which *mathematical fields* do the applications belong? (categories as in A1).

To illustrate parts of this schema, we use the same conceptions:

(*Ad* 4*a*): Lietzmann's applications use mathematics mostly routinely or intelligently, but there are some examples with parts of mathematizations. With Glatfeld one can see a difference from his theoretical requirements. The examples use nearly exclusively mathematics routinely or recipe-like, mathematizations are very seldom. This is different with Schupp, who develops, as theoretically required, mathematizations besides examples with an intelligent use of mathematics.

(*Ad* 4*b*): Lietzmann's applications don't discriminate in this way; which Lietzmann indeed does not intend. Within Glatfeld's examples there exist only a few applications which are based on this discrimination. Schupp, in the contrary, explicates this distinction in his examples very often and carefully.

(*Ad* 4*c*): This tendency is continued with the consideration of the modelbuilding process.

4.3. For analysing *single applications*, not (necessarily) included in teaching conceptions, we firstly use the *situational* dimensions. Among the other dimensions, the dimension '*function* of an application' is also applicable to single applications. It makes sense to base the evaluation upon this dimension too, because one can mostly recognize which function a single application can perform or not perform, e.g. from the context or the type of presentation.

(*B*) *Dimensions for analysing single applications*
(1) Which (possible) *function* has the application? (categories as in A1).
(2a) How is *mathematics used*? (categories as in A1).
(2b) Is a *distinction* between *model and reality* made? If so, how precise is it?
(2c) If need be, in what way is the *process of modelbuilding* taken into account and what steps does it contain?
(3) Which type of *connection to the real world* does the application have? (categories as in A1).
(4a) To which *extra-mathematical fields* does the application belong? (categories as in A1).
(4b) To which *mathematical fields* does the application belong? (categories as in A1).

5. SELECTED EXAMPLES OF PERFORMED EVALUATIONS

5.1. First we illustrate *schema A* by the complete evaluation of Lietzmann. (See p. 210/211).

Up to now about 40 teaching conceptions and accompanying applications from the German, English and French language area have been analysed in this way.

5.2. In the framework of a *bibliography* of literature for an application-oriented mathematics instruction (Kaiser, Blum, Schober, 1982), *schema B* was used with slightly modified categories, because the emphasis lay on a systematical analysis of 'relevant' single applications whereas, with regard to teaching conceptions, only a rough survey was intended.

The following dimensions were used:

—type of use of mathematics (in the following example under 2): (categories as in A1).
—type of connection to the real world (in the following example under 3):
  —situations which are close to real world with data which are close to real world
  —'intentional' unreal situations
—function of the application (in the example under 4): (categories as in A1)
—extra-mathematical fields:

—arts	—everyday life	—military science	—sociology
—biology	—geography	—music	—sports

—chemistry —linguistics —political science —technology
—economics —medicine —psychology

—mathematical fields:
- —algebra/number theory
- —arithmetic/weights and measures
- —calculus
- —finite mathematics
- —foundations of mathematics
- —functions
- —geometry
- —linear algebra/analytic geometry
- —numerical methods
- —stochastics

The two last-mentioned dimensions were differentiated with 75 headings in all, to which *indexes* were produced to make a systematic research of literature possible. The bibliography contains 350 selected publications, 235 of these are reported in detail.

Here is an *example* with extracts from the text and the indexes:

Example of a summary:

HENN, H. W.
Preisindices in der Schule.
In: Didaktik der Mathematik, 8(1980a)3, S. 222–238.

ZIEL
Mathematische Analyse und Hinweise zur unterrichtlichen Behandlung von Preisindices.

INHALTE
1. Ausgangspunkt: Verwendung von Preisindizes in tarifpolitischen Auseinandersetzungen und öffentlichen Diskussionen als scheinbar objektive Bestandsaufnahme wirtschaftlicher Verhältnisse.
   Mathematische Präzisierung der Preisindizes: Darstellung der gebräuchlichen Preisindizes, mathematische Forderungen an einen Preisindex, Entwicklung verschiedener mittlerer Preissteigerungsraten wie Inflationsrate, lokale Preissteigerungsrate, spezielle Preisindexfunktionen.
   Vorschläge zur unterrichtlichen Umsetzung:
   —Sekundarstufe I: Zunächst Preisindex für Einzelwaren, dann Preisindex für Warenkorb, Vergleich verschiedener Warenkörbe, Inflationsrate, graphische Darstellung über Jahrzehnte.
   —Sekundarstufe II: Präzisierung, Behandlung weiterer mittlerer Preissteigerungsraten und der lokalen prozentualen Preissteigerungsrate.
   Diskussion der mathematischen Verfahren ausgehend von Tagesproblemen; ständige Rückübersetzung aus der mathematischen Beschreibung in die Realität.
2. Verständige Anwendung mathematischer Verfahren.
3. Realitätsnahe Situationen mit realitätsnahen Daten.
4. Mathematik soll zum Verständnis der Situation beitragen.

WERTUNG
Darstellung mathematischer Zusammenhänge mit unterrichtlichen Hinweisen zur Behandlung aktueller, relevanter Anwendungen und zur Vermittlung grundlegender Übersetzungsqualifikationen. Unterrichtsvorschlag für die Sekundarstufe I direkt umsetzbar, beim Unterrichtsvorschlag für die Sekundarstufe II sind method. Überlegungen nötig. Kenntnis wirtschaftl. Zusammenhänge hilfreich. Unterrichtliche Verwendbarkeit: In fächerübergreifenden Unterrichtseinheiten ab Jg. 7, bei Behandlung von Folgen in Jg. 10, in Analysiskursen ab Jg. 11/II.

	Analysis of Lietzmann's teaching conception	Analysis of Lietzmann's applications
Goals of mathematics teaching	Formal goals should be to train the scientific way of thinking, to become familiar with weights and measures, and to develop an intuitive faculty. Practical goals should be to train the students to apply mathematics to reality and to recognize mathematical problems in reality. Furthermore aesthetic and ethical goals such as education to objectivity.	
Relevance of applications	In the 'traditional practical mathematics' (Sachrechnen), especially in the last year of the 'low level' (Hauptschule), there should be high relevance of the applications by organizing instruction in extra-mathematically structured teaching units; besides applications should be considered by linking the different school subjects.	
Participation	Independent work should be considered in homework, in projects e.g. during stays in country boarding schools.	
Functions of applications	Applications should illustrate mathematical concepts, promote general goals and help to organize mathematical areas; besides, mathematics should contribute to the solution of real problems and to the development of translation skills.	Applications illustrate mathematical concepts and organize mathematical areas; furthermore, mathematics can contribute to the solution of real problems.
Role of math. theory	Emphasis should be on mathematics, even in the 'traditional practical mathematics'.	Explicit, broad treatment in mathematically structured applications.
Role extra-math. context	As a rule, applications should be avoided in which explaining the context takes longer than the arithmetical operations.	Explicit, broad treatment, especially in extra-mathematically structured applications.

Connections	Arrangement in extra-mathematically structured teaching units.	Co-ordination with other sciences and school subjects.
Type of use	Placing the emphasis on intelligent use by avoiding schemas, sometimes situations should be mathematized, formulated as 'functional thinking'.	Frequent routine use of mathematical methods, partly mathematizations with reference to necessary simplifications.
Model and real world	The discrimination between model and real world remains implicit, within the practical goals, especially with reference to necessary simplifications.	No discrimination is made.
Process of model-building	Only few steps should be distinguished: picking out the mathematical problem from the environment, problem-solving, translating the results in real world.	The process of model-building is not considered.
Connection to real world	Emphasis should be on problems which are close to real world, besides imaginative problems also have a right to be included (so-called recreational mathematics).	A three-part division: the largest consisting of applications which are close to real world, a smaller part are remote from real world (mostly because of imaginary problem-setting) and a third part are unreal problems.
Extra-math. fields	Applications should be taken from everyday life, from the wider environment and other school subjects; overly specific relations to vocation and business should be avoided.	Applications belong to everyday life, to the wider environment and to other school subjects.
Mathemat. fields	Applications should be taken from foundations/arithmetic/weights, measures; algebra; geometry; functions/calculus; applied mathematics.	Applications belong to foundations/arithmetic/weights, measures; algebra; geometry; functions/calculus; applied mathematics.

Extract from the indexes:

Volkswirtschaftslehre

BAUMANN 1977b	HENN 1980a	LIETZMANN 1943
BAUMANN 1977c	*HERGET	LING
BENDER, E. A.	HESSISCHES INSTITUT . . .	SCHNEIDER 1968
BOSCH; WITTMANN	INEICHEN	*STONE
. . . .	. . . .	. . . .

Differentialrechnung

. . . .	. . . .	. . . .
DREETZ	*MONTGOMERY	*STOCK
*DRENCKHAHN; SCHNEIDER	*NOACK	THEODORE
FREUDENTHAL 1973	NOBLE 1967	*TIETZE
HENN 1980a	OPEN UNIVERSITY COURSE TEAM	TIMISCHL 1980
HENN 1980b	OTTE; STEINBRING; STOWASSER	TWERSKY
*HERGET	*RIEBESELL	*WEYGANG 1976
HOGBEN	DE SAPIO	WODE

Folgen

BOARD OF EDUCATION . . .	LIETH	SCHNEIDER 1968
DEUTSCHES INSTITUT . . .	LIETZMANN 1924	SCHOOLS COUNCIL SIXTH FORM . . .
ENGEL 1968	LING	*SLOYER
HENN 1980a	MENNINGER	STREHL
*HIRSTEIN	MEYER	TANNERT
JACOBS	OPEN UNIVERSITY COURSE TEAM	THEODORE
*KAISER u.a.	ROHRBERG	*WEYGANG 1976
. . . .	. . . .	. . . .

6. SOME PRELIMINARY FIRST RESULTS OF THE ANALYSIS OF TEACHING CONCEPTIONS

Finally, we present some results—still preliminary ones—which G. Kaiser has attained through a detailed analysis of familiar *German* conceptions for an application-oriented mathematics instruction (further analyses show that there are relevant differences between the English and the German language area; we have to omit details here).

At the beginning of this century very precise differentiations were developed according to the connection of the applications to the real world. At that time no distinction was made, however, between model and the real world. Around 1930 there was a vague talk about 'schema' or 'model'. Only after World War II do we find the beginning of a distinction between model and real world, and a development of ideas resembling today's conceptions of modelbuilding. But all this remained rather imprecise. Only *in the seventies*, in connection with a strong trend toward applications in mathematics instruction, *precise notions* about the *distinction between model and real world* and about *possible procedures in the mathematizing process* were developed. These notions, according to our analysis, have met with *general agreement* since the end of the seventies in the discussion of an application-oriented mathematics instruction. This consensus has been relatively independent of the different tendencies within application-oriented mathematics instruction. Some writers set up a great variety of ambitious goals, but their suggestions about teaching have little to do with these goals. *Gaps between claims and realization* also appear on other levels, such as the connection of the applications to the real world, the function of the applications, and the proportion of extra-mathematical context to mathematical theory. But these gaps are not as large as those in the realm of *modelbuilding activities*. So, for example, at the present time more modest demands are being made for the connection of the applications to the real world. In recent years, however, increasing numbers of instructional suggestions have been developed which indeed bring in the real world, but which are not based on mathematizations.

After these rather general remarks, we want to add some preliminary results of the analysis, which are more closely related to the process of modelbuilding and to the relation between mathematics and the real world. It became clear that the *demand for realistic examples* on the one hand and for *considerations of modelbuilding* on the other are only *distantly related* to each other. The process of modelbuilding can be carried out using examples unrelated to the real world. In such cases the phases of idealization of the real situation and the interpretation of the results in the initial problem are missing, since the real situation had already been radically simplified. It seems questionable whether any genuine mathematization abilities or metaknowledge about applying mathematics can be conveyed using such abbreviated examples.

Furthermore, the *degree of preciseness* of the *distinction between model and real world* has *consequences for the process of modelbuilding*. So, for example, when we start with situations which have already been idealized, the resulting mathematization appears almost compulsory, which is practically a falsification of any genuine process of modelbuilding.

We are well aware that some of the results just presented may already seem familiar. As we said at the outset, the function and purpose of our analysis-schema is—among other things—to provide a more precise and more carefully differentiated grasp of the gaps between claim and realization, as well as of shortcomings, trends, changes, etc. Thus, already supposed statements can be supported more precisely or, if necessary, can be refuted.

REFERENCES

Becker, G. *et al.*, 1979, *Anwendungsorientierter Mathematikunterricht in der Sekundarstufe* I, Klinkhardt.

Burkhardt, H., 1981, *The Real World and Mathematics*, Blackie and Son.

Giles, G., 1966, The Place of the Problem, in *The Development of Mathematical Activity in Children—The Place of the Problem in this Development,* Ed. Research and Development Panel of the Association of Teachers of Mathematics, p. 28–32.

Glatfeld, M. (Ed.), 1983, *Anwendungsprobleme im Mathematikunterricht der Sekundarstufe* I, Vieweg.

Jäger, J., Schupp. H., 1983, *Curriculum Stochastik in der Hauptschule,* Schöningh.

Kaiser, G., Blum, W., Schober, M., 1982, *Dokumentation ausgewählter Literatur zum anwendungsorientierten Mathematikunterricht*, Fachinformationszentrum Energie, Physik, Mathematik.

Kruckenberg, A., 1935, *Die Welt der Zahl im Unterricht,* Schroedel.

Kühnel, J., 1916, *Neubau des Rechenunterrichts*, Vol. 2, Klinkhardt.

Lietzmann, W., 1916, 1919, 1924, *Methodik des mathematischen Unterrichts,* Vols. 1–3, Quelle & Meyer.

Lörcher, G. A., 1974, Realitätsbezogenheit im Mathematikunterricht, in *Schulwarte,* **27,** 74–83.

Pollak, H. O., 1969, How can we Teach Applications of Mathematics? in *Educational Studies in Mathematics,* **2,** 393–404.

Modelling: What do we really want students to learn?

J. H. Mason
The Open University

1. INTRODUCTION

Since we have so many modellers gathered here, perhaps I should follow a framework and begin by Recognizing a Problem:

> In the last six years or so there has been a spate of books on mathematical modelling, presenting polished models which answer a wide variety of questions. What is the student to make of it all? What do we really want students to learn?

I shall approach this question from three mutually supporting directions. In brief, the three claims which I shall develop are as follows:

1. There is a fundamental error in believing that relevance (ie modelling) is a complete answer to getting students engaged and participating in mathematics. Involvement and relevance are opposite sides of the same coin, and they require an action inside the student. I shall argue that such an action is most clearly manifested in asking questions, and that one of the most important things we can do for ourselves and for our students is to foster questioning. A necessary condition for real questioning seems to be seeing and recognizing generality when looking at particularities.
2. Demonstrating completed or well-begun models is a useful component in teaching modelling, but it is only relevant once the inner action of point 1 has already begun. We might learn a great deal from using and studying the quintessential modelling process, namely metaphor. The particular lesson to be learned from metaphor is a glimpse of the inner activities that give rise to the outer events of modelling, shedding light on 'entering a question and making it my own'. In the context of modelling I shall develop one

metaphor in particular which seems to me to offer pertinent and specific suggestions for the teaching of any craft.

3. Modelling is a complex process. Even when the initial 'getting-started' hurdle has been surmounted, producing effective and useful models is very much an art. The expert may feel no need for an overview, but the novice is very often lost in detail, unable to appreciate the modelling process as a whole—particularly if the modelling has been done by someone else! Why then are people so reluctant to employ a framework to structure their activity? Some people fear a straitjacket, others fear that students will use it as a crutch to avoid thinking. But frameworks (possibly fragmentary) are present whether they are acknowledged or not. I claim that they can and should be used positively, with suitable weaning where necessary.

Before treating these three claims in detail, it is necessary to set the scene by examining typical responses to the question

'What do we really want students to learn?'.

Typical responses are usually along the lines of

to apply mathematical techniques and skills to obtain useful answers to practical questions;

to see mathematics in action, explaining and predicting;

to become more interested and involved in mathematics, and so to pursue it further;

to become better at using mathematics to help resolve questions.

These are the sorts of reasons that are listed in modelling books, in those brief introductory chapters that talk sparingly about the act of modelling and about why it is worth doing with students, before getting down to the case studies. As justifications, these are really very superficial. They may be a delayed response to students of the 1960s who cried out for relevance, but they remain at that same level of vagueness. They may express a utilitarian philosophy, but they are not the result of looking closely at the question in the manner that utilitarians espouse. When students ask

'Why are we doing this?'

they are rarely satisfied by a reply of the form

'It is important in the steel industry'.

In fact they are usually not asking a question at all, but rather making a plea for help, because they have lost contact with the content of the class. It is not 'relevance' in the utilitarian sense that is sought, but rather a statement that they are no longer coping. I claim that most attempts to 'make mathematics relevant to students'

are misconceived and doomed to failure, because relevance is not a property of mathematics, nor of its application to a particular physical context. Relevance is a property of an appropriate correspondence between qualities of some mathematical topic and qualities of the perceiver. Relevance is a relative notion: it describes a 'ratio' between aspects of the content and of the student.

Relevance is often used as a buzz word to signal that students will become involved, as if all or even most students will become involved when presented with certain material. Involvement like relevance is descriptive of a relationship between content and student, and does not reside in the content itself. I suggest that if a student becomes involved in some topic or question, then they see what they are doing as relevant to them at that time, and conversely, if something seems relevant, then the way is open for involvement. In other words, relevance and involvement are different ways of describing the same experience. I cannot by my efforts alone involve a student, I can only work for conditions which facilitate involvement for students who wish it. I can for example make it easier for students to become involved by reducing pressures to get correct answers, and listening to what students have to say in a conjecturing atmosphere. I can even institute commercial-like pressures as described by Borelli and Busenberg (1980), and by Henrikson (1977) and hope that the students respond. Whatever approach is taken, it is bound to be more successful if attention is placed on the relationship between student and content, and not on the content alone. Care must be taken not to confuse mathematical involvement with physical activity. It is easy to get students 'doing' something, but often they are actively involved only in accumulating grades, and avoiding real thinking. Often physical activity actually precludes mathematical thinking!

The belief that 'I can involve my students' is rather like picturing students as pianos, as passive instruments waiting for me as pianist to strike a chord. It does great injustice to our students, and wastes a great deal of our own time and energy. Students of all ages and all topics are active. They are constantly construing meaning from salient fragments of lectures and texts (and unavoidably ignoring parts that fail to spawn meaning). When we present a fully developed model like Newtonian mechanics, or our own version of a case study taken from a book, what fragments of our careful exposition can we truly expect students to pick up? For example, the following extract is taken from a comparatively sensitive and carefully constructed book on modelling by Saaty and Alexander (1981):

Strategy for a Mailman

A mailman must deliver mail to each house on both sides of a straight street of length L and width W. Suppose that all N houses have the same street frontage of length D. The houses on both sides may be considered as points in the plane and start precisely at the beginning of the street, and end at its end. The houses on one side are a mirror image of those on the other. If the mailman crosses the street he does so on a perpendicular to its length. Compare the strategy of delivering mail to all houses on one side, crossing the street, and returning to his starting point by delivering to all houses on the other side, with the strategy of crossing the street from one house to that opposite it on the other side, walking to its next door neighbor and crossing the street again.

Note that $\frac{N}{2} - 1 = \frac{L}{D}$.

And so it continues for 6 more lines of packed algebra. The exposition of the model is the content. The modelling is totally obscured, as is the original question. Only exceptional students are likely to learn anything about modelling. From this sort of presentation, most students are likely to construe modelling as something that 'they' do, and mathematical models as just another source of awkward assessment questions on mathematical theory which has to be learned. Indeed it is doubtful if many students would be able to distinguish between this sort of modelling and theorems of pure mathematics. This cannot be what we really want students to learn.

The next section develops my first claim: that relevance and involvement are intimately connected with being aware of generality when focusing on particularities.

2. SEEING THE GENERAL IN THE PARTICULAR

There is a fallacy being promulgated currently that questions to do with the 'real world', the world of material objects, are by their very nature relevant and hence involving. As I argued in the last section, I maintain that questions themselves are simply questions. When they become MY questions in the sense that they are a part of me and I of them, then relevance and involvement are non-issues, because the question-tension is inside me, and the only response apart from intentional avoidance is involvement.

Relevance/involvement begin when a person experiences an inner tension of surprise at something unexpected or curious, and then tries to make use of that tension rather than avoid it. It may begin with a question posed by someone else, but the question itself is not the source of tension. The question is more like a catalyst or seed crystal. The tension is a result of the student construing meaning. I refer here to a positive form of cognitive tension which unfortunately is often confused pejoratively with emotional tension, perhaps because of the singular lack of vocabulary for such things in Indo-European languages.

It does not matter whether the question is practical:

> What is the most efficient route for a postman on foot delivering letters in a street?

or abstract:

> Which integers can be written as the sum of consecutive integers and in how many different ways?

Once a tension appears, and providing that the emotional context is supportive, mathematical thinking can begin. In order that the tension arise, the student must do some work, so that some sort of action, some bed of energy is essential in a student before a question can spark off activity.

It is true that an energetic and emotive lecturer can carry a large number of students along, sweeping them into at least a partial appreciation of a question and its resolution, but even then students must supply something, must experience in themselves some of the tension. Lecturers who have a charismatic quality in their style of presenting questions or in the way they conduct themselves mathematically have an energy or absorbing interest which can be infectious. Even so, the student must bring something to the event, and it is the nature of that something which I would like to understand. When that 'something' is present, mere 'questions' can become involving and relevant. The image which comes to my mind is of waves on a pond—their size depends on the depth of the pond as well as the strength of the wind. Perhaps a more apposite image is of people asleep. No matter what you say (without shouting!), little will come of talking to them. Even shouting will only rouse them from slumber, not equip them to take in what you have to say. So too a certain energy level is necessary in students.

To test the hypothesis that power does not reside in the question, one need only try recording oneself posing a question to a class (preferably one that takes off), transcribing it, and then presenting the written version to various people under various conditions. The result will be that 'interest' in the question will be unpredictable. It will be taken up most readily by colleagues who are already mathematically active (modulo time available), who have an attitude of mathematical (if not broader) questioning.

We now have a new (and possibly self-referent) question:

> How does a question become involvingly mine?

What to me is most surprising about this question is that it should need to be asked at all. As purveyors of models and modelling are aware, there is a plethora of potential problems around us all the time. Perhaps the best explanation for failure to question is as a defence against the overwhelming quantity of difficult questions. One of the most comprehensive collections of questions I know is Walker's *The Flying Circus of Physics* (1977) and it illustrates my point perfectly. Walker has assembled a vast array of questions about everyday events from thunder to vacuum-cleaners. Despite my predilection for seeking questions, there is for me something initially exciting but ultimately debilitating about the Flying Circus. His questions are not my questions, and furthermore I know that he has (mostly) resolved his questions at the back of the book. A book full of questions seems to me to run entirely opposite to a classroom atmosphere of enquiry, of asking questions, since much of the force of a question lies in the asking. This is why Burkhardt (1982) recommends eliciting students' concerns and massaging them into questions, though the chances of getting something mathematically meaty is small. A supply of seed questions can be helpful to get going, but they must be used very sparingly because a whole catalogue of them can be off-putting. The gap between a question being someone else's and being mine seems to widen when I know that there is a profusion of questions already written down, waiting to be set.

We wish to stimulate enquiry, because that can be followed by investigation and modelling. Usually the question source will be from someone else, but to stimulate appreciation of, and adoption of externally provided questions, it may be (and I claim it is) essential to help students develop their ability to recognize and pose questions for themselves. This brings us to a third set of questions:

> How can we foster an atmosphere of enquiry, investigation and modelling?
> How can we re-awaken questioning in ourselves and our students?

Having pondered these questions for a long time myself, it has recently come to me that there is one feature about the noticing of interesting (to me) questions which is fundamental to mathematical thinking. To show you what I mean, have a look at the following pictures. Several of them evoke in me questions whose resolution could easily be aided by mathematical modelling. I would like to draw your attention to what you do when you look at these pictures. I am interested at this point *not in the questions*, but in *how the questions arise.*

Each of the pictures is either a particular object or a particular scene. Yet in order to make sense of them, to construe some sort of meaning, we each must surely relate what we see to things or scenes we have seen before. When we see the particular, we relate it to other similar instances. We automatically and subconsciously stress some features and ignore others. Without contrast there are no distinctions, and so no tension. With the awareness of contrast between what is seen and what is not, between the scene as it is and as it might be, tension arises. When our eyes are open and we look at a particular scene, we can only *see* what we can generalize or fit into a broad schema. For example most of us simply cannot see distinctions in snow which are important to an eskimo.

The noticing of a question requires the focus of our attention to be seen *as* a specific instance of a general phenomenon. It seems to me that awareness of

generality is essential. Only then does it make sense to ask a general question such as

What is happening?
Does that always happen?
What is the maximum/minimum . . .?
. . .

Many if not most mathematical questions are to do with explaining or predicting a repeating phenomenon, and so are inherently general. Do they make sense if there is little or no prior experience of particular instances, or if students' attention is totally absorbed by the particular to the exclusion of the generality?

These observations suggest to me that we can be of direct benefit to ourselves and our students by taking every opportunity to indicate when we are 'seeing the general in the particular', because it is highly likely that our students are engrossed in the particular. Certainly this is true at the macro level of modelling when we present students with specific models. To the presenter, the model is only a specific instance of a general process called modelling. To the student, the model is the content to be made sense of and learned. Case studies become content, and not particular cases of anything more general like the process of modelling.

The same discrepancy between a tutor's perception of generality when talking about a particular example, and a student's perception of the particular as the totality takes place at all levels of mathematical presentation. (See Mason and Pimm (1983) for elaboration).

Having developed the notion of 'seeing the general in the particular' by pursuing the theme of relevance and involvement into the realm of recognizing and posing questions, and thus learning to make questions my own as well as making my own questions, the next section invokes metaphor as an aid to distinguishing between

'entering a problem and making it your own'

and the more usual student activity:

'keeping a problem at arms length, failing to make contact with it'.

Metaphor is closely akin to modelling, and appreciation of the richness of a metaphor is closely connected with being involved.

3. METAPHOR

The previous section has indicated that it is important to distinguish between

learning A model

and

learning TO model.

I invite you to reflect on that difference a moment before rushing on.

STOP NOW AND REFLECT

As I was contemplating the distinction between the noun 'model' and the verb 'to model', and the import of the word learning, an image of pots and pottery classes came to me:

I wouldn't attempt to teach someone to throw pots simply by showing them a lot of pots on a shelf.

Similarly, no-one would attempt to teach modelling simply by showing students lots of mathematical models. Or would they? I recently approached

some twenty or more books on modelling to find out what they had to say, and found very little in the way of comment on the process of modelling. (One notable exception was Burkhardt (1982).) They consist largely of worked out models, and exposition of techniques needed to solve the mathematical problems posed in the models. Perhaps this is not surprising, since books are often seen as the repository of knowledge, and models are often confused with knowledge in the same way that pots are often confused with pottery.

The image of pots which came to me has already begun to develop itself into a metaphor, but just as someone else's model is less immediately engaging than your own, so it is with metaphors. Therefore I invite you to stop reading and to construct your own metaphor for the process of teaching modelling to students.

CRYSTALLIZE YOUR METAPHOR NOW!

A metaphor is apt if it captures an elusive essence, evoking mental images or associations. In that sense, the notion of metaphor as quintessential model (see Leatherdale (1974) for a full discussion) leads me to conjecture that modelling can be seen as constructing and developing a metaphor, with all the richness of inner images and kinaesthetic associations that are associated with metaphors. A model is not just a collection of variables and relationships stated in qualitative or quantitative symbolism. It is not the clay, but the pot. It is a way of looking at, of being present in a question, full of metaphoric richness and imagery. The ubiquity of metaphor is nicely put by George Eliot:

> O Aristotle! If you had had the advantage of being 'the freshest modern' instead of the greatest ancient, would you not have mingled your praise of

metaphorical speech, as a sign of high intelligence, with a lamentation that intelligence so rarely shows itself in speech without metaphor—that we can so seldom declare what a thing is, except by saying it is something else?

The Mill on the Floss
quoted in Leatherdale, 1974)

In the next section I will try to draw together the threads of metaphoric imagery, questioning, and the early stages of modelling, but before doing that I want to explore the image of the potter working with clay because it does suggest specific areas where we could learn from teachers of pottery, as well as from the metaphor itself.

Metaphors like models are not exact. Exploring the details of a metaphor is like exploring the details of a model, examining its assumptions, checking accuracy and appropriatness of fit, and making predictions. We don't abandon models because they are approximate, and neither should we abandon metaphors because they are not perfect. Each model is a way of preceiving an event, and so is each metaphor. Exploring where a model/metaphor is accurate, and where not can shed light.

The physical elements of pottery include the clay, wedging to prepare the clay, throwing and centring the clay on the wheel, fashioning the pot itself, letting the pot green, glazing and firing.

I see the clay as the mass of variables and relationships. When a question is first posed, the variables are only vaguely sensed—they need preparing, getting the 'air' out of them. As wedging proceeds my hands become familiar with the clay and a relationship develops between me and it. I wonder how many students could say that about the variables and relationships in the Newtonian model of mechanics, or in any other model to which they are exposed?

The act of choosing a wedge of clay, throwing it on the wheel and centring it corresponds to pruning the feature list and making all sorts of assumptions so that work can begin. How often have you seen students unable to proceed because they are wedded to complexity—too large a lump of clay to work?

The potter usually has a sense of the sort of pot that is wanted, though perhaps not in fine detail. This corresponds to the unspoken sense of what sort of model is wanted and for what purpose—features that are rarely made explicit in case studies.

The pot begins to rise and take shape, corresponding to the formulation of the mathematical model. The model is the pot itself, not the clay of which it is made. Its basic form is quickly determined, and all the remaining work is in dealing with details—coating with slip, clarifying lip and base, and incising decoration. It may easily be that something goes wrong, so the clay is scrapped and a new wedge selected.

The finished pot must be allowed to dry slowly—to green—and in the process it may crack, split or sag, just as a model may do when it is checked for mathematical errors and for agreement with the situation being modelled. Glazing corresponds to minor (decorative) details, Firing corresponds to writing up the model in terms understandable by the people who want to use the model, and putting the pot up for sale is like trying to get them to use the model.

There may be many weaknesses in the analogy as I have laid it out—it depends both on how you view modelling and pottery, and the extent to which you are willing to try to enter this metaphor and try to learn from it. The pottery metaphor seems to me to be sufficiently accurate to invite further exploration to see if it brings about any perceptual shifts either in the way we practise modelling, or in the way we teach modelling to students. I recommend highly that you pursue your own metaphor in depth, and note differences and agreements.

A metaphor is only helpful if it actually informs me about my questions. So far all I have done is sketch some details of an analogy (which by the way was originally a mathematical term meaning proportion). Now I shall turn to some of the details in connection with learning.

Students of pottery pick up the language—wheel, slip, throwing, green, firing, glaze, . . . by hearing it used in context. There is no need for it to be 'taught'.

The teacher acts as a model of behaviour, demonstrating how tools and clay are to be treated, how calmness and inner quiet are more effective than rush and tension. The potter and student are usually facing in the same direction in the sense that they are both working on the pot, whereas mathematics teachers are very often the medium, the intermediary between student and mathematics. When teacher and student face each other, the student is more oriented to 'guessing what is in teacher's mind' than to the common task. So too the teacher who is bound and determined that the class must end up with THE model must almost inevitably push students in that direction. Not so the potter who acts as aide and guide, rather than locomotion.

Aesthetic appreciation of a fine pot (or model) can be gained from books on the design of pots, and from handling well-thrown pots, in conjunction with practical work with clay. These can provide a flavour that can resonate later when students are looking for ideas, helping to guide their hands as they fashion a pot or model. Case studies can play a similar role, once students have worked some clay themselves. The difference between pots and models is that in order to handle someone else's model a good deal of mathematical sophistication is required, whereas a pot can be appreciated at what ever level of experience the student brings.

It helps to make coiled pots and other pottery objects in order to become familiar with clay and with textures which can be produced. So too familiarity with mathematical skills is essential. There is a meta-heuristic which says that a modeller is unlikely to be able to use a mathematical technique effectively unless it was first encountered at least two years previously. Obviously 'two years' must be interpreted as a gestation period which is a function of mathematical sophistication, but for undergraduates, two years seems to be pretty accurate.

There is nothing mysterious about throwing pots, but it does depend on being centred oneself—in a calm state and in a harmonious relationship with the clay. Something similar may be true for modellers as well. Certainly a skilful mathematician can drive a route through to a model if the question is not too challenging, but a really successful model which is going to be used frequently requires considerable sensitivity to the real questions being tackled, and to the nature and role of models in helping resolve those questions.

When a potter is demonstrating, it can be very helpful to hear a running commentary—it gives only a glimpse of what the potter is experiencing, but it is a good deal better than simply watching. In mathematical demonstrations we all tend to mimic the textbooks, wary of wasting time by showing students dead ends along the way. Yet despite the difficulties I am struck by the disparity in modelling books which do offer some sort of framework for modelling, between sometimes sensitive exposition of the framework and the mechanical, superficial use of that framework in expounding the models. Particular attention is drawn to recognizing, becoming familiar with, and formulating the problem—at least in the introduction—but little is done about it in the case studies.

Textbooks may correspond to books about pottery, and completed models to pots on the shelf, but it is still possible to let students in on the really significant choices and awarenesses at the beginning of the modelling, which get the whole thing going.

4. FRAMEWORKS AND STRAITJACKETS

According to the experts, teaching modelling is difficult:

> The only way we have so far to train an individual in modelling is to expose him to a wide variety of problems and to a corresponding variety of models which provide representations of those problems. This establishes a need for a methodological framework for problem formulation.
>
> Saaty and Alexander (1981)

> Once the significant features have been identified, the next stage is to translate these into mathematical entities. This is generally the most difficult stage, and one in which it is impossible to give formal instruction!
>
> Andrews and McLone (1976)

Behind all the discussions and case studies it is not hard to detect fragments of models and metaphors for the process of mathematical modelling. Ocassionally these are made explicit—for example many books refer to the two-state diagram:

REAL WORLD	MATHEMATICAL WORLD

It draws attention to a separation, but it gives little assistance as to how to go about modelling. A more detailed seven-box diagram is used in some courses at the Open University (1981) which expands the two states and tries to indicate what to do at each stage:

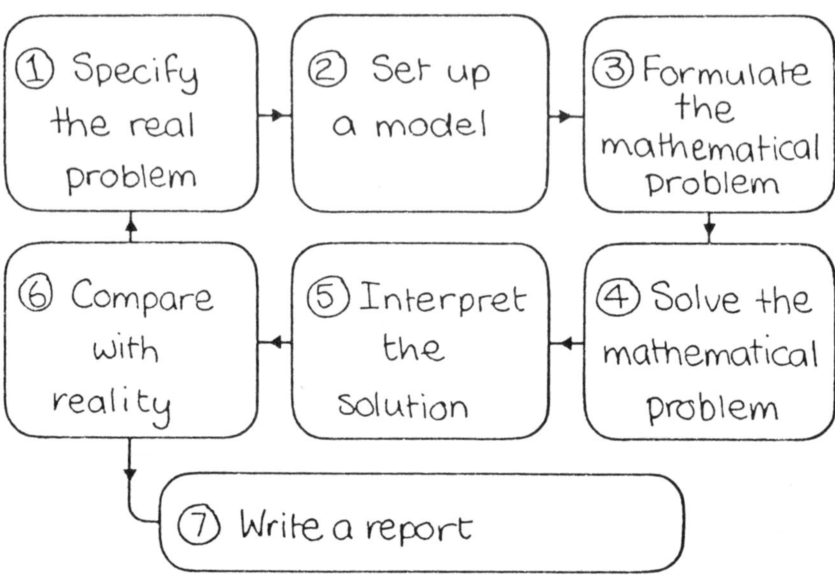

It is a refinement of the first diagram, the left column belonging to the real-world domain, the right column to the mathematical world, and the centre column offers a hint of the metaphoric quality of modelling which is cognitive by nature and belongs neither to the real-world nor to the mathematical. The conceptualizing is often not easy to talk about because it involves mental imagery, modeller-predilections, resonance with past experience and sensitivity to the purpose of the exercise. Rather than following most books and abandoning all attempts to try to communicate the metaphoric aspects of modelling because it is not easy to talk about, it seems to me sensible to grasp the nettle and invoke metaphors when ever possible. Describe what you are doing when modelling by using several different metaphors. Show that there *are* ways of talking about mathematical activities that we perform in our heads—they may be partial and idiosyncratic, but at least acknowledge that they exist!

James and McDonald (1981) offer nine stages in list form:

Recognition	Analysing
Familiarity	Interpreting
Formulating	Implementing
Constructing model	Monitoring
Validating	

which correspond closely to the seven boxes but have the disadvantage of profusion (nine is too many to remember) and lack of any supporting mental images. The only way that a student could try to make use of them would be as a flowchart/algorithm. Burkhardt (1982) offers both a complex flowchart and also a list of modelling skills:

Generating variables or features
Selecting variables or pruning features
Formulating questions
Generating relations between variables
Selecting relations

Frameworks such as these arise as a result of reflection, in response to a wish to be able to encompass the whole and to be able to offer advice when the going gets sticky. They are of no use to students if they are mistaken for content, for another layer of things to be learned. In the words of St. Paul, they are to be

read, learned, and inwardly digested

or as Halmos (1960) said in his beautiful book on set theory,

read it, absorb it, forget it

by which he meant, let it descend so deeply into your awareness that it lies below the surface, subconsciously structuring what you are doing. We all have bits and pieces of frameworks operating subconsciously. By making them explicit we can actually modify them in the light of experience.

A lecturer in mathematical modelling who refuses to expose a framework explicitly to students is like the potter who throws magnificent pots and expects students to discover the inner activities of mental imagery connected with aesthetic sense which guide the hands in the outer task of fashioning a pot, simply by watching and occasionally trying. Students under such circumstances naturally focus on technical details, and soon lose sight of the overall purpose.

Students need help to keep the woods in mind as well as the trees, and even some experts might just be able to improve their own performance a little bit here and there by adopting some structure or framework. Even the 'doers' that Halmos (1981) distinguishes from 'reflectors' could learn by a little reflecting.

One frequently offered justification for concentrating on case studies of mathematical models without talking about the modelling process is that it seems so artificial to try to force what I want to say into someone else's framework. The clue lies in the 'someone else's', which suggests that Halmos's advice has not been taken. When a framework becomes one of my ways of perceiving the world, there is no sense of trying to force things, because the framework determines what is perceived. This positive feature is often turned around and used as a reason for dispensing with frameworks, because people are afraid that they will be trapped or controlled.

Many people claim they reject frameworks because they feel straitjacketed, that it banishes creativity, and does not accurately reflect what really happens. It seems curious that people who claim to be experts at modelling and who espouse

making a simple model to start
then modify it as necessary,

are unwilling to do the same thing consciously at a meta-level. By being explicit, the framework can be modified. A framework is only dangerous when it is implicit and fragmentary. Multiple frameworks, like multiple metaphors add richness and flexibility, neither of which are available when the whole process takes place subconsciously.

The most substantial objection to promulgating frameworks to students is that process, when talked about and dwelt upon, can easily turn into a combination of content and empty ritual. The seven-boxes or nine stages, when perceived as further content to be learned, get memorized and trotted out when a suitable stimulus (examination question) is applied. Once classified as 'learnable content', a framework is consigned to its own compartment and has no impact on student behaviour. Students can also latch onto a framework as a crutch to avoid thinking. By treating it as an algorithm, they can avoid contact with the problem, hoping that by going through the motions the answer will emerge. In this way, frameworks can turn into empty ritual.

Frameworks can certainly be misused, or rather fail to be used, but this is not a valid reason for ignoring them altogether. On the contrary, the real challenge, if either of the previous two sections have made any sense, is to exploit frameworks for helping students to become involved and to appreciate what it is like to construct mathematical models. Even in the modelling books which I sampled and which do offer some framework, the framework is most often used mechanically, even superficially by the authors, thus giving students an impression that it is only frill. The framework is usually used only to label stages, rather than to afford entry into the experience. No wonder students come away with an algorithmic, mechanical approach to mathematics.

So what could be done? The main thing, building on the observations of the previous sections, must be to develop and become imbued with some coherent way of going about modelling, and sharing that explicitly in detail with students. For example, taking the seven-box approach:

Specify the real problem

The initial stages of coming to grips with a problem—called variously Recognition and Specify the Problem—are represented in most case studies simply by a statement of the question, but with no remarks about how that formulation of the question was reached, of what was considered and rejected and why.

Set up a model

This is a particularly subtle stage because it is by nature conceptual and metaphoric, and so students need most help. The phase of Setting up a Model (Constructing a Model) is usually presented by stating certain equations or inequalities. Even if the model is being studied as content to be learned, a student wants to know where these came from, and if the student is intended to be studying the model as a particular example of the modelling process, then it is vital to make process remarks. Even where detailed exercises are offered which indicate how to go about setting up a model by writing down all the features that seem relevant, and then discarding complex ones, the case studies rarely indicate *why* certain features were considered and rejected. Nowhere have I seen assistance given on the conceptual aspects of modelling, which involve forming and entering some sort of mental image of the situation being modelled, trying to experience it mentally. We all do this in some form or other, but we give students the impression that one simply 'thinks' and then writes down beautiful feature lists or completed models.

Formulate the Mathematical Problem

Translation of verbal relationship into symbols is not as easy as it seems, particularly when the relationships are someone else's. It is much easier when I have struggled to articulate the relationships myself. One author, Aris, dismisses formulation as

> 'rational accounting for the various factors that enter the picture in accordance with the hypotheses that have been laid down.'

If only students could 'enter the picture', perhaps then they could confidently and competently undertake 'rational accounting'!

Careful attention to what is known—to which symbols stand for parameters or data—and to what is wanted, will assist students to formulate the mathematical problem.

Solve the mathematical problem

This is frequently the domain of assessment, so naturally attention is focused on it, beyond its importance. Frequently in practice the problem cannot be solved exactly, so further assumptions or approximations are required. Unfortunately many case studies present the work of experienced modellers who have anticipated mathematical difficulties and so built in simplifying assumptions at an early stage. These mystify students who lack experience. Many modelling books focus on models built around a particular mathematical technique or theory (and usually this is the students' first introduction to the theory!). Tackling a question when you know what the technique is going to be is more like doing exercises at the end of the chapter than it is like modelling. Modelling should not be confused with theory and techniques. It is a separate and difficult art to learn. Students will only too happily focus on techniques which can be memorized and mastered, and avoid the conceptual challenge, the metaphoric thinking required by modelling.

Interpreting the solution

Interpreting the mathematical solution involves retracing the conceptual-metaphoric leap, and can only be done by someone who appreciates what was in the modeller's mind to begin with. Routine mechanical exercises at this stage do nothing to help students appreciate the model. For stimulating assessment ideas see Saaty and Alexander (1981).

Compare with reality

Comparing with reality (implementing) is usually barely even mentioned. Never have I found discussions about the difficulties of measuring quantities whose values are required in the model. Each measurement will introduce error, and may cancel out mathematical refinements in the model.

Off-hand remarks are often made about the cyclical, refining, multi-pass aspect of modelling in which the first model is almost guaranteed to be too simplistic and too rough, but rarely does the exposition actually traverse the cycle more than once. I suspect this is because the author presents the best, most refined model found to date, and omits the early versions. Students construe from the paradigm that it may be necessary to make several passes, but that one ought to get a pretty good one the first or second time.

Writing a report

The natural assumption is that the case study itself constitutes the report, but it is very rare to find any mention of difficulties in writing a report. It is important for the modeller to pretend to be the user so that the report speaks in the user's language. Many of the comments suggested for previous stages would actually help the user to understand what the modeller set out to model, and how the model is meant to operate.

A framework is only a straitjacket if it is employed as a ritual algorithm in order to avoid thinking. Remember the Euclidean geometry classes in school? All proofs had to begin with Given, To Prove, Proof. These were meant to help get into the question, to feel internally and to get inside the tension between Given and To Prove, the proof being a bridge constructed between them. And how did most students respond? By quickly writing down the rubric, then asking themselves

'What do I do now?',

or in the language of schools,

'I've read the workcard miss, but is it add or multiply?'.

The framework was treated as empty ritual. Why? I submit that it was because the teacher *never* spoke about or acknowledged the inner work when writing down what was Given and To Prove. Most students never even realized that there was more to be done than writing down some words. (I can assert 'never realized' because clearly they never real-ized or made real for themselves.) You can avoid this by Being a Modeller in front of, and with your students, by voicing your inner thoughts as you metaphorically sit at the wheel and throw a pot, and by joining with your students, facing a problem together, listening to what they say, trying to enter their perceptions and helping them build on those rather than rushing along in your own direction. After all, what we really want students to learn, is to *become modellers*, do we not?

5. SOME SUGGESTIONS, BY WAY OF SUMMARY

1. Stimulate students to ask their own questions.
2. Practise exploring and developing metaphors, both mathematical and non-mathematical.
3. Work *with* students, genuinely. Supress your own ideas and focus on helping students articulate and modify their own ideas. Don't push your own model. Give students time to express their thoughts and images in a mathematical conjecturing atmosphere—everything that is said is intended to be modified.
4. Present completed models, but stress the generic aspects—the modelling process, and play down technical details. Dwell on the choices and decisions made in the early stages. Give voice to the mental imagery (the implicit metaphors) which accompany the written statements.
5. Present incomplete models and invite development and criticism. For example, students could role play a panel who have to select one proposal to be 'funded', and justify their choice.
6. Become aware of how often you use a particular example to illustrate a general point, and make sure that your students are aware of what you are doing. Take opportunities to draw explicit attention to the movement from particular to general and vice versa, at all levels, from modelling as a process to specific mathematical techniques.

REFERENCES

Modelling processes

Borelli R. L., and Busenberg S. N., 1980, Undergraduate Classroom, Experiences In Applied Mathematics, *UMAP*, **1** (3), 17–26.

Burkhardt H., 1981, *The Real World and Mathematics*, Blackie.
Henriksen M., 1977, Applying Mathematics without a License, *Am. Math. Monthly*, **84** (8), 648–650.
Halmos P. R., 1960, *Naive Set Theory*, Van Nostrand, Princeton.
Halmos P. R., 1981, Applied Mathematics is Bad Mathematics, *Mathematics Tomorrow*, Ed. L. Steen, Springer-Verlag, New York.
James D. J. and McDonald J. J., 1981, *Case Studies in Mathematical Modelling*, Halsted Press.
Leatherdale W. H., 1974, *The Role of Analogy, Model and Metaphor in Science*, North-Holland, Oxford.
Mason J. and Pimm D., 1983, Generic Examples: Seeing the General in the particular, preprint.
Spanier J., 1980, Thoughts about the essentials of modelling, *Math. Modelling*, **1,** 98–108.
Spanier J., 1981, Solving Equations is not Solving Problems *Mathematics Tomorrow*, Ed L. Steen, Springer-Verlag, New York.
Witten M., 1981, Brief Note: Thoughts about the essentials of simulation modelling, *Math. Modelling*, **2,** 393–397.
Open University, 1981, MST204 Project Guide for Mathematical Modelling and Methods, Open University, Milton Keynes.

Case Studies

Aris R., 1978, *Mathematical modelling techniques, Research Notes in Maths* 24, Pitman, London.
Bender E. A., 1978, *An Introduction to Mathematical Modelling*, Wiley, New York.
Olinick M., 1981, Mathematical models in the social and life sciences: a selected bibliography, *Math. Modelling*, **2,** 237–258.
Saaty S. L. and Alexander J. M., 1981, *Mathematical Models and Applications*, Pergamon, Oxford.
Walker J., 1977, *The Flying Circus of Physics*, Wiley, New York.

Mathematical Modelling and the Preparation of Industrial Mathematicians

J. G. Sekhon, A. G. Shannon, C. Chiarella,
New South Wales Institute of Technology
and
A. F. Horadam
University of New England

SUMMARY

This paper defines mathematical modelling, delineates an approach to teaching it, observes its importance in industry in three areas, and explains that many mathematical graduates, even at the doctoral level, are unequal to the tasks required by industry.

This chapter pleads that experience in mathematics *and* experience in the field being modelled are critical. The paper maintains that the philosophy of assisting students to become successful modellers implies the need for decisively different emphases, methodologies and assessment techniques.

The authors advance the view that this philosophy is admirably caught in student projects, in both undergraduate and graduate studies, which may arise out of the consultancy activities of academics or from the work-environment of students themselves. The difficulties of client communication and in the recognition and formulation of these problems will parallel those met in future professional practice.

1. INTRODUCTION

The purpose of this chapter is to consider the role of mathematical modelling in the preparation of mathematicians for industry, commerce and other incidental uses of quantitative skills. Mathematical modelling aims to represent the behaviour of a real world situation by means of a carefully contrived mathematical description. Aris (1978) defines a mathematical model as

> any complete and consistent set of mathematical equations which are designed to correspond to some other entity, its prototype. The prototype

may be a physical, biological, social, psychological or conceptual entity, perhaps even another mathematical model.

In some senses, then, any mathematics course involves modelling. There are also modelling courses which are specifically designed to introduce the student to the steps of recognition, formulation, solution, computation, interpretation and testing by means of well-documented cases such as *Case Studies in Mathematical Modelling*, edited by James and McDonald (1981).

The qualitative analysis involved in modelling is an important ingredient in the development of that conceptual framework which is essential for the industrial mathematician. A good example of this *qualitative analysis* can be provided by the model of population growth with a variable growth rate and we shall use this model to illustrate the mathematical and educational implications of our major points. The model, of course, is not new, but the use we make of it seems to be rarely advocated.

2. AIMS OF MATHEMATICAL MODELLING

There are many purposes for the construction of a mathematical model. Davis and Hersh (1981) list some of these as

> (1) to obtain answers about what will happen in the physical world (2) to influence further experimentation or observation (3) to foster conceptual progress and understanding (4) to assist the axiomatization of the physical situation (5) to foster mathematics and the art of making mathematical models.

Providing we do not restrict *physical* to the relatively narrow scope of the physical sciences, this list also includes the goals of learning more mathematics and learning about mathematics without which the project may be commercially valuable but educationally void.

The physical growth model to which we referred occurs most naturally in a biological setting, and in this context, it is convenient to envisage a fish population which is growing in a limited environment such as a large lake. The population P of fish grows according to

$$\mathrm{d}P/\mathrm{d}t = k_0(1 - k_1 P)P, \ (P_0 \text{ given}), \tag{1}$$

where k_0, k_1 are both positive. When $k_1 = 0$, equation (1) reduces to the classic population growth model studied in all standard courses. When $k_1 > 0$, the $(1 - k_1 P)$ term has the effect of reducing the growth rate as the population rises. In this simple way the effect of the limited environment can be modelled, at least to a first approximation.

The benefits to be derived from the effective modelling of a complex system are, as Osborne and Watts (1977) observe, three:

- gaining an insight into the mechanism underlying the operation of a system;
- predicting changes in the system due to changes in the input parametric variables;
- optimizing the performance of a system in accordance with some defined *criterion function.* More and more in recent times, industrial problems of ever-increasing complexity have to be optimized with regard to such factors as cost, reliability, maintenance and design.

Industry is not always aware of the extent to which mathematics and mathematical modelling can assist in the solution of a wide range of problems. However, the importance of mathematical modelling in industry becomes plain when it is considered that:

- mathematical modelling is the only alternative for optimal utilisation of large and complex plant which cannot be modelled experimentally, as in the design and construction of a blast furnace;
- the use of mathematical models often avoids the necessity to perform long and/or expensive experiments, as in the examination of properties of new materials.

3. PERCEIVED INADEQUACIES IN MATHEMATICS GRADUATES

Employers in industry are often disenchanted with mathematics graduates from institutions of higher education. They complain that students learn to solve well-defined problems leading to unique solutions—problems which are expected to be complete in themselves, to contain all the relevant information to involve the routine application of one or other of the mathematical techniques they have learnt and to have correct answers. They perceive that students lack confidence in their ability to use the mathematics they learn in a constructive manner to assist in solving real-life problems.

Many mathematics graduates, even at the doctoral level, experience genuine difficulties in the art of applying mathematics, expressed in its quintessence in mathematical modelling. Among the tasks to which they feel unequal may be mentioned:

- the abstraction of an industrial problem;
- the formulation of a mathematical model;
- the development of a mathematical discretion that leads to effective decisions in the simplification of a model;
- the selection of an optimal solution method;
- the critical evaluation of the solution so as to extract the maximum of interpretation and inference.

The contributory causes of the difficulties experienced by mathematical modellers may be found, in the main, to be:

- educators who have little clear perception of the priorities of the industrial world where the motive is profit and where the problems are unstructured, incompletely defined and hedged with many constraints;
- the content and methods of teaching and examining mathematics courses inimical to effective modelling.

For example, let us analyse equation (1) with the approach and methodology by which students are currently taught. Thus, separating variables the student would write

$$dP/(1-k_1P)P = k_0\,dt. \tag{2}$$

Then, after finding the partial fractions,

$$\int\left(\frac{1}{P}+\frac{k_1}{1-k_1P}\right)dP = \int k_0\,dt. \tag{3}$$

Integrating and rearranging, the student obtains

$$P/(1-k_1P) = Ae^{k_0t}, \tag{4}$$

where A is the arbitrary constant. Finally our student would manipulate (4) to obtain

$$P(t) = P_0/(k_1P_0+(1-k_1P_0)e^{-k_0t}). \tag{5}$$

Our student will then interpret equation (5) by drawing Fig. 1 to see that $P(t)$ tends to the steady state level $1/k_1$ from above or below depending on the initial value.

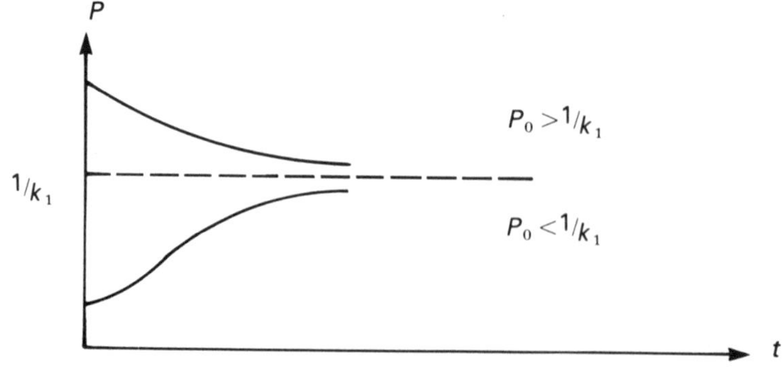

Fig. 1.

Whilst the above-average student will proceed smoothly to Fig. 1, the fair-to-middling student is likely to make some manipulative errors at equations (3) and (4) and might leave equation (5) in a form which makes the sketching more difficult for him. For example if the student leaves (5) in the form

$$P(t) = P_0e^{k_0t}/(k_1P_0e^{k_0t}+(1-k_1P_0)), \tag{6}$$

then the average student is unlikely to deduce Fig. 1 from this expression. We shall return to the modelling approach to this shortly.

4. ACADEMIC AND INDUSTRIAL ENVIRONMENTS

It is of value to pause on some of the differences that exist between academic and industrial environments.

Mathematics courses, as indeed courses in physics, chemistry, and engineering, tend to be dominated by theory in higher education institutions. The output of institutions of higher education is largely in the form of papers where theory plays a dominant role. The function of industry, however, is not to produce theory or papers but to sell devices, products and systems and to make a profit.

Academic problems have reasonably simple geometries and, by and large, the solution of differential equations which are linear. Fairly accurate solutions are possible. On the contrary, the industrial world seldom, if ever, presents a well-defined problem. Most of the industrial problems are usually non-linear and involve geometries which are complex. They appear to be completely intractable and to have no rational solution. The problems of industry appear to be quite unrelated to the skills of being able to invert matrices, manipulate Laplace transforms or understand group theory.

A certain haziness characterizes the majority of industrial problems: in one manner or another the problems are incompletely defined and part of the problem is the state of ignorance in which mathematicians in industry must often work. The use of such expressions as 'engineering judgement', 'factors of safety' and 'allowable limits' not only reveal that engineers and mathematicians are accustomed to, and aware of, imprecision but also they represent attempts by engineers, mathematicians and scientists to cope with this lack of determinism. The framework within which the design and planning of engineering systems must be taken is hedged with many restrictions such as limits on size, weight, shape, demands for a certain measure of reliability, optimization of one variable or another, and economic constraints. Although modern scientific methods have done much to eliminate uncertainty fostered by ignorance, the inescapable reality is that (Singh, 1980)

> uncertainty is still the environment in which an industrial mathematician does his work.

5. THE FUNCTIONS OF INDUSTRIAL MATHEMATICIANS

The engineer or mathematician in industry has three main tasks. The first is the idealization of the physical or engineering situation and its formulation in mathematical terms. The second is the manipulation of the mathematical symbolism to obtain a solution. And the third is the interpretation of the solution in physical terms.

The first of these, frequently referred to as the making of a mathematical model, is often the most difficult task. The raw material for this is information which will almost certainly be incomplete or redundant, and occasionally even inconsistent.

There should be a range of mathematics taught, but an *early* and *intensive* cultivation of the power of thinking about real things and the application of mathematical symbolism to physical ideas is very desirable for the preparation of industrial mathematicians. Flegg (1974) echoes this:

> The philosophy and the practicalities of mathematical modelling should be an integral part of the education of scientists and technologists.

It is a commonplace in educational circles that it is comparatively easy to teach the methods of solutions of standard mathematical equations, but much harder to communicate the ability to formulate the equations adequately and economically. The Secretary and Registrar of the Institute of Mathematics and its Applications (Clarke, 1978) shares this view:

> To explain the methods that have been devised to bring mathematics into some aspects of the everyday world is a more difficult task than to explain mathematics itself.

We return now to the qualitative or geometric approach to our population growth model. The first step is to plot $\dot{P}(\equiv \mathrm{d}P/\mathrm{d}t)$ as a function of P as shown in Fig. 2. From equation (1) the student can easily deduce that

$$\begin{cases} \dot{P} > 0 \text{ for } 0 < P < 1/k_1, \\ \dot{P} < 0 \text{ for } \quad\quad P > 1/k_1. \end{cases} \tag{7}$$

Hence the increase or decrease of P must be as shown by the arrows on the parabola in Fig. 2. In fact the combinations of P and $\dot{P}$ that satisfy (1) must lie

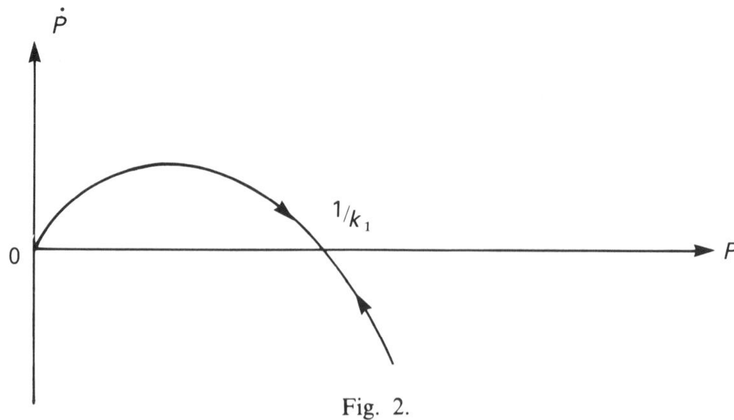

Fig. 2.

on the parabola. The student sees immediately that for $P_0 < 1/k_1$, $P(t)$ rises to $1/k_1$ and the reverse for $P_0 > 1/k_1$. In other words our student has the picture in Fig. 1 without having to go through a lot of algebraic manipulations.

6. TEAM APPROACHES

The solutions to unstructured, complex industrial problems demand team approaches. Industry tends to organize itself in multidisciplinary teams of scientists, engineers, economists, mathematicians and statisticians.

We believe that set exercises in mathematical modelling should be attempted by students working in small groups. This way of working helps the student to understand, and to be understood by, the other members of the group and simulates future professional practice. Further, we feel that, on some occasions, an identical problem should be given to three or four groups. This procedure would help to reinforce the view that, as different models make different types of simplifying assumptions, usually there is no single best model for describing a given situation and hence no single best answer. It is worth observing that mathematical modelling is an art and that there are no correct single answers to the problems one seeks to solve in industry: often there are many alternative answers and each one of them is a compromise between conflicting advantages and disadvantages.

7. COURSE IN MATHEMATICAL MODELLING

The subject of mathematical modelling is now so fundamentally important that we might well introduce into the mathematics curriculum a section *Mathematical Modelling Techniques*, where particular and careful attention should be paid to the derivation of mathematical equations as well as to the interpretation of the solutions obtained.

The topic of problem-formulation could possibly be introduced by a *case-history* approach, with the discussion group replacing the formal lecture. Case studies have been successfully and extensively used in teaching law, medicine and business. They not only help to inject a note of realism in professional education, but the solution of the problem presented by a case study generally requires a synthesis of discrete packages of knowledge. There are a few books coming on to the market now that pay attention to the little things that the experienced modeller does—almost by instinct. Amongst these may be mentioned the source book edited by Andrews and McLone (1976). We have also referred to some others elsewhere in this paper.

These case studies lead naturally to numerous extensions. Thus, for example, once examples like equation (1) have been digested, the student can be led on to analyse the differential equation

$$dP/dt = \phi(P) \tag{8}$$

by sketching $\dot{P} = \phi(P)$ and considering the sign of $\phi(P)$ at points where $\dot{P} = 0$ as in Fig. 3.

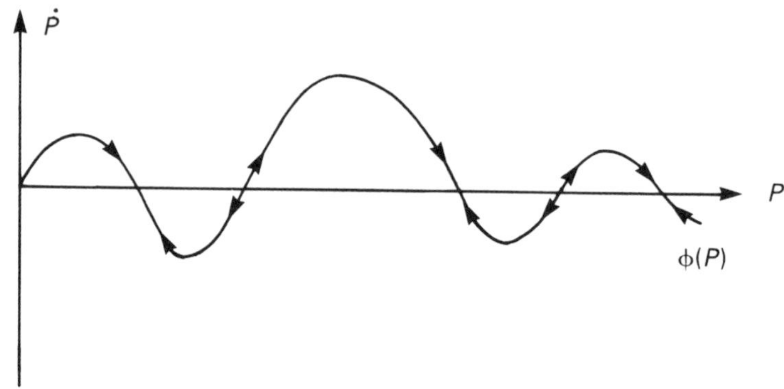

Fig. 3.

The philosophy of assisting students to become successful modellers implies the need for decisively different emphases, methodologies and assessment techniques. The constraints of the conventional time-constrained examination paper should not deflect us from preparing students for the type of environment in which they will practise the art of their calling. A student appearing for the conventional type of examination is usually asked, in the words of Deare (1974),

> to consider a task which is more or less circumscribed, is shorn of most of its ambiguous fringe, and leads to an answer which is tightly defined.

Many a successful examinee has painfully to unlearn the attitudes and skills garnered while preparing for examinations. They have to:

- learn that in problem-solving one needs time to ponder over the problem and hence assimilate a new time scale of perhaps months instead of minutes;
- realize that at times they may have to work without the knowledge that their problem is capable of resolution;
- appreciate that practical industrial problems contain much irrelevant and redundant information;
- become accustomed that in a practical situation there may not exist one correct answer;
- understand that it is their function to choose a compromise solution from a set of different answers produced by a number of conflicting criteria.

No examination can hope to reproduce completely the actual conditions of professional practice. However, we do believe that the more general student project has a rule to play in the acquisition of the heuristics of mathematical modelling, particularly in the case of the part-time students whose employment situation will sometimes (more often than the employer might realize) contain suitable topics for such projects. For the full-time student project topics can arise from the consultation work of academics. Some recent examples of student projects which arose out of academic research activities include:

- Credit creation in banks and non-bank financial intermediaries (Morath and Shannon, 1978);
- An analog computational application of a differential model of diabetes mellitus (Rozsasi and Shannon, 1982);
- A macro-educational simulation model of the Sydney Catholic educational system (Shannon, 1983);
- Mathematical attitudes of technology students (Shannon and Clark, 1979);
- Scaling and adjustment of weighted average marks (Kwan and Shannon, 1982).

These few are mentioned because they are accessible in the literature. The essential mathematics in each case was, in turn, progressions, first order differential equations, simulation techniques, factor analysis and cumulative frequencies.

Projects provide one of the more promising aspects of professional preparation of mathematicians with destinations in industry. For, in effect, by carrying out a project, the student assumes the mantle of a researcher or an industrial mathematician. The general abilities and attitudes developed in the use of projects as a form of professional simulation can include:

- the confidence to make judgements and take decisions;
- the ability to generate and consider several possible avenues to approach a problem;
- the capacity to integrate discrete packages of knowledge and skills acquired during the course of study;
- the need to bring a measure of organization, resourcefulness and intelligence to the undirected structure and open-ended content that characterize his (her) undertakings in industry;
- the ability to express oneself clearly and cogently in documented form.

As a project confronts a student with a realistic situation which calls for him to appraise it, appreciate its implications in mathematical terms, plan its investigation, pursue any leads which subsequently arise and interpret conclusions, the context of the examination, and education, is brought closer to the work content of the life of the industrial mathematician and/or engineer. An approach to the assessment of projects is described in Shannon (1981).

8. CONCLUSION

There are many institutions which may find it difficult to design and implement a formal course called Mathematical Modelling. However, by a judicious choice of problems and methods of teaching and examining, mathematical modelling can be made to be central in undergraduate programmes. This chapter has attempted to show that there is another sense in which mathematical modelling can be taught incidentally. This is via student projects which can be an alternative to some couse work in the undergraduate degree studies but a compulsory component for higher degrees.

Too many mathematics courses are presented as though computers did not exist, as though the students did not have calculators, as though we were trying to satisfy our nineteenth-century forebears, and developments in the teaching of mathematical modelling thus have general curriculum reform implications. This is so because the context of modelling provides an opportunity to modernize the content and methodology of mathematics syllabuses, and our population growth model also attempted to illustrate this.

Similar implications for curriculum reform can be drawn from Burghes and Wood (1980) whose chapters are based on mathematical topics, except for the first chapter which is an account of the aims and philosophy of mathematical modelling. An elaboration of the latter in terms of the learning needs of students may be found in Freudenthal (1961, 1973).

Mathematical modelling in the preparation of tomorrow's industrial mathematicians, while not a panacea for all ills, is a realistic means of accommodating the interests and skills (or lack of them) that the students of today bring with them to tertiary study.

REFERENCES

Andrews J. G., McLone R. R., 1976, *Mathematical Modelling*, London, Butterworths.

Aris R., 1978, *Mathematical Modelling Techniques*, San Francisco, Pitman.

Burghes D. N., Wood A. D., 1980, *Mathematical Models in the Social, Management and Life Sciences*, Chichester, Ellis Horwood.

Clarke N., 1978, Foreword in J. Lighthill (ed.), *Newer Uses of Mathematics*, Harmondsworth, Middlesex, England, Penguin Books.

Davis, Philip J., Hersh R., 1981, *The Mathematical Experience*, Boston, Birkhäuser.

Deare N. T., 1974, The Assessment of Project Work in H. G. McKintosh (ed.), *Techniques and Problems of Assessment*, London, Edward Arnold.

Flegg H. G., 1974, The Mathematical Education of Students and Technologists—a Personal View, *International Journal of Mathematical Education in Science and Technology*, **5**, 1, 65–74.

Freudenthal H., 1961, *The Concept and Role of the Model in Mathematics and Social Sciences*, Dordrecht, Reidel.

Freudenthal H., 1973, *Mathematics as an Educational Task*, Dordrecht, Reidel.

James D. J. G., McDonald J. J., 1981, *Case Studies in Mathematical Modelling*, Cheltenham, England, Stanley Thornes.

Kwan P. Y. K., Shannon A. G., 1982, A comparison of three methods of scaling measurements for aggregation, *International Journal of Mathematical Education in Science and Technology*, **3**, 3 299–310.

Moráth R. L., Shannon A. G., 1978, An extension of the credit creation multiplier, *Transactions of the Institute of Actuaries of Australia and New Zealand*, 523–535.

Osborne M. R., Watts R. O., 1977, Model Construction and Implementation, in M. R. Osborne and R. O. Watts, *Simulation and Modelling*, Brisbane, University of Queensland Press, 3–29.

Rozsasi R. I., Shannon A. G., 1982, An analog computer simulation of diabetes mellitus, *Reflections: Journal of the Mathematical Association of New South Wales*, **7**, 3, 9–11.

Shannon A. G., 1981, Student Projects and Mathematical Modelling, *Newsletter on Teaching Mathematical Modelling*, **2**, 2, 6–7.

Shannon A. G., 1983, Pupil Population Progressions, Teaching Mathematics and its Applications, 2, 2, 82–84.

Shannon A. G., Clark B. E., 1979, Mathematical attitudes of some polytechnic students, *British Journal of Educational Technology*, **10**, 1, 59–68.

Singh J. G., 1980, An investigation into The Mathematical Education of Engineering Undergraduates at Australian Colleges of Advanced Education, M.Sc. Thesis, Loughborough University of Technology.

Experience with Team Projects in Mathematical Modelling

A. M. Gadian, P. C. Hudson, M. J. O'Carroll and W. P. Willers

Teesside Polytechnic

SUMMARY

A new course in mathematical modelling was introduced to the first year of the HND course in Mathematics, Statistics and Computing at Teesside Polytechnic in 1981/82. The programme of studies aims to reflect the needs of employers of mathematicians, that they should be better equipped to apply their knowledge, and able to communicate with non-mathematicians and work cooperatively with colleagues within a prescribed schedule.

An underlying principle is that students should enjoy the course. They learn, not by instruction, but by active participation, the work being creative and socially interactive, thus complementing the traditional lecture/tutorial arrangement encountered in other parts of the course.

The ice is well and truly broken in a two-week introductory period during which the organization and objectives of the course are explained and lecturer-led teams take on a simple mini-project. The students are then divided into small teams to work on their own initiative on the first of three projects. The teams are given a statement of a real problem which they work on for a period of five weeks, with a strict deadline for producing a written team report. The problem is deliberately defined in an open-ended way to make the teams think about its definition and interpretation. They are free to solve the problem as they see fit. Written and oral communication skills are practised, and each team agrees and prepares a written project report. Finally a representative from each team presents a verbal report at an often lively post mortem session, and then is cross-examined by the other teams.

No marks are awarded for this course; a student passes the course if he is judged to have participated and gained the intended experience.

Reactions have been very favourable. Staff have found the course enjoyable and several visitors from other institutions have been favourably impressed. Student response is recorded systematically and indicates significant success in meeting the objectives. The course received an award of the Royal Society of Arts in its Education for Capability Recognition Scheme in 1982.

1. INTRODUCTION

This chapter describes a new mathematical modelling course currently being presented to students on the first year of the HND course in Mathematics, Statistics and Computing at Teesside Polytechnic. The course aims to provide experience both of the processes of communication and teamwork (as prescribed in the BEC/TEC Committee for Mathematics and Statistics Guidelines), and the practices of mathematical modelling relevant to mathematicians working in industry and commerce. There are three main objectives of the course:

FUN—There is no pressure of preparing for examination nor grasping new theory. The work is creative and social and should complement the more usual lecture/tutorial activities. This is an active course in which the students are required to participate and be involved in the experience.

COMMUNICATION—Students gain experience in written and oral communication, being involved in informal and formal meetings, working in teams to strict deadlines, and cooperating in the preparation of written reports.

MODELLING—Mathematical modelling is a skill acquired by experience rather than by learning about it, and the course aims to provide the experience. The student is required to understand a real problem (often deliberately presented in vague terms), formulate the problem mathematically, solve it, and finally interpret and validate the results.

2. COURSE ORGANIZATION

2.1 Introduction and mini-project

During the first two weeks the students are introduced to the course. The organisation and method of assessment is described and each student is issued with a Course Guide and some notes on the preparation of written reports. The law is laid down firmly about attendance requirements and it is made clear that participation is the keynote.

The students spend part of this time working on a mini-project in teams led by a lecturer. One purpose of the mini-project is to introduce the students to the main stages of mathematical modelling, i.e. formulation, solution, interpretation and validation, as well as the idea of a feature list. At the conclusion of the mini-project each student is required to submit a brief written report of the progress made by his team.

2.2 Team projects

After the initial two weeks of the course, teams are formed, usually of five students, although this depends on the number of students on the course and staff and space resources. Each team works on the same project over a five-week period. The students undertake three projects in all, as a member of a different team in each case.

A lecturer is assigned to each team, acting mainly in an advisory capacity, although he may contribute more fully in the first project. It is important that the lecturer leaves the modelling initiative to the students, and only advises occasionally on such things as relevance of approach or need for validation or admissibility of data. The idea is that the lecturer acts as an industrial manager or client might, and not as a superior modeller. This requires some self-restraint from lecturers at first. Sometimes additional information is available on request, but is not volunteered by the lecturers.

In the first week the teams are given a project specification which they are asked to consider in general terms. The final week is mainly devoted to producing a joint team project report, and in the final session of each project period a nominee from each team presents a brief verbal report (usually lasting about two minutes), and each team answers questions from the rest of the class.

2.3 Procedural aspects

Three class hours per week are assigned to team project work. Two hours are devoted to informal meetings and the remaining hour to a formal meeting.

The purpose of the formal meeting is to record progress made and plan the work for the following week. An agenda for the meeting is provided in the course guide and includes

1. Recording of attendance
2. Confirmation of minutes of the previous meeting
3. Receipt of individual team members' reports and technical reports
4. Discussion of progress to date
5. Work schedule for following week, and assignment of tasks to team members
6. Any other business

The position of chairman and secretary rotates between the students in the team from week to week (five students per team ensures that each team member acts as chairman and secretary once throughout a five-week project period), and the secretary is responsible for recording and writing the minutes of the meeting. In the first project the lecturer assigned to a team acts as chairman of the first formal meeting, thus providing the students with procedural information. Each team is required to submit a full set of minutes with its project report, and any technical reports are usually included as appendices.

The informal meetings are used for informal discussion and completion of individual tasks. Lecturers do not normally attend informal meetings for the full hour. They call in to see if consultation or advice are needed and to keep a check on attendance.

Each student is also required to complete a weekly report sheet. This usually consists of a few lines mentioning in brief outline his project activities during that week. Students give a fuller report of their weeks activities at the formal meeting item 3 of the agenda. Each student is issued with a sheet covering the five-week project period, the weekly entries being approved by the lecturer at the formal meeting. The completed report sheets are submitted with the team project report.

3. ASSESSMENT

Assessment is entirely continuous, the purpose being to assess whether a student has participated and gained the appropriate experience. The extent of participation is not quantified in the form of an assessment mark. For each project assessment is made by the lecturer in charge of each team, considering whether each student has participated satisfactorily each week.

It is implicit in this method of assessment that attendance at all meetings, formal and informal, is obligatory. Students are expected to give advance notice of any absence for special appointments etc., and to give a satisfactory explanation of any unforeseen absence from any meeting. It is accepted that students will not always be present in the room set aside for project work; they may, for example, be using library or computing facilities.

The normal assessment for each project is satisfactory, although exceptionally an assessment of unsatisfactory is given, usually when a student fails to attend without good reason. A credit is sometimes awarded if a student has made exceptionally good contributions both in modelling and in promoting group activity. Students are informed if their assessment is unsatisfactory at the end of the project period. Sometimes students are informed that they will be expected to participate more fully in future projects if they have merely met the attendance requirement. To pass the course students must obtain a satisfactory assessment in at least three projects, including the mini-project.

4. CHOICE OF PROJECTS

At the time of presentation of this paper there will have been two student intakes to the modelling course and two mini-projects and six five-week projects will have been used. One of the most time consuming and difficult aspects of the course is the selection of suitable projects. The team of lecturers involved must be convinced of the suitability of a particular project and totally committed to making the project a success, otherwise the lack of enthusiasm is soon detected by the students and they are reluctant to become totally involved in the work. So far no project has been repeated, although it is envisaged that some of the projects which have been particularly successful might be repeated in future years. Dozens of projects ideas have been scrutinized and found wanting in some respect. A list of projects already undertaken is given in Appendix 1.

All students on the course have A-level mathematics or an equivalent qualification and no additional mathematical expertise is assumed.

Further, a conscious attempt is made to make the projects as relevant and attractive to the students as is possible, in order to promote a general desire on the students' part to want to obtain a solution to the problem. Whilst it is important that each project contains an objective which is easily achievable within the five-week period it is also desirable that the projects are sufficiently open-ended so as to provide for more competent and enquiring student.

There is also a deliberate policy to select projects based on different aspects of mathematics in order to dispel the idea that modelling is specifically related to one particular area of mathematics. Many models considered are empirically based on some prescribed data, or data to be accumulated by the students themselves, although an attempt is made to include a theoretical model involving some mathematical analysis. This highlights some of the difficulties students have in communicating mathematical analysis, and the use of technical reports becomes more important.

Where possible a feature lecture is associated with the project. For example, during the project on predicting the time of evacuation for the new Computer and Mathematical Sciences building at the Polytechnic, the Polytechnic safety officer talked to the students and answered queries concerning fire regulations and the conduct of fire drills (a source of data for comparable buildings). External speakers from industry and commerce are particularly valuable.

5. COURSE REVIEW AND FUTURE DEVELOPMENTS

The course has now operated successfully for two years. The keynote of the whole programme of studies is participation and it is important that this is established at the start of the course. By firmly enforcing the one hundred per cent attendance rule few problems have been encountered as regards this aspect of the course. Occasionally a student will let his team down by failing to give his particular task due attention, or by being absent without good reason when he was expected to turn up with some data or draft section of their report. Not only does he then risk failing assessment, but he often receives sharp criticism from the rest of the team and often extra tasks to do.

In 1981/82 there were thirty students on the course and five teams, each of six students were formed. It was generally felt that smaller teams would ensure fuller participation by all members of the team, but as the number of teams increases more staff are required for supervision and the course becomes very expensive of staff resources. This year there are fifty-two students on the course and ten teams of five or six students have been formed. The course team consists of five members of staff, each supervising two student teams simultaneously. While this is easy as far as informal meetings are concerned, it has meant that formal meetings are restricted to half-an-hour. Although there were reservations about this, it has worked well and the formal meetings are more business-like and less discursive on the technical work. Although stimulating, the course is a little expensive to run, and the above measures were introduced in an effort to keep down the staff commitment. For a class of 50 on the HND first year the course uses about 0.7 FTE staff and accounts for about 6 FTE students.

Initially students were allocated to teams in an arbitrary manner. However, two other methods have been tried in the third project. In 1981/82 students were invited to form their own groups, the philosophy being that students would tend to associate themselves with students who they felt would make a significant contribution to the project. This year an attempt has been made to ensure that each team contains at least one student with good organizational abilities. On balance it is felt that the arbitrary allocation of students to teams is to be recommended.

The method of assessment has posed few problems. The course team meets regularly to discuss the progress of students, and students who are making exceptionally good contributions to the work are identified. Students who are not considered to be participating adequately, in the sense that they are merely attending, are also identified and warned that they are likely to fail the project. Normally absence without satisfactory explanation has resulted in failure of the project. During the first two years of operation no team has failed to hand in its project report by the specified deadline.

The last session of a project period when each group presents a verbal report is considered an important part of the course. In the first year of operation with just five teams this resulted in a very successful and lively session lasting for one hour. Student spokesmen had the disciplined exercise of presenting a summary in two minutes, usually using at most two OHP transparencies, and then answering lively questions or directing them to colleagues, for ten minutes. There was often a spirit of competition between the project teams. This year there were ten simultaneous teams and their reports were squeezed into a one-hour session. Groups of three or four reports were taken successively before questions on them jointly. This led to too much repetition of ideas and too much màterial presented in total, and discussion was cut short. The lesson here is that five team reports are about as many as can be handled well in an hour's meeting, and if there are ten teams then it would be better to split into parallel sessions.

The response of students relative to the objectives described in Section 1 was tested by issuing the questionnaire given in Appendix 2 for anonymous completion at the end of the course. Results are summarized in Table 1. The advice of one student was taken up by acquiring and showing the John Cleese film 'Meetings Bloody Meetings' as part of the course. Among the comments received were:

Table 1

SCORE / QUESTION	1	2	3	4	5	6	7	8	9	10
1	3					1		1	3	8
2	1				5	1	6			3
3	2				1		1	2	1	9
4	1				8			2		5

Frequency of scores for 1981/82
(Note: some students thought they had to score exactly 0, 5 or 10, so these numbers have higher frequency as a result)

SCORE / QUESTION	1	2	3	4	5	6	7	8	9	10
1			1		1	2	4	4	3	3
2				1	4	2	3	2	4	2
3					1	1	5	3		8
4					5	6	2	2	2	1

Frequency of scores for 1982/3

'One foreign student did not understand what taking minutes meant, and kept checking his watch during one of the formal meetings'—comment made in calling for more instruction about meetings.

'Competition in each project will make the question more interesting. Announce the champion team on the notice board, or award a £50 prize.'

'Every maths course should have a modelling component as students tend to think that every problem has one solution and can be solved in the time of an exam question.'

'The main problem for me was having the confidence to put forward my views. The more confident students took over.'

> 'I liked the lemonade selling one best, probably because the group got wired in from the start and made the most profit.'

> 'I could really relate to the length of the toilet roll. My dad got into that one as well and unrolled one to measure it.'

The pass rate for this course has been higher than normal and students have indicated that they have enjoyed the experience and gained confidence in tackling real problems. It is intended to develop this kind of programme further in furture course planning operations, including the difficult step of integrating this methodology/communication experience with lecturer-led case studies and concept learning in mechanics, business and society.

APPENDIX 1

Mini-projects

1. Electricity tariffs:
   To decide whether it is more economical to pay for electricity consumed by quarterly account or by having a meter installed.
2. Length of a roll of toilet paper:
   To estimate the length of a roll of toilet paper given an unwrapped toilet roll and one sheet of toilet paper.

Team projects

1. Building evacuation times:
   To predict the evacuation time for the new Computer and Mathematical Sciences building given fire drill reports for other buildings, fire precaution act 1971, plans of the new building.
2. Population prediction:
   To forecast the population of the UK for the next decade using data from Annual Abstracts of Statistics, 1981 edition.
3. Conversion of a rugby try:
   Where should the kicker place the ball to maximize his chance of converting a try?
4. Running times:
   Given a jogger's diary for 1982 estimate the jogger's time for arbitrary distances and routes over varied terrain from 3 miles to marathon distance.
5. Soft drinks stall:
   Each day the owner of a soft drinks stall must decide (a) how many drinks to make (b) how many advertising signs to display (c) the price to charge per drink. Given weather forecasts and report of recent sales devise an optimal strategy for the decisions to be made.

6. Manufacture of soft drinks cans:
   To determine the dimensions of a cylindrical can to minimize the surface area, and the best method of constructing a prescribed number of cans of specific capacities from rectangular sheets in order to minimize the wastage of metal.

APPENDIX 2: HNDI MODELLING QUESTIONNAIRE

I would be most grateful if you would indicate your reflections on the modelling projects this year by completing the form below and returning it to me by if possible. This is of course completely voluntary on your part and without prejudice to any present or future assessment of your performance—it is simply to help in adapting the course in future. Please enter scores from 0 to 10 in the boxes as you feel would be most appropriate in representing your personal experience on the course. You will recall the course objectives briefly outlined in Section 1 of the project guide under the headings of Fun, Communication and Modelling. If you wish to add any comments please use the back of this form.

1. How did you enjoy this type of activity? SCORE: ☐

Calibration: 0 = very tedious, boring and irritating; would have much preferred lectures or a general studies type option lecture course.

5 = it would have been much the same to me to have had a lecture course or general studies option instead.

10 = it was a welcome change in the week's pattern of lectures and tutorials, the activities were so stimulating that I actually looked forward to them, much preferred to having another lecture course instead.

NOTE: You may enter intermediate numbers, 3, 7, etc, as appropriate:

2. How is your confidence in team work and communication? SCORE: ☐

0 = I dread it; as a result of the projects I have become much more nervous and ill-at-ease in meetings, afraid of being given tasks to do on time, and totally confused about writing a report.

5 = Much as before.

10 = No problem; I have learned how to behave in meetings and how to get my point across while also picking up key points from others; I would have no difficulty in arranging or cooperating with allocation of tasks and reporting on time.

3. How is your confidence in mathematical modelling? SCORE: □

0 = Hopeless; I now realise how useless I am at it; I hope I never have to solve any real problems for I wouldn't know where to start and certainly couldn't get any useful result.
5 = Much as before.
10 = Well, at least I have now learned how to set about it, clarifying the question and making a features list etc. I am sure I could get some useful result from most real problems, although I expect my maths isn't (yet) up to producing astounding, sophisticated results. I would always be prepared to have a go, may be even eager.

4. Did the exercise reinforce other parts of the course? SCORE: □

0 = The modelling projects confused and distracted me, making the whole course confusing and contradictory as a result.
5 = No influence either way.
10 = It helped to give meaning and examples reinforcing ideas such as data collection/analysis, programming, forecasting, hence much improving my motivation and performance on the course as a whole.

ADDENDUM: THE POLYMODEL GROUP AND CONFERENCES

The Polymodel Conferences are organised annually by the North East Polytechnics Mathematical Modelling and Computer Simulation Group.

The Group is a non-profit organization based on the mathematics departments of the three polytechnics in North East England and with membership drawn equally from polytechnics and industry. Its objective is to promote research and collaboration with industry in these subjects.

Special meetings are held in addition to the annual conferences. Annual Conference Proceedings are:

1. *Modelling and Simulation in Practice*
   Editors: M. Cross, *et al.*, Pentech Press
   (held at Sunderland Polytechnic, May 1978)
2. *Modelling and Simulation in Practice*, 2
   Editors: M. J. O'Carroll *et al.*, Emjoc Press
   (held at Teesside Polytechnic, May 1979)
3. *Cases Studies in Mathematical Modelling*
   Editors: R. Bradley *et al.*, Pentech Press
   (held at Newcastle upon Tyne Polytechnic, May 1980)
4. *Numerical Modelling in Diffusion Convection*
   Editors: J. Caldwell and A. O. Moscardini, Pentech Press
   (held at Sunderland Polytechnic, May 1981)
5. *Mathematical Modelling of Industrial Processes*
   Editors: P. C. Hudson and M. J. O'Carroll, Emjoc Press
   (held at Teesside Polytechnic, May 1982)
6. *Industrial Electromagnetics Modelling*
   (to be published)
   (held at Newcastle upon Tyne Polytechnic, May 1983)
7. The seventh conference is to be on ocean dynamics and offshore engineering and will be held at Sunderland Polytechnic in May 1984

Further information may be obtained from the group's secretary, Dr. A. W. Bush at Teesside Polytechnic.

Price Index: Mathematization without a Happy Ending

Wilfried Herget
Technical University of Clausthal (F.R.G.)

SUMMARY

This chapter shows how the problem of the definition of the price index can be used in mathematical lectures and seminars at universities and colleges.

It will be shown how the starting problem almost invariably leads to an axiomatic investigation—and the few axioms turn out to be incompatible. The usual mathematical happy end fails to come—but even this (in usual mathematics very unusual) situation turns out to be extremely fruitful.

Why always only 'happy–ending–mathematics'?

1. INTRODUCTION

Usually 'one' (that is: students at colleges, universities and so on), learns mathematics in order to gain a powerful tool to solve problems dealing with the real world, and by applying this device these problems always turn out to be small and harmless ones (Fig. 1).

But I feel that, to some extent, this sounds like a fairy-tale. Without any doubt mathematics *is* powerful, indeed—but I believe that students also must learn that sometimes a problem may turn out to be even more complicated, more powerful than the mathematical methods used.

In this paper I shall confront you with such an unusual situation—there will be a big problem, and we shall see nothing but a little bit of mathematics (Fig. 2).

That problem is: *How to work out a formula for a really good price index, for example for a consumer price index.*

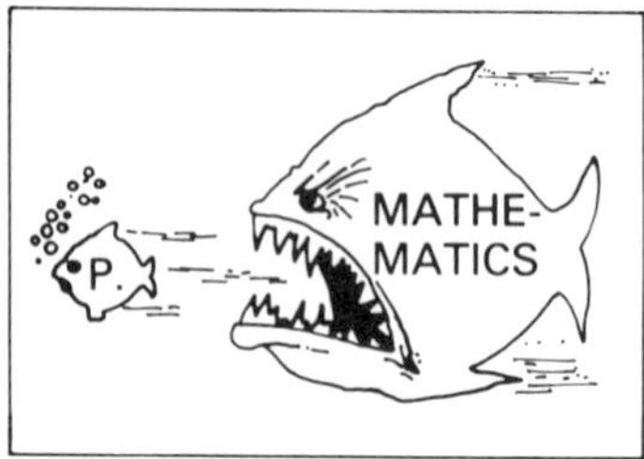

Fig. 1.

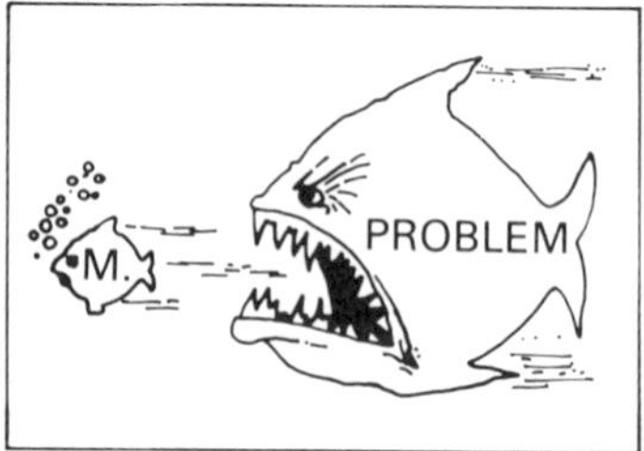

Fig. 2.

Well, nearly everybody is concerned about money, so we need not worry about motivation. And it is fairly easy to find a newspaper article concerning the price index.

But I feel that people know little or nothing of the computation and the meaning of such an index number—and this is reason enough to study the computation of price index numbers.

2. FIRST APPROACH: CHECKING SOME KNOWN FUNCTIONS

To make the matter easier for us, let's just concentrate on one single commodity. The price index is then to measure the changing of the price (per unit) of this commodity between a base year and the current year. In this case it is rather obvious to define the price index by

$$\mathrm{I}(p^0, p) = \frac{p}{p^0},$$

where p^0, p denote the price (per unit) of that commodity of the base year and the current year, respectively. This gives a function I: $\mathbb{R}_+^2 \to \mathbb{R}_+$, where $\mathbb{R}_+$ is the set of positive real numbers.

Perhaps one might suggest the difference $p - p^0$ for $I(p, p^0)$—but the function above is a better one, as you will see right away: a dimensional change in the weight (as from grams to pounds) or in currency (as from £ to p) would change the value of the difference, but not that of the quotient—and the dimension of a good might be chosen rather arbitrarily.

But, unfortunately, you can hardly live on a single article, so we have to consider the case of two (or even more) commodities. But the prices of different commodities may have different rates of change, hence the price index has to be a suitable mean of these rates.

To make the matter a little bit easier for us, we shall at first consider only two commodities and try to collect some possible formulas for a suitable price index function, where p_i^0, p_i denote the price (per unit) of the commodity i of the base year and the current year, respectively (see Fig. 3).

$$\frac{\frac{1}{2}(p_1+p_2)}{\frac{1}{2}(p_1^0+p_2^0)}$$ ratio of the arithmetic means

$$\frac{1}{2}\left(\frac{p_1}{p_1^0}+\frac{p_2}{p_2^0}\right)$$ arithmetic mean of the ratios

$$\sqrt{\frac{p_1 \cdot p_2}{p_1^0 \cdot p_2^0}}$$ ratio of the geometric means = geometric mean of the ratios

$$\left(\frac{p_1 \cdot p_2}{p_1^0 \cdot p_2^0}\right) \cdot \left(\frac{p_1^0+p_2^0}{p_1^0+p_2^0}\right)$$ ratio of the harmonic means

$$\frac{2p_1p_2}{p_1p_2^0+p_1^0p_2}$$ harmonic mean of the ratios

Fig. 3. Price index formulas for 2 commodities

By considering nothing but the usual meanvalue-functions (and it is easy to continue this list) the trouble begins: what is the 'best', the 'right' price index formula? A ratio of meanvalues of prices? A meanvalue of price ratios? And which mean should we select? And we have not yet mentioned a further question which will cause further trouble: how should we weigh the different commodities? Above we made no distinction between any given commodities, we treated them in the same way—but in general this procedure certainly is *not* suitable. For example, let's take the two commodities 'sugar' and 'housing': If the price of sugar rises by 5 per cent and the cost of living in a house decreases by 5 per cent in the same period, would this really be 'balanced', that is, should our price index actually rest unchanged in this situation? I feel the answer must be 'no', for these two goods have a very different effect on the usual cost of living.

Again: what is the 'right' price index?

Now our analysis has reached a very important stage. We learnt that looking only into the infinite list of mean value formulas is not very successful as long as we do not exactly know what we should demand of a 'right' price index.

What exactly are our expectations? We have to state precisely our demands!

3. SECOND APPROACH: AXIOMATIC INVESTIGATION

At this point, I feel, you can see how the problem we are confronted with nearly inevitably leads to a more axiomatic investigation. We define those demands which we want of a 'right' price index. I developed the following list of demands (or a similar one) in several seminars and courses together with student teachers. Special knowledge in economics is not required for that.

Demands of a price index

(1) If some price rises, then the index shall rise, too.

(2) If no price changes, then the index shall not change, either.

(3) If all prices double (change k-fold), then the index shall double (change k-fold), too.
(4) The index shall always be an intermediate value, that is shall lie between the fewest and the greatest of the ratios of the corresponding prices.
(5) The index must be independent of the unit of money. (In other words, if two economies are identical except for the definition of the unit of money (for example £ and p, resp.), then the respective values of the price indexes shall be equal.)
(6) The index must be independent of the units of measurement of the quantities (of the corresponding commodities).
(In other words, if two economies are identical except for the definition of the unit of some good (for example inch or centimeter), then the respective values of the price indexes shall be equal.)
(7) The 'index between two years (or dates)' must be independent of the chosen base year. (To explain this demand let us take the base years to be 1972 and 1978, respectively, and consider index numbers which use 1979 and 1980 as a current year. The following table gives some (authentic) index numbers from the U.S.A.

	1979	1980
Index I (1972 = 100%)	166.0%	184.3%
Index II (1978 = 100%)	109.3%	120.9%

And here demand (7) means $\frac{184.3}{166.0} = \frac{120.9}{109.3}$, which measures the price change in the 1979–1980 period.)

Up to now these demands have only been stated in colloquial language. The next step will be to precisely transform them into mathematical notation. As a matter of fact, this is an ambitious, exacting task for students—unfortunately such a task usually is very seldom, too seldom given to them. I feel that teachers and students together should treat such mathematizing tasks more often—it is not very easy, but important and interesting.

Mathematizing these demands, there can be different versions, different translations by the students, and this deserves further interest.

In Fig. 4, such different versions are marked by (.a), (.b), (.c) and the following notations are used: n is the number of commodities in our 'market basket' ($n > 1$), $i \in \{1, 2, \ldots, n\}$, p_i^0 and p_i denote the price (per unit) of the commodity i of the base year and the current year, respectively, q_i^0 and q_i denote the weight (the quantity) of the commodity i of the base year and the current year, respectively,

$$\mathbb{R}_+ = \{x | x \in \mathbb{R},\ x > 0\},\ \Lambda = \begin{pmatrix} \lambda_1 & & 0 \\ & \ddots & \\ 0 & & \lambda_n \end{pmatrix},\ \Lambda^{-1} = \begin{pmatrix} \lambda_1^{-1} & & 0 \\ & \ddots & \\ 0 & & \lambda_n^{-1} \end{pmatrix}.$$

We compile the prices and weights to price vectors $p^0 = (p_1^0, \ldots, p_n^0)$, $p = (p_1, \ldots, p_n)$ and weight vectors ('market basket') $q^0 = (q_1^0, \ldots, q_n^0)$, $q = (q_1, \ldots, q_n)$, and we shall write $I(p^0, q^0 | p, q)$ for the value of the index function. That is, we are searching for a function $I\colon \mathbb{R}_+^{4n} \to \mathbb{R}$ which fulfils the demands (or 'axioms') as follows, and that for all $p, \bar{p}, p^0, \overline{p^0}, q, \bar{q}, q^0, \overline{q^0} \in \mathbb{R}_+^n$ and all $\lambda, \lambda_1, \ldots, \lambda_n \in \mathbb{R}_+$:

($\bar{p} \geqslant p$ means $\bar{p}_i \geqslant p_i$ for all i, and $\bar{p} > p$ means $\bar{p} \geqslant p$ but $\bar{p} \neq p$.)

(1) $\bar{p} > p \Rightarrow I(p^0, q^0 | \bar{p}, q) > I(p^0, q^0 | p, q)$

(1a) $\bar{p} \geqslant p \Rightarrow I(p^0, q^0 | \bar{p}, q) \geqslant I(p^0, q^0 | p, q)$

(2) $I(p^0, q^0 | p^0, q) = 1$

(2a) $I(p^0, q^0 | p^0, q^0) = 1$

(2b) $I(p^0, q^0 | p^0, q^0) = I(p, q | p, q)$

(3) $I(p^0, q^0 | \lambda p, q) = \lambda I(p^0, q^0 | p, q)$

(3a) $I(p^0, q^0 | \lambda p^0, q) = \lambda I(p^0, q^0 | p^0, q)$

(3b) $I(p^0, q^0 | \lambda p^0, q^0) = \lambda I(p^0, q^0 | p^0, q^0)$

(4) $\min\limits_i \dfrac{p_i}{p_i^0} \leqslant I(p^0, q^0 | p, q) \leqslant \max\limits_i \dfrac{p_i}{p_i^0}$

(5) $I(\lambda p^0, q^0 | \lambda p, q) = I(p^0, q^0 | p, q)$

(6) $I(,p^0,, {}^1 q^0 | ,p,, {}^1 q) = I(p^0, q^0 | p, q)$

(7) $\dfrac{I(p^0, q^0 | p, q)}{I(p^0, q^0 | \bar{p}, \bar{q})} = \dfrac{I(\bar{p}^0, \bar{q}^0 | p, q)}{I(\bar{p}^0, \bar{q}^0 | \bar{p}, \bar{q})}$

(8) $\lim\limits_{x_i \to 0+} I(\ldots, x_i, \ldots) \in \mathbb{R}_+$

Fig. 4. Demands of a price index—transformed into mathematical notation

Obviously we have:

(1) $\Rightarrow$ (1a), (2) $\Rightarrow$ (2a) $\Rightarrow$ (2b); (3) $\Rightarrow$ (3a) $\Rightarrow$ (3b).

For technical reasons, we have always considered the positive reals as the domain of price indexes. Otherwise even simple ratios of prices are not defined and cannot appropriately be defined for certain points, that is if some price is zero.

But it may happen that certain commodities disappear out of our 'market basket' which collect those things an average consumer usually needs, as you know. (For example: do you still need a slide rule?) On the other hand some new products may enter (as the microcomputers have done in the recent years). All these events should not cause severe troubles (I mean: troubles for our price index function—I do not consider the troubles referring to this for maths education . . .) . . . thus we demand at least a 'good' behaviour of a price index as any scalar argument (price or quantity) tends to zero, namely

(8) $\lim\limits_{x_i \to 0+} I(\ldots, x_i, \ldots) \in \mathbb{R}_+$.

In mathematics, axioms used to be the base of a theory, naturally of a perfect theory with wonderful theorems and propositions. Here all these things look as good as always—still.

3.1 Again: possible price index functions

Before working with this handsome collection of demands, let us first consider some possible and generally used price index functions:

$$I_1(p^0, q^0 | p, q) = \frac{q^0 p}{q^0 p^0} = \frac{q_1^0 p_1 + \ldots + q_n^0 p_n}{q_1^0 p_1^0 + \ldots + q_n^0 p_n^0} \quad \text{(Laspeyres, 1871).}$$

This index function measures price change by using the weights of the base year. Thus it is designed to provide the answer to the question 'How much would it cost in the current period to buy the average living standard of the base period?' Most of the commonly used indexes are Laspeyres indexes.

$$I_2(p^0, q^0 | p, q) = \frac{qp}{qp^0} \quad \text{(Paasche, 1874).}$$

In contrast to the Laspeyres index the Paasche index uses the weights of the current year. Thus the corresponding 'market basket' must be updated in each year, which is very expensive.

Taking the geometric mean of (1) and (2) we get

$$I_3(p^0, q^0 | p, q) = \left[\frac{q^0 p}{q^0 p^0} \cdot \frac{qp}{qp^0}\right]^{\frac{1}{2}} \quad \text{(Fischer, 1911).}$$

The following index

$$I_4(p^0, q^0 | p, q) = \left(\frac{p_1}{p_1^0}\right)^{\alpha_1} \left(\frac{p_2}{p_2^0}\right)^{\alpha_2} \cdots \left(\frac{p_n}{p_n^0}\right)^{\alpha_n}, \quad \alpha_i > 0, \sum_{j=1}^{n} \alpha_j = 1$$

uses only prices and (at least explicitly) no quantities—but, of course, the weights are hidden in the α_i. To express the α_i explicitly through the q_i there are several possibilities, for example:

$$I_{4.1}(p^0, q^0 | p, q)\colon \ \alpha_i = \frac{1}{s} q_i^0 \text{ where } s = \sum_{j=1}^{n} q_j^0,$$

$$I_{4.2}(p^0, q^0 | p, q)\colon \ \alpha_i = \frac{1}{s} q_i \text{ where } s = \sum_{j=1}^{n} q_j,$$

$$I_{4.3}(p^0, q^0 | p, q)\colon \ \alpha_i = \frac{1}{s} p_i^0 q_i^0 \text{ where } s = \sum_{j=1}^{n} p_j^0 q_j^0,$$

$$I_{4.4}(p^0, q^0 | p, q)\colon \ \alpha_i = \frac{1}{s} p_i q_i \text{ where } s = \sum_{j=1}^{n} p_j q_j.$$

Now it is obvious to check whether these functions fulfil all the demands. Well, here is the result of the check (see Fig. 5). None of our candidates satisfies all axioms—hence we must go on hunting for the best index function!

	(1)	(2)	(3)	(4)	(5)	(6)	(7)	(8)
I_1	+	+	+	+	+	+	−	+
I_2	+	+	+	+	+	+	−	+
I_3	+	+	+	+	+	+	−	+
I_4	+	+	+	+	+	+	+	−
$I_{4.1,\,4.2}$	+	+	+	+	+	−	−	−
$I_{4.3,\,4.4}$	+	+	+	+	+	−	−	−

Fig. 5. None of the functions satisfies all axioms

3.2 Examination of the system of axioms

It is self-evident to make that 'hunt' as easy as possible. In order not to check any demand more often than is necessary we examine each axiom to see whether it can be derived from the others—for then the test of this axiom can be omitted and we can restrict ourselves to testing the other axioms. It is obvious that students like methods which make labour a little bit easier—there is no problem to motivate them. If a certain axiom could not be omitted then there must be a function which does not satisfy this special axiom but all the other ones. Fig. 6 gives a list of such functions for the axioms (1), (2), (3), (5), (6) (cf. Eichhorn, 1978, p. 161).

$I(q^0, p^0 \mid q, p)$	(1)	(2)	(3)	(5)	(6)
$\left(\frac{p_1}{p_1^0}\right)^{-\alpha_1}\left(\frac{p_2}{p_2^0}\right)^{\alpha_2}\cdots\left(\frac{p_n}{p_n^0}\right)^{\alpha_n}$, $\alpha_1, \ldots, \alpha_n > 0$, $-\alpha_1+\alpha_2+\ldots+\alpha_n = 1$	−	+	+	+	+
$\frac{qp}{q^0p^0}$, $q \neq q^0$	+	−	+	+	+
$\sqrt{\frac{qp}{q^0p^0}}$	+	+	−	+	+
$\frac{1}{n}\cdot\frac{qp^0}{qp^0+1}\sum_{i=1}^{n}\frac{p_i}{p_i^0}+\frac{1}{qp^0+1}\max\left\{\frac{p_1}{p_1^0}, \ldots, \frac{p_n}{p_n^0}\right\}$	+	+	+	−	+
$\frac{ap}{ap^0}$, $a_1, \ldots, a_n > 0$	+	+	+	+	−

Fig. 6. Independence of the axioms (1), (2), (3), (5), (6)

By 'hunting' for such examples to prove the independence of an axiom (functions with . . . , but not with . . .) we would gain further confidence in our demands—for we can look at such a 'monster function' and say, 'This terrible function for price index?—No, never!'.

But the hunting does not go on in this easy way. We might be on the look-out for 'monsters' for axiom (4) as long as we want to—there will never be a function satisfying (1), (2), (3), but not (4)—for (1), (2), (3) always implies (4)! And, now hunting for similar implications, we could get for example the following theorems (where $n \in \mathbb{N}$, $n > 1$, I: $\mathbb{R}_+^{4n} \to \mathbb{R}_+$):

Theorem 1: I satisfies (4) $\Rightarrow I$ satisfies (2) and (3a).
Theorem 2: I satisfies (1a), (2), (3a) $\Rightarrow I$ satisfies (4).
Corollary: I satisfies (1), (2), (3) $\Rightarrow I$ satisfies (4).
Theorem 3: I satisfies (2b), (3b), (7) $\Rightarrow I$ satisfies (5).
Corollary: I satisfies (2), (3), (7) $\Rightarrow I$ satisfies (5).

The proofs of the theorems are listed at the end of this article. If you like to, you are invited to recover some more theorems, perhaps by considering different versions of axioms. It is still an open problem to give explicitly the set of all functions satisfying (1) through (6) and (1) through (7), respectively (Eichhorn, 1978, p. 154). But it is not hard to prove:

Theorem 4: $I_1, \ldots, I_k$ satisfy (1), . . . , (6) $\Rightarrow$

$$I = (\beta_1 I_1^\delta + \ldots + \beta_k I_k^\delta)^{\frac{1}{\delta}} \text{ where } \delta \neq 0,\ \beta_i > 0,\ \sum_{i=1}^{k} \beta_i = 1$$

satisfies (1), . . . , (6).

Theorem 5: $I_1, \ldots, I_k$ satisfy (1), . . . , (6) or (1), . . . (7), resp. $\Rightarrow$

$$I = I_1^{\alpha_1} I_2^{\alpha_2} \ldots I_k^{\alpha_k} \text{ where } \alpha_i > 0,\ \sum_{i=1}^{k} \alpha_i = 1$$

satisfies (1), . . . , (6) or (1), . . . , (7), resp.

But after all, at the end of all our efforts, there will always be deep disappointment: we shall never find the desired best price index function! For, as it can be proved by elementary mathematics, there is no function which meets all requirements:

Theorem 6: There is no function I, which satisfies (2), (3a), (6), (7), (8) at the same time.
Corollary: There is no function I, which satisfies (2), (3), (6), (7), (8) at the same time.

3.3 The moral of this story

The last theorem makes our painfully constructed system of axioms collapse. We have learnt—and this may lead to resignation—that there will never be any index meeting all our requirements at the same time!

Here you might feel like Goethe's Faust:

Da steh, ich nun, ich armer Tor,
und bin so klug als wie zuvor!

(All my efforts did not really get me anywhere.)

But, is it right to assume that we are not any wiser than before? I definitely don't think so!

- We have considered a motivating problem 'out of the real world'.
- The axioms did not do what the French would call 'parachuter', that is come like parachutes out of the blue (mathematicians easily yield to the temptation of the latter method, as Freudenthal (1973, p. 101) stated.
- The axioms have turned out to be incompatible.

Of course, this contradiction was contained in the axioms—or, figuratively speaking, was hidden in the axioms—for only our deductions have brought it forward! (And, by the way, we have learnt a little bit of the independence and the consistency of a system of axioms.)

The usual mathematical happy ending fails to come—and this is the very reason why this example differs from others! This is why we do not succumb to the temptation to lean back contentedly in our armchair. We are forced

—to think over our problem again,
—to reflect on the mathematizing procedure,
—to interpret the mathematical result.

The feedback cannot be forgotten!

And this will lead to both understanding the situation and to understanding mathematics:

We learn that price index numbers can only give but a weak picture of the real facts. For they are intended to aggregate the price changes of very different commodities to one single number—and this cannot be done without a loss of pieces of information.

3.4 Blind trusting in index numbers is reduced

In particular the fact that (7) is not satisfied by both Laspeyres and Paasche index functions leads to a more critical view of index numbers: (7) is that property which allows us to build index chains, chain-index numbers by connecting indexes, for example, from 1978 to 1979 and from 1979 to 1980 to a new index number for 1978 to 1980—and this very axiom is *not* fulfilled by all the index formulas used by the statistical bureaux nearly all over the world!

It is quite certain that this lack would only give way to a small effect—but a half per cent or even one per cent difference would easily occur, and this would sometimes be of some interest.

Furthermore, please remember that we have dropped any discussion of those problems concerning the making of the weights and those to determine the 'average prices' of a commodity in a certain period.

It's obvious, or, better, should be obvious, that similar problems would arise for all such index numbers which are intended to aggregate various data to one single number—we should have learnt that this can never be done without any loss of information!

And finally there is an overlapping aspect: we get to know that in mathematizing we should concern both the mathematical logic and the 'logic' of the object concerned:

—if forgetting the mathematical logic it might happen that the system of the axioms is inconsistent;
—and if we ignore the logic of the object concerned the constructed mathematical model will not properly describe that situation and our mathematical results will not solve that original problem.

Sometimes it is necessary to make a compromise because there is no 'best' solution—as it is here, where there is no price index function 'filling the bill'.

This insight in the limits of mathematical modelling is the main purpose of the example I have presented to you. I wished to demonstrate, that for some given situation there is (may be) no mathematical model with all the desired properties and that sometimes we have to accept a compromise.

I hope you will see it now: mathematics is not always a big fish hunting for a ridiculously little problem fish (as you remember, Fig. 1), but sometimes there is a big problem fish which is laughing about the very little mathematical tools we have to 'hunt' it with.

Why always only look for 'happy-ending-mathematics'? Let's do mathematics in another way, at least sometimes!

'. . . it is a hard task to look for those axioms which describe a given complicated situation in the right way: But—there would be no mathematics, if everything therein were easy (Revuz, 1965).

4. APPENDIX: PROOFS OF THE THEOREMS 1, 2, 3, 6

Theorem 1: I satisfies (4) $\Rightarrow$ I satisfies (2) and (3a).
Proof: From (4) with $p = \lambda p^0$ we have:

$$\lambda = \min\left(\frac{\lambda p_i^0}{p_i^0}\right) \leqslant I(p^0, q^0 \mid \lambda p^0, q) \leqslant \max\left(\frac{\lambda p_i^0}{p_i^0}\right) = \lambda,$$

hence $I(p^0, q^0 \mid \lambda p^0, q) = \lambda$ for all $p^0, q^0, q \in \mathbb{R}_+^n$, $\lambda \in \mathbb{R}_+$.
For $\lambda = 1$ we get (2) and finally (3a).

Theorem 2: I satisfies (1a), (2), (3a) $\Rightarrow I$ satisfies (4).
Proof: (cf. Eichorn (1978, p. 155)): For all $p^0, p \in \mathbb{R}^n_+$ we have:

$m \cdot p^0 \leqslant p \leqslant M \cdot p^\circ$, where $m = \min\left(\frac{p_1}{p_1^0}, \ldots, \frac{p_n}{p_n^0}\right)$ and

$M = \max\left(\frac{p_1}{p_1^0}, \ldots, \frac{p_n}{p_n^0}\right)$.

Hence $m = m \,.\, I(p^0, q^0 | p^0, q)$ with (2)
$= I(p^0, q^0 | m \cdot p^0, q)$ with (3a)
$\leqslant I(p^0, q^0 | p, q)$ with (1a) because of $m \cdot p^0 \leqslant p$.
Analogously we get $I(p^0, q^0 | p, q) \leqslant M$.

Theorem 3: I satisfies (2b), (3b), (7) $\Rightarrow I$ satisfies (5).
Proof: For all p^0, p, q^0, $q \in \mathbb{R}^n_+$, $\lambda \in \mathbb{R}_+$ we get:

$$\frac{I(\lambda p^0, q^0 | \lambda p, q)}{I(p^0, q^0 | p, q)} = \frac{I(\lambda p^0, q^0 | \lambda p, q)}{I(p^0, q^0 | p, q)} \cdot \frac{I(p^0, q^0 | p^0, q^0)}{I(\lambda p^0, q^0 | \lambda p^0, q^0)} \quad \text{with (2b)}$$

$$= \frac{I(\lambda p^0, q^0 | \lambda p, q)}{I(\lambda p^0, q^0 | \lambda p^0, q^0)} \cdot \frac{I(p^0, q^0 | p^0, q^0)}{I(p^0, q^0 | p, q)}$$

$$= \frac{I(p^0, q^0 | \lambda p, q)}{I(p^0, q^0 | \lambda p^0, q^0)} \cdot \frac{I(p^0, q^0 | p^0, q^0)}{I(p^0, q^0 | p, q)} \quad \text{with (7)}$$

$$= \frac{I(p^0, q^0 | \lambda p, q)}{I(p^0, q^0 | p, q)} \cdot \frac{I(p^0, q^0 | p^0, q^0)}{\lambda I(p^0, q^0 | p^0, q^0)} \quad \text{with (3b)}$$

$$= \frac{I(p, q | \lambda p, q)}{I(p, q | p, q)} \cdot \frac{1}{\lambda} \quad \text{with (7)}$$

$$= \frac{1}{\lambda} \cdot \lambda = 1 \quad \text{with (3b).}$$

Theorem 6: There is no function I: $\mathbb{R}^{4n}_+ \to \mathbb{R}_+$, which satisfies (2), (3a), (6), (7), (8) at the same time.
Proof: (cf. Eichorn & Voeller, 1976, p. 85/88, Satz 4/6, and Eichorn, 1978, p. 168, Theorem 8.5.27):
Suppose there is a function I satisfying the axioms (2), (3a), (6), (7), (8). With $e = (1, \ldots, 1) \in \mathbb{R}^2_+$ we get

$$\frac{I(e, e | \Lambda M e, \Lambda^{-1} M^{-1} e)}{I(e, e | e, e)} = \frac{I(e, e | \Lambda M e, \Lambda^{-1} M^{-1} e)}{I(e, e | \Lambda e, \Lambda^{-1} e)} \cdot \frac{I(e, e | \Lambda e, \Lambda^{-1} e)}{I(e, e | e, e)}$$

$$= \frac{I(\Lambda e, \Lambda^{-1} e | \Lambda M e, \Lambda^{-1} M^{-1} e)}{I(\Lambda e, \Lambda^{-1} e | \Lambda e, \Lambda^{-1} e)} \cdot \frac{I(e, e | \Lambda e, \Lambda^{-1} e)}{I(e, e | e, e)} \quad \text{with (7)}$$

$$= \frac{I(e, e | M e, M^{-1} e)}{I(e, e | e, e)} \cdot \frac{I(e, e | \Lambda e, \Lambda^{-1} e)}{I(e, e | e, e)} \quad \text{with (6), hence with (2a)}$$

(*) $I(e, e|\Lambda Me, \Lambda^{-1}M^{-1}e) = I(e, e|Me, M^{-1}e) \cdot I(e, e|\Lambda e, \Lambda^{-1}e)$

for all matrixes $\Lambda = \begin{pmatrix} \lambda_1 & & & 0 \\ & \cdot & & \\ & & \cdot & \\ & & & \cdot \\ 0 & & & \lambda_n \end{pmatrix}$, $M = \begin{pmatrix} \mu_1 & & & 0 \\ & \cdot & & \\ & & \cdot & \\ & & & \cdot \\ 0 & & & \mu_n \end{pmatrix}$ where $\lambda_1, \ldots, \lambda_n$, $\mu_1, \ldots, \mu_n \in \mathbb{R}_+$. Furthermore, by using (6) and (7), we have

$$\frac{I(e,e|e,p)}{I(e,e|e,e)} = \frac{I\left(p; \frac{1}{p_1}, \ldots, \frac{1}{p_n} | p, e\right)}{I\left(p; \frac{1}{p_1}, \ldots, \frac{1}{p_n} | p; \frac{1}{p_1}, \ldots, \frac{1}{p_n}\right)} = \frac{I(e,e|p,e)}{I\left(e, e|p; \frac{1}{p_1}, \ldots, \frac{1}{p_n}\right)},$$

hence with (2a)

(**) $I(e, e|p, e) = I(e, e|e, p) \cdot I\left(e, e|p; \frac{1}{p_1}, \ldots, \frac{1}{p_n}\right)$ for all $p \in \mathbb{R}^n_+$.

We consider now $P = I(e, e|\lambda, \ldots, 1; e) \cdot \ldots \cdot I(e, e|1, \ldots, \lambda; e)$, where $\lambda \in \mathbb{R}_+$. From (**) we get

$P = I(e, e|e; \lambda, \ldots, 1) \cdot \ldots \cdot I(e, e,|e; 1, \ldots, \lambda) \cdot Q$ where

$Q = I\left(e, e|\lambda, \ldots, 1; \frac{1}{\lambda}, \ldots, 1\right) \cdot \ldots \cdot I\left(e, e|1, \ldots, \lambda; 1, \ldots, \frac{1}{\lambda}\right)$. From (*) we get

$Q = I\left(e, e|\lambda, \ldots, \lambda; \frac{1}{\lambda}, \ldots, \frac{1}{\lambda}\right) = I\left(e, e|\lambda \cdot e, \frac{1}{\lambda} \cdot e\right)$ and finally

$Q = \lambda I\left(e, e|. e, \frac{1}{\lambda} e\right) = \lambda$ with (3a) and (2).

So we have $P = \lambda I(e, e|e;\lambda, \ldots, 1) \cdot \ldots \cdot I(e, e|e; 1, \ldots, \lambda)$.
(8) implies that the right-hand side tends to zero for $\lambda \to 0$, hence at least one of the factors of P tends to zero for $\lambda \to 0$. But this contradicts (8).

REFERENCES

Eichhorn, W., 1978, *Functional Equations in Economics* (Applied mathematics and computation; no. 11), Addison Wesley, Reading, Mass.

Eichhorn, W. and Voeller, J., 1976, Preisindextheorie, in Fuchssteiner u.a. (Hrsg.), *Jahrbuch Überblicke Mathematik*. BI, Mannheim.

Freudenthal, H., 1973, *Mathematics as an Educational Task*, D. Reidel, Dordrecht, Holland.

Revuz, A., 1965, *Moderne Mathematik im Schulunterricht*, Herder, Freiburg.

Teaching Mathematical Modelling in O.R.

P. B. Taylor
City of London Polytechnic

SUMMARY

The process of *creating* a mathematical model has always been recognized as an activity of central importance in the practice of O.R. It is therefore surprising that so little attention has been paid to teaching mathematical modelling, as distinct from mathematical models, in undergraduate and postgraduate courses. This chapter describes a short course, presented to graduate mathematicians, specifically designed to teach modelling. The principles on which the course is based are discussed, together with illustrations of the teaching method and assessment material used. The chapter concludes with a pedagogic hypothesis concerning the development of modelling skills which if accepted, has important implications for the design of O.R. courses at all levels. The material should be of interest to all those involved in the education and training of O.R. practitioners.

1. IS TEACHING MODELLING FEASIBLE?

There now exists a substantial body of literature concerning education *in* or *for* Operational Research: a useful bibliography is given by Hicks (1973) and more recent contributors include Miser (1976) and Tate (1977). What is perhaps surprising about these contributions is that so little attention has been given to mathematical modelling, an activity which many would hold to be central to the practice of O.R.

Let us be clear about what is meant by mathematical modelling. If we adopt the Open University course team's (1977) definition of a model as a representation of some aspect of reality 'created for a specific purpose or purposes' then mathematical modelling must have a broader definition than merely converting a verbal problem into a symbolic one. To some extent it must include aspects of problem formulation, since the model-builder's perceptions of reality will condition the form of the symbolic representation used. In addition, some manipulation of the model, perhaps amounting to a solution, is necessary in order to determine the model's fitness for purpose.

Why is the literature so sparse on the teaching of mathematical modelling? The answer is not difficult to find. Morris (1967), in the only substantial paper on the process of modelling known to the author, describes modelling as 'an art in the sense that it must remain largely intuitive'. He continues, 'Any set of rules for obtaining models could only have the most limited usefulness at best, and at worst, might seriously impede the required intuition'. Bonder (1973), commenting on Morris's paper, adds:

> I do not believe that experience in modelling can be obtained from lectures that present O.R. techniques or models. Textbooks give a misleading experience, since, when the models are described, the descriptions are usually justifications of the model and not the process by which one creates the model, as students often assume.

So modelling poses some formidable pedagogical problems; worse still, it can be argued that those who do the teaching and the learning (mathematicians—in fact, or by any other name) have a predisposition towards the mathematical manipulation, rather than the creation, of the model. As Richards (1973) put it: 'They are happiest when they have minimised the descriptive setting of their problems and have retreated rapidly into the rather safer harbour of mathematical jargon and manipulation of equations'.

With these strictures in mind it is as well to tread warily when designing courses in mathematical modelling. What then should such a course attempt to do? Here is one proposal:

(a) Give the students both first- and second-hand experience of constructing models.
(b) Set the modelling activity within a methodological framework, gradually developing such principles of the model-building process as can be recognized.

Both (a) and (b) are judged to be essential: if (a) is attempted without (b) the course lacks cohesion and progression, like a menu made up of tasty but unrelated tit-bits; if (b) is emphasized to the exclusion of (a) then it becomes indigestible and is likely to be rejected by the students. Perhaps the latter point explains why White's (1975) excellent book, *Decision Methodology* is so often found unpalatable by students when used as course material.

If these general aims are accepted one is left with selecting the models to be used for (a), and then sequencing them in such a way that (b) may be achieved. Before launching into the compilation of a list of models it appears prudent to consider the criteria for selection. Three maxims suggest themselves: avoid complexity; avoid rigour; avoid techniques. The complexity issue is not difficult to justify—with a complex problem at hand any methodological issues designed to be highlighted by the model tend to be swamped by the sheer weight of material. One should not be ashamed of a model which consists of a pair of linear simultaneous equations if it can be used as a vehicle for illustrating some fundamental methodological point. On rigour, Hammersley (1973) put it nicely: 'Avoid rigour like the plague: it only leads to rigor mortis at the manipulative stage'. Besides, it is contrary to the spirit of modelling, in its initial stages, at least. The question of techniques is a little more difficult—short of total originality, not possessed by the author, they seem a little difficult to avoid, and indeed to define: we all accept linear programming as a technique, but what of the calculus? The essential point is not to give the opportunity to the students (or, to be honest, onself) to go off into reveries on the beauty of the technique. One stimulant is the provision of computer packages for techniques which dispense with 'solution' as a problem and give an important opportunity for solution-formulation feedback, thus increasing the creative element.

The author's selection of models, following these maxims, is listed in Table 1: this has been used as the basis for a course on mathematical modelling given to graduate mathematicians. The list *does* include a linear programming formulation—Model 9, Product Mix and Politics—chosen with the aim of illustrating how a model, normally associated with a solution technique, can be employed equally powerfully to demonstrate alternative formulations of the same problem. The essence of any such list of models is variety in context and spread of type. The contexts vary from the homely (Model 5, Planning a Student Party), through environmental (Model 4, Road Tunnel Speed) to the more traditional industrially based (Model 2, Shall we use a Depot?). The models differ in type, being deterministic (most), stochastic (Model 7), static (most), dynamic (Model 6), *a priori* (Model 4), empirical (Model 1), functional (most), logical (Model 7). Like the superstitious advice to the English bride there is something old (Model 3), something new (Model 5), something borrowed (Model 4—from Pollak (1968)), something blue (Model 9).

However carefully the models are selected to accord with our maxims, whatever skill is employed in putting them in a sequence which allows sensitive methodological development, what determines whether a course has a chance of being successful in teaching modelling (as distinct from models) is largely the teaching technique employed, some principles of which must next be discussed.

Table 1

MODEL	*AIM*	*SELF ASSESSMENT*	*READING*
1 Petrol Sales	Basic ideas. Problems of measurement and model form	Critique of model	Rivett (1975) The Structure of league football
2 Shall We Use a Depot?	Ideas of Sensitivity Analysis	Hire or buy a car	
3 Simple Stock Control	Use of influence diagram. Optimisation	Bank withdrawals	Eilon (1974) Mathematical modelling for management
4 Recommended Road Tunnel Speed	Mixing physical principles with assumptions on human behaviour	Right hand turn at a T-junction	Burkhardt (1978) Temporary traffic lights
5 Planning a Student Party	Estimating model forms and parameters. Multiple criteria.	Allocating demand between competing retail outlets.	
6 Fleet Size for Seasonal Demand	Notation Conditional averages.	Newsboy problem.	
7 Appointments System at Doctor's Surgery	Logical and functional model comparison	Reserve bus crew. Analytical vs simulation solution.	
8 Bus Scheduling	Philosophy of reduction to a smaller problem	Aircraft scheduling	Lampkin and Saalmans (1967) The design of routes etc.
9 Product Mix and Politics	Alternative formulations of same problem	Comparative analysis of the models	

2. PRINCIPLES OF TEACHING METHOD

2.1 Start from the beginning

An obvious enough injunction—the difficulty lies in determining where the beginning is. It is the author's experience that the teaching of O.R. is moving in a direction where the starting point for any modelling activity is getting further and further away from the reality which is being represented. If this trend continues O.R. will come to be regarded in much the same way as what used to be known as 'applied mathematics', with its inelastic balls on smooth planes, without it even occurring to the students whether there exist any real world counterparts to the abstractions used.

Consider simple stock control (Model 3) as an example. One starting point is shown in Fig. 1 with its familiar saw-tooth pattern. Having explained the context and the graph the first step in the modelling process and the students are likely to hear is: let Q represent the order quantity. Note the unstated assumption that the order quantity is constant. Thus using Fig. 1 as a starting point deprives the students of the opportunity of *deducing* that the order quantity should be constant, under certain conditions, and determining, what those conditions are.

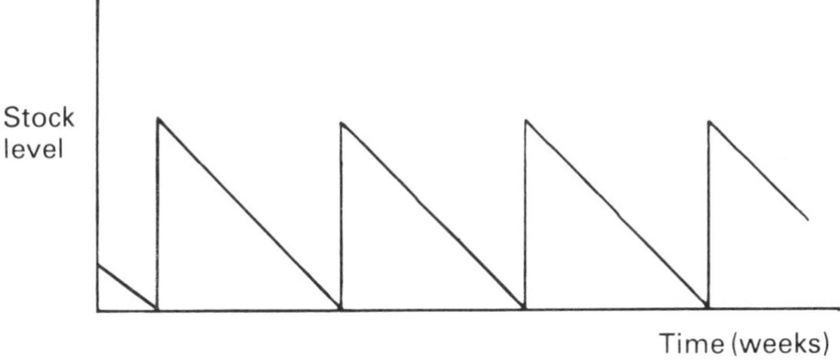

Fig. 1. An abstracted graph

By contrast, starting from Fig. 2, much closer to reality, it is not at all immediately obvious that the order quantity should be constant, nor that by an appropriate ordering policy the stock–time graph can be given a saw-tooth pattern, the abstraction of which is Fig. 1. The opportunity can thus be given for much modelling activity prior to the introduction of the symbol Q.

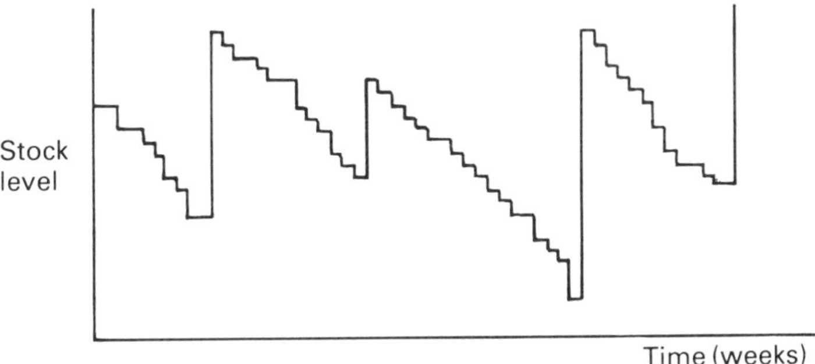

Fig. 2. A more realistic representation

One device that has been found useful when attempting to start closer to reality is for the lecturer to outline very briefly the setting of the problem and then invite students to discover more about it by questioning him as the 'man on the spot'. This will lead students to suggest modelling assumptions, and gauge reaction to them, as the enquiry proceeds. A typical dialogue, referring to Fig. 2, might be as follows:

Q. Why did you place an order in week *n* when you already had plenty of stock?
A. Because I was going on a fortnight's leave and didn't want to risk running out.
Q. Did you place a particularly large order in week *m*?
A. I see what you mean, but I wasn't aware that I had at the time.
Q. Why don't you save yourself trouble and order enough for the whole year at once?
A. Lots of reasons. It wouldn't be fair on our supplier, we haven't got enough space and it would incur the wrath of our accountant.

And so on.

2.2 Make assumptions explicit

Another definition of a model, given by Tate and Jones (1975), is a 'set of organized assumptions about a particular aspect of the world and the way it works'. Modelling activity, then, is principally concerned with the generation, testing and modification of assumptions; and so to teach modelling it is necessary to make these processes explicit.

One method of focusing attention on assumptions is by means of an 'influence diagram' (alternatively called a conceptual or qualitative model) as we now illustrate for Model 9 (product mix and politics). A simple analysis of the influence diagram of Fig. 3 shows that the model has the following inputs and outputs:

Inputs	*Outputs*
Unit profit	Machine loading
Demand	Unsold goods/unsatisfied demand
Machine hour availability	Total profit
Product mix	
Machine times per unit	

A model with five inputs and three outputs is particularly difficult to handle: we may cast it into a more manageable form by making a series of assumptions. First for the inputs:

Assumption 1 We assume that the company has (or chooses to have) no

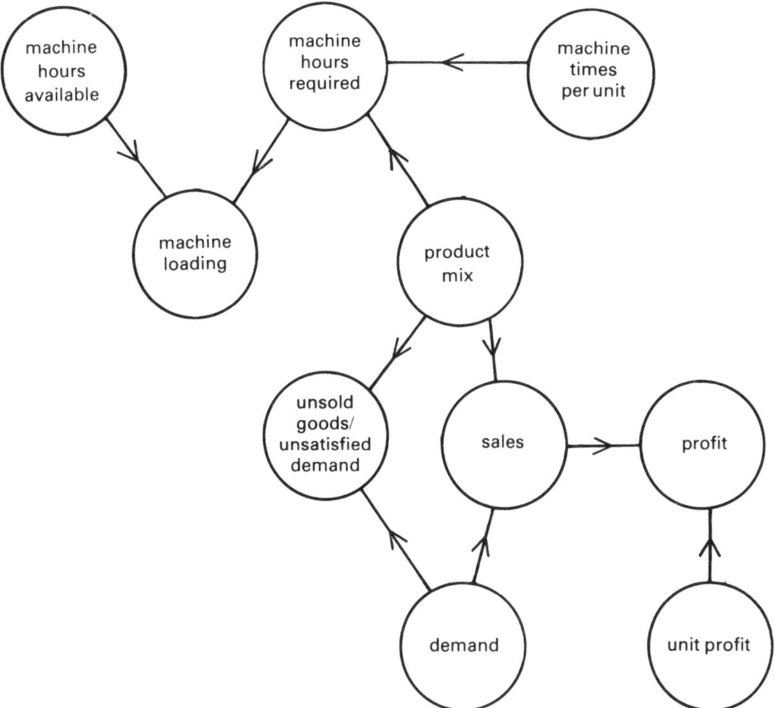

Fig. 3. Influence diagram for Model 9

influence on the unit profit, demand, machine hours availability and machine times per unit.

This reduces the problem to one where there is only one factor over which the company, in this context, has control—the product mix.

Assumption 2 We assume that the unit profit, demand and machine hours availability are fixed and known.

This further simplifies the representation to a deterministic model. At this stage it is worth exploring the implications of this assumption (no machine breakdowns etc.). We then proceed to a more technical assumption which results in a considerable algebraic saving:

Assumption 3 Machine change-over times (set-ups) are negligible.

So far our assumptions have been neutral—posited for reasons of simplicity, crude approximation to reality and mathematical convenience. From now on, in connexion with the outputs, they take on a distinctly political flavour:

Assumption 4 No shortfall in capacity is allowed (e.g. no overtime working or subcontracting); no cost is incurred for spare capacity.

Assumption 5 No unsold goods are allowed; no cost is incurred for unsatisfied demand.

This leaves profit as the only output factor of concern and the model, called by the author the *right wing version*, becomes obvious:

Maximize profit,

subject to

Machine hour requirement $\leqslant$ Machine hour availability

and

Production quantities $\leqslant$ Demand

However, with a small change in perspective, one could have:

Assumption 4′ Overtime is allowed under full working conditions; idle working is to be reduced as far as possible.

This leads to the (not very) *left wing* formulation of:

Minimize idle time

subject to

Total profit $\geqslant$ Acceptable minimum profit

and

Production quantities $\leqslant$ Demand

More consumer orientated formulations may be arrived at by modifying assumption 5.

2.3 Emphasize validation

Validation is the linch-pin linking abstraction to reality; without it one is confined to 'building models in the air', an exercise which, however intellectually satisfying, falls short in terms of purposeful activity. A distinction is often made between *pedagogic modelling* and *professional modelling*, the former being where one is content to use reality as a form of backdrop against which mathematical symbolism can be imaginatively employed, and the latter which demands time-consuming attention to the detail of the reality. Such a distinction, it may be argued, fails to recognize that validation takes many forms—and that there is considerable pedagogic value in finding that a model's behaviour does (or does not) conform to one's expectations or observations, i.e. in the validation process itself. Professional modelling, in turn, becomes seriously impoverished if the ideas which underlie it become fixed too soon, followed by extensive checks on the model for verisimilitude—it is, after all, probably more important to explore what will be or could be, rather than what is.

In the classroom, perhaps more so than in a professional capacity—dare it be said—one is severely limited by time constraints. It behoves one to employ those forms of validation which lend conviction, and achieve this by a wide variety of means, in the shortest possible time.

Without attaching labels to various forms of validation (as do some authors), the following is a crude checklist of possible questions to be asked of the model:

(a) Does the model behave in special or limiting cases as you would expect?
(b) Do the outputs of the model change as one would expect to changes in the imputs?
(c) How sensitive is the model to estimates in parameter values and functional forms?
(d) How does any recommended course of action from the model compare with observed decisions?
(e) How well does the model fit observed data?

Questions (a) and (e) are those familiar to mathematicians and statisticians, respectively, in the normal course of their work. Validation along the lines of question (b) can be a more complex affair involving, for instance, a detailed examination of how outputs correlate with each other in response to changing inputs. However, for teaching purposes, the idea can be simplified to an examination of the form of an output variable as one input is changed (e.g. does it have a turning point?) Questions (c) and (d) relate specifically to those models built for decision-making purposes and it may be useful to give examples of the teaching technique employed.

Sensitivity analysis may be introduced in the following manner. For Model 2 (Shall we use a depot?), reduced to its barest essentials, the models for the daily cost of operations for the case of no depot, CND, and with a depot, CD, are respectively:

$$CND = ax$$

and,

$$CD = bx + c,$$

where x represents some measure of the level of business. If we denote estimates by a circumflex accent, the extra cost of using a depot may be expressed in the following ways, in preparation for a sensitivity analysis with respect to a, b, c and x:

$$CD - CND = [\hat{b}\hat{x} + \hat{c}] - [\hat{a}\hat{x}](a/\hat{a})$$
$$CD - CND = [\hat{c} - \hat{a}\hat{x}] + [\hat{b}\hat{x}](b/\hat{b})$$
$$CD - CND = [(\hat{b} - \hat{a})\hat{x}] + [\hat{c}](c/\hat{c})$$
$$CD - CND = [\hat{c}] + [\hat{x}(\hat{b} - \hat{a})](x/\hat{x})$$

In each case the square brackets are, for the purpose of the analysis, constant. Denoting the ratio of actual to estimate by α, Fig. 4 shows graphically how much more critical is the estimation of a and b than c and x, given typical values of the estimates.

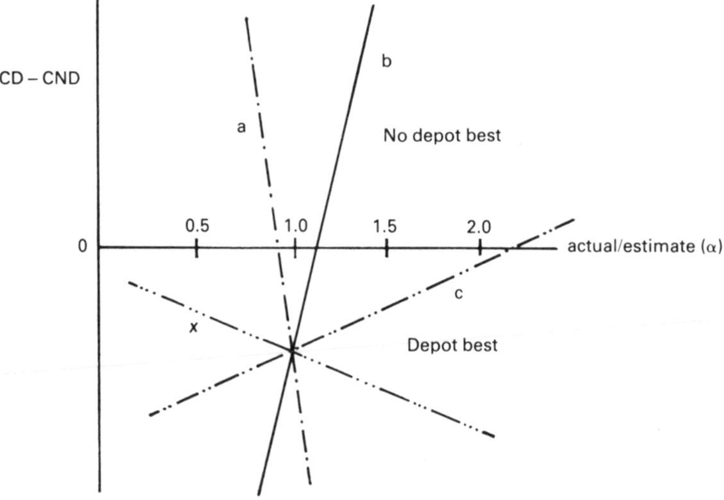

Fig. 4. Sensitivity analysis for Model 2

Model 5 (planning a student party) uses two decision criteria—the number of students attending, N, and the profit made by the party, P. From the simple model an 'efficient set' of decisions is obtained which involve a trade-off between these two criteria. Essentially the model prescribes a low entry-ticket price (to get students in) and a reasonable mark-up on drinks (to cover the fixed costs and protect against loss). Now many party organizers, in the author's experience, do just the opposite—they set a high ticket price and use the door receipts to subsidize the drink. This casts doubt on the whole basis of the model until one searches for a decision criterion not used in the original model; the one that seems appropriate here is that of the rate of alcohol consumption, held by many to be an essential ingredient to make the party 'go'. This element of surprise—whether against general experience, observation or intuition—has been found to be a useful pedagogic device for stimulating further and improved modelling suggestions.

2.4 Support by self-assessment questions

A self-assessment question gives the students the opportunity of testing their own understanding of the principles developed by constructing and validating their own model and comparing their own efforts with a suggested answer. It is, of course, of paramount importance, to emphasize that there is no one 'solution' to a problem of this kind. The extent to which one can demand of students a sizable conceptual leap from classroom material to the self-assessment question will no doubt never be resolved satisfactorily; however, it is the author's experience that, in terms of frequently employed self-assessment questions, the leap should be small, concentrating on one or, at most, two issues. Such questions can be interspersed with a few more free-ranging problems, designed to develop creative skills rather than to reinforce principles.

Throughout, it is important that not too much contextual knowledge is required of the students and, for this reason, the self-assessment questions hinted at in Table 1 are of a more homely kind than the presentation material itself. Each of these requires a rather different kind of conceptual leap and despite the fact that many of these may appear undemanding to seasoned model-builders, students starting out on the path of model building find them intellectually taxing. As a particular example, the classic newsboy problem, following on from Model 6, causes considerable difficulty: not only is the notion of conditional expectation found troublesome by many, the transition from a deterministic dynamic model to the (formally identical) stochastic static model proves particularly troublesome.

To give an idea of the additional material (not listed in Table 1), unrelated to the presentation models, the following two questions have been used and generated very considerable interest. Such questions are perhaps best attempted as group exercises with guidance hints being given from time to time to steer the students towards more productive ideas.

Question 1 Suppose it is autumn and a thick layer of leaves cover your large lawn. You have a lawn rake and a barrow to clear the lawn of leaves and transfer them to a compost heap just beyond the lawn. Consider alternative ways of organizing this and assess these by means of models.

Question 2 We may categorize books on an undergraduate reading list as being (a) useful for a whole year, (b) useful for a topic covered in lectures in six weeks or (c) useful in support of an individual lecture. Suppose that m copies of a particular book are on the shelves and n students ($> m$) attempt to borrow (or reserve) the book at the same time, for the standard loan period of L weeks. Suggest a model for the total value of these m copies to the students. How could your model be used?

3. A PEDAGOGIC HYPOTHESIS

It has been suggested cynically that no paper concerned with education should be without an untested (and possibly untestable) pedagogic hypothesis. This chapter conforms. The hypothesis concerns an abstraction-reality axis for modelling activity and movements along this axis: it may be stated as follows:

Hypothesis Skills in modelling can be accelerated by encouraging movement back and forth along the abstraction–reality axis. Movements in the direction of abstraction allow greater freedom in mathematical manipulation and deduction; movements towards reality develop critical skills in assessing the utility and validity of models and suggest reformulations. Taken together, the skills, in Morris's terms, assist in the development of the required intuition for modelling.

The following summarizes moves in the direction of *abstraction*:

- Writing down a statement of the problem.
- Defining suitable variables (and parameters) and suggesting a notation.
- Making assumptions.
- Defining the form of functional relationships.
- Considering special cases (e.g. the deterministic solution of a stochastic problem).
- Finding a particular formulation intractable and making simplifying assumptions.

Similarly the following are moves in the direction of *reality*.

- Data collection.
- Making intelligent guesses at parameter values.
- Examining whether solution values look intuitively reasonable and whether the solution has the right properties.
- Testing parameter values and functional forms for sensitivity.
- Comparing solution values with reality.
- Comparing solutions with different assumptions.

4. CONCLUDING REMARKS

The course we have described is short (one hour a week for a term) and elementary. It is followed by more advanced courses on modelling within particular technique areas, namely mathematical programming and simulation, based on a similar approach.

Whether we have been successful in teaching modelling is very difficult to assess in the short term; and in the long term, when such skills become apparent (or not), we no longer have access to the students. Perhaps instead of teaching modelling we have simply taught modelling—but this seems one stage better than teaching about models. One particularly telling remark by a student was that he felt that he did not possess the necessary *emotional* make-up to be able to make the broad, sweeping assumptions in mathematical modelling.

REFERENCES

Bonder, S., 1973, Operational Research eduation: some requirements and deficiencies. *Ops. Res.*, **21**, 796–809.

Burkhardt, H., 1978, Temporary traffic lights. *J. Mathematical Modelling for Teachers*, **1**, 1, 11–19.

Eilon, S., 1974, Mathematical modelling for management. *Interfaces*, **4**, 2, 32–38.

Hammersley, J. M., 1973, Maxims for manipulators. *Bull. I. M. A.*, **9**, 276–280.

Hicks, D. 1973, Education for Operation Research. *Omega*, **1**, 1, 107–116.

Lampkin, W. and Saalmans, P. D., 1967, The design of routes, service frequencies and schedules for a municipal bus undertaking. *Opl Res. Q.*, **18**, 4, 375–398.

Miser, H. J., 1976, Introducing Operational Research. *Opl. Res. Q.*, **27**, 655–670.

Morris, W. T., 1967, On the art of modelling. *Mgt. Sci.*, **13**, B707–B717.

Open University course team, 1977, Modelling by mathematics. TM 281. Unit 1.

Pollack, H. O., 1968, On some of the problems of teaching applications of mathematics. *Ed. Studies in Maths*, **1**, 24–30.

Richards, E. J., 1973, Some thoughts on mathematical education. *Int. J. Math. Ed, Sci. Technol.*, **4**., 377–395.

Rivett, P., 1975, The structure of league football *Opl Res. Q.*, **26**, 4, 801–812.

Tate, T. B. and Jones, L., 1975, Systems, Models and Decision. Open University, Systems Modelling Course, T314 1/2. Sections 1 and 2.

Tate, T. B., 1977, Teaching Operational Research Technique. *Opl. Res. Q.*, **4**, 765–779.

White, D. J., 1975, Decision Methodology. Wiley, Chichester.

Modelling a Games Programme at a Tenth Birthday Party

C. R. Haines
The City University

SUMMARY

It is often difficult to devise novel problems which are suitable for use with students of diverse mathematical backgrounds. The model described in this chapter was devised in making arrangements for a tenth birthday party. It concerns the setting up of a league system for games for which the following objectives are borne in mind.

(1) Each child would play every other child once and only once.
(2) In a given session, different games would be played by each pair as only one set of each game was available.
(3) Each child would play as many of the available games as possible.

It turns out that objectives (1) and (2) are easily attainable but that (3) is only partially so. The main features of the model involve the existence of a many–many map from a Cyclic group or a Klein group $(S_n, *)$ to the set $S_n = \{1, 2, 3, \ldots, n\}$ on which a particularly uninteresting operation $\circ$ is defined. For this application it appears that such a map does not exist for all groups $(S_{4n-2}, *)$, $n \varepsilon N$, so do not have six children at the party!

1. INTRODUCTION

The creation of a successful mathematical model is usually a very personal experience, even unique to an individual student and is very difficult to cater for within a traditional lecture course. It is not surprising, therefore, that a great deal of time and effort has been put into the development of teaching programmes in Mathematical Modelling, several of which are well documented in the literature, see for example the *I. M. A. Bulletin*, Volume 19, March/April 1983. Whatever approach is adopted, these programmes usually promote an enthusiasm for mathematics and motivate students to think independently. The attendant skills of the formulation of the mathematical model, the co-operation with others in a similar field and the communication of the model and its results are similar to those skills successfully achieved by many engineering and other students in pursuing a major project as part of their undergraduate course. (Cunningham-Greene, 1983; Tayler, 1983).

Modelling and achieving these skills is to some extent a group activity. There are many examples of research groups which flourish from the interaction of ideas within the group and conversely some mathematicians work in splendid isolation. The great Indian mathematician Ramanujan spent the greater part of his life isolated from other mathematicians and much of his original work was rediscovery. Confidence levels at creating models can be increased by interaction and communication with others, Haines (1980), but one of the major problems is that any analysis of models should use those mathematical techniques which are already familiar to the student. For this reason it is usually much easier to propose novel problems and to illustrate the main ideas at the graduate level, although, in particular, the Open University has made invaluable contributions in this field at first-year undergraduate level (for example: Summer School Investigations as part of the course M101: A Foundation Course in Mathematics and also the course MST204 Mathematical Models and Methods).

This chapter does not seek to advance any new theories on mathematical modelling but simply puts forward a model which the author believes will prove a useful addition to those novel problems appropriate to a very wide range of 'students', and could even include pre 'O'-level candidates.

2. THE MODEL

In trying to arrange suitable activities for my son's tenth birthday party, little did I realize that two years later, the party would still be on my mind and not for the inevitable disasters which usually occur at such functions.

Like many families with young children, we had available several different boxed games for two players; we therefore decided to arrange a 'league' system in which the following objectives would be borne in mind.

(1) Each child would play every other child once and once only.
(2) In a given session, different games would be played by each pair as only one set of each game was available.
(3) Each child would play as many of the available games as possible.

As a relative newcomer to this sort of modelling exercise and as a mere 'concrete' applied mathematician, it seemed, at first sight that having fixed the structure it would be a simple matter to make the appropriate arrangements for the eight children at the party. A full description of a first attempt at solving this problem can be found in a short paper by Haines, 1983; however it will be useful to look at some of the results of that paper in order to illustrate the structure.

Consider the case of four players and the objective (1) above. We will come to the other two objectives later. Table 1 gives the position, numbering the players 1, 2, 3 and 4 and the sessions in which the games will take place as 1, 2, 3 and 4 also. The table as a whole represents a fuller system in which each player plays each player (including himself) once, but for our purposes only the six entries above the leading diagonal are relevant. Consequently, we shall only need to use the sessions numbered 2, 3 and 4.

Table 1

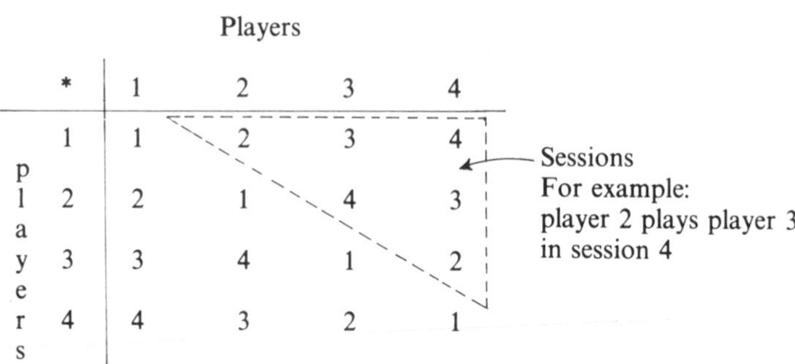

		Players			
	*	1	2	3	4
Players	1	1	2	3	4
	2	2	1	4	3
	3	3	4	1	2
	4	4	3	2	1

Whilst the arrangement put forward in Table 1 is not unique, it does ensure that all of the children are occupied all of the time, an essential prerequisite for a successful party. The second objective (2) was to ensure that a given game was used only once during each of the sessions outlined in Table 1. Games 1, 2, 3 and 4 have to be allocated to the sessions of Table 1 so that, for example, the four games in progress during session 2 consist of one pair playing game 1, one pair playing game 2, one pair playing game 3 and one pair playing game 4. With an eye to the third objective too, it was necessary to allocate games in each session so that there were not repeats in any row or column. The result is shown in Table 2.

The information given in Tables 1 and 2 can be summarized as shown in Table 3 and suggests that, whilst the objectives (1) and (2) have been satisfied completely, the attempt to meet the third objective (3) has not been entirely successful. The model and its limited solution so far proposed is certainly within the grasp of students at a very elementary level and it is only through a closer examination of the problem that a far deeper and interesting mathematical structure emerges, suitable for study, up to undergraduate level.

Table 2

Players

Players	$\circ$	1	2	3	4
	1	1	3	4	2
	2	4	2	1	3
	3	2	4	3	1
	4	3	1	2	4

games

Table 3

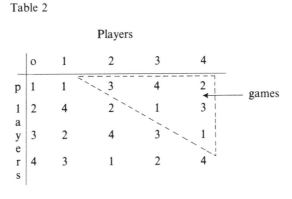

Session 2	1v2 game 3	3v4 game 1
Session 3	1v3 game 4	2v4 game 3
Session 4	1v4 game 2	2v3 game 1

Table 1 can be regarded as defining a binary operation $*$ on the set $S = \{1, 2, 3, 4\}$ and it is easy to see that $(S, *)$ is an Abelian group which is in fact a manifestation of the non-cyclic group K_4. Table 2, on the other hand, defines a particularly uninteresting operation $\circ$ on the set S, which has the following properties

(i) $\circ$ is closed on S
(ii) $\circ$ is self-distributive on S
(iii) all cosets of S are S.

This last property is a direct consequence of objective (2).

This simple case of four children illustrates the main features of the model which involve the construction of the operations $*$ and $\circ$ in such a way that the map

$$h: (S, *) \to (S, \circ)$$

exists in a form which satisfies the objectives (1)–(3) as far as possible. Unfortunately there is a great deal of flexibility in defining $*$ and the difficulties arise in finding a suitable operation, $*$, such that a map h can be found.

The original problem involved eight children at the birthday party, and a solution in this case can be deduced from the group structure mentioned above.

If we let $S_n = \{1, 2, 3, \ldots, n\}$ then the required map

$$h: (S_8, *) \to (S_8, \circ)$$

can be defined such that $*$ and $\circ$ are given by Table 4.

Table 4

*	1 2	3 4	5 6	7 8
1	1 2	3 4	5 6	7 8
2	2 1	4 3	6 5	8 7
3	3 4	1 2	7 8	5 6
4	4 3	2 1	8 7	6 5
5	5 6	7 8	1 2	3 4
6	6 5	8 7	2 1	4 3
7	7 8	5 6	3 4	1 2
8	8 7	6 5	4 3	2 1

o	1 2	3 4	5 6	7 8
1	1 3	6 8	7 5	4 2
2	4 2	7 5	6 8	1 3
3	8 6	3 1	2 4	5 7
4	5 7	2 4	3 1	8 6
5	3 1	8 6	5 7	2 4
6	2 4	5 7	8 6	3 1
7	6 8	1 3	4 2	7 5
8	7 5	4 2	1 3	6 8

Having solved the problem, to some extent, for the particular cases of four and eight children, the immediate needs of the birthday party were satisfied. But, of course, in setting up the model, the aim is usually to try to achieve a general result. Indeed, the first objective can be satisfied for all n by the construction of cyclic groups $(S_n, *)$ defined by Table 5.

Table 5

*	1	2	3	4	. . .	$n-1$	n
1	1	2	3	4	. . .	$n-1$	n
2	2	3	4	5	. . .	n	1
3	3	4	5	6	. . .	1	2
4	4						.
.	.						.
.	.						.
$n-1$	$n-1$	n	1	.		$n-3$	$n-2$
n	n	1	2	3		$n-2$	$n-1$

It turns out that if n is prime and $n \neq 2$ and if $(S_n, *)$ is formed in the manner of Table 5 then the map

$$h: (S_n, *) \to (S_n, \circ)$$

exists in which the operation o may be defined by Table 6. This construction can be extended by employing some well-known theorems on groups and is illustrated by the following example.

Table 6

o	1	2	3	4	5	. . .	n
1	1	2	3	4	5	. . .	n
2	n	1	2	3	4	. . .	$n-1$
3	$n-1$	n	1	2	3	. . .	$n-2$
4	$n-2$	$n-1$	n	1	2	. . .	$n-3$
5	$n-3$	$n-2$	$n-1$	n	1	. . .	$n-4$
.							
.							
.							
n	2	3	4	5	6	. . .	1

Example 1 Since $h: (S_3, *) \to (S_3, \circ)$ is given by Table 7 and $h: (S_5, *) \to (S_5, \circ)$ is given by Table 8 then $h: (S_{15}, *) \to (S_{15}, \circ)$ may be constructed by considering cosets of order 3 or 5. One such definition of h in this case is given by Table 9.

Table 7

*	1	2	3
1	1	2	3
2	2	3	1
3	3	1	2

o	1	2	3
1	1	2	3
2	3	1	2
3	2	3	1

Table 8

*	1	2	3	4	5
1	1	2	3	4	5
2	2	3	4	5	1
3	3	4	5	1	2
4	4	5	1	2	3
5	5	1	2	3	4

o	1	2	3	4	5
1	1	2	3	4	5
2	5	1	2	3	4
3	4	5	1	2	3
4	3	4	5	1	2
5	2	3	4	5	1

Table 9

*	1	2	3	4	5	6	7	8	9	10	11	12	13	14	15
1	1	2	3	4	5	6	7	8	9	10	11	12	13	14	15
2	2	3	4	5	1	7	8	9	10	6	12	13	14	15	11
3	3	4	5	1	2	8	9	10	6	7	13	14	15	11	12
4	4	5	1	2	3	9	10	6	7	8	14	15	11	12	13
5	5	1	2	3	4	10	6	7	8	9	15	11	12	13	14
6	6	7	8	9	10	11	12	13	14	15	1	2	3	4	5
7	7	8	9	10	6	12	13	14	15	11	2	3	4	5	1
8	8	9	10	6	7	13	14	15	11	12	3	4	5	1	2
9	9	10	6	7	8	14	15	11	12	13	4	5	1	2	3
10	10	6	7	8	9	15	11	12	13	14	5	1	2	3	4
11	11	12	13	14	15	1	2	3	4	5	6	7	8	9	10
12	12	13	14	15	11	2	3	4	5	1	7	8	9	10	6
13	13	14	15	11	12	3	4	5	1	2	8	9	10	6	7
14	14	15	11	12	13	4	5	1	2	3	9	10	6	7	8
15	15	11	12	13	14	5	1	2	3	4	10	6	7	8	9

o	1	2	3	4	5	6	7	8	9	10	11	12	13	14	15
1	1	2	3	4	5	6	7	8	9	10	11	12	13	14	15
2	5	1	2	3	4	10	6	7	8	9	15	11	12	13	14
3	4	5	1	2	3	9	10	6	7	8	14	15	11	12	13
4	3	4	5	1	2	8	9	10	6	7	13	14	15	11	12
5	2	3	4	5	1	7	8	9	10	6	12	13	14	15	11
6	11	12	13	14	15	1	2	3	4	5	6	7	8	9	10
7	15	11	12	13	14	5	1	2	3	4	10	6	7	8	9
8	14	15	11	12	13	4	5	1	2	3	9	10	6	7	8
9	13	14	15	11	12	3	4	5	1	2	8	9	10	6	7
10	12	13	14	15	11	2	3	4	5	1	7	8	9	10	6
11	6	7	8	9	10	11	12	13	14	15	1	2	3	4	5
12	10	6	7	8	9	15	11	12	13	14	5	1	2	3	4
13	9	10	6	7	8	14	15	11	12	13	4	5	1	2	3
14	8	9	10	6	7	13	14	15	11	12	3	4	5	1	2
15	7	8	9	10	6	12	13	14	15	11	2	3	4	5	1

Evidently maps h can be found for groups $(S_{4n}, *)$ using appropriate cosets based on the cases $n = 1, 2$ given in Tables 1, 2 and 4 together with Tables 5 and 6. For example, it is easy to find

$$h: (S_{12}, *) \to (S_{12}, \circ)$$

using cosets of order 3 or 4.

The above discussion covers all cases except those relating to $4n$-2 children. The case of two children at the party is easy to resolve without any sophisticated modelling, but so far it has not been possible to suggest ways of accommodating say six children within the stated objectives.

An interesting variation has been suggested by Lockwood (1983) in which objective (3) is interpreted very strictly to mean that for eight children only seven different games are available. The model he puts forward depends upon pairings given by tangents and chords of circles and gives a complete solution for the cases

8 children with 7 games
12 children with 11 games
16 children with 15 games

His method appears to be good for $4n$ children but he too cannot deal with the cases of $4n-2$ children.

REFERENCES

Cunningham-Greene, R. A., 1983, *I. M. A. Bulletin*, **19,** 3/4, 56.
Haines, C. R., 1980, *Teaching at a Distance*, **18,** 34.
Haines, C. R., 1983, *Mathematical Gazette*, **67,** 439, 1.
Lockwood, E. H., 1983, private communication.
Tayler, A. B., 1983, *I. M. A. Bulletin*, **19,** 3/4, 53.

Modelling Growth and Development

D. E. Prior and A. O. Moscardini
Sunderland Polytechnic

1. CURRENT MODELS OF GROWTH

Growth is a concept that lends itself very readily to mathematical modelling. Models abound that describe the growth of single species and competing (predator–prey) species. Here the concept to which the term 'growth' is applied is that of the changeover time in some volume measure of an amorphous aggregate of similar individual entities. A broad word description of the types of functions involved in this sort of growth situation might be 'more breeds more in proportion up to some environmental limiting factor'. Many methods exist for dealing with these, encompassing Monte Carlo techniques, Leslie matrices, Lotka–Volterra type equations and system dynamics.

Monte Carlo methods use the random number generator of a computer and a sampling distribution. Many assumptions have to be made to determine the sampling distribution which usually is taken to be one of the standard ones, i.e. Poisson, normal or negative exponential. If Leslie matricies are used, then there are two difficulties, i.e. the breakdown into sub-classes and the estimation of the transition probabilities. If the growth of population of deer is being modelled then the groupings have to be decided, e.g. male and female or year groups. Then the probability of say a three-year-old male contributing to a foal has to be estimated. Controlled experiments can be conducted on animal populations and using these results, reasonable estimates can be obtained for such parameters but the best one can obtain is some sort of average value. The system dynamics approach is useful for distinguishing the individual components and their interrelationship but again is very dependent on the choice of various constants.

If the concept to which the term 'growth' is applied is that of the morphological change of an organism over time, then use of the familiar growth methods above may be inappropriate. One or two have been applied to models of differentiated growth and it is enlightening to consider what assumptions lie behind the application of these methods in such cases. Consider, for instance, modelling the growth of a plant. The simplest model, Maynard-Smith (1968), will consist of some type of logistic equation i.e.

$$\frac{\mathrm{d}m}{\mathrm{d}t} = \alpha m\left(1 - \frac{m}{M}\right)$$

where $m(t)$ is the mass of the plant at any time t and M is the maximum size of the plant. What has been assumed here is that the shape of the plant is unimportant, in fact it is generally taken to be a cylinder! Also all parts of the plant are assumed to be growing at the same rate. In reality, of course, different parts of the plant have different functions and do not grow at the same rate. To try and build this into the model is difficult. Even splitting the plant into two parts, say roots and shoots, results in six differential equations Moscardini (1981) instead of one and a host of parameters, whose values have to be estimated. As some of these parameters are theoretical rather than physical, this is a difficult exercise and often inspired guesswork is the only answer. The solution to the logistic equation or the more complicated models, will usually involve negative exponential terms which approach zero as t approaches infinity. In reality, of course, maximum growth is attained after a finite time, T, so usually 99 % of the final value is deemed acceptable. The logistic equation or the continuous approach has several limitations but can still be useful. For example, if a farmer was wanting to model the growth of wheat in his fields, the important attribute would be the mass of wheat he grows for sale. The shape, size and behaviour of individual wheat stalks is unimportant and this model is therefore a reasonable one. But this is because the modeller is now in the first type of growth situation discussed—that of the changing size of an amorphous aggregate. Among other analyses of growth phenomena are those based upon streak photographs obtained by marking the organism with particles of carbon and photographing them through a slit onto a moving strip of film, Erickson (1976). Obviously this technique can only be used when the organism concerned lacks means of locomotion and has therefore been used to study root growth and the like. The resulting analysis is based upon growth rates obtained from estimates of the slopes of the photographic streaks, $\mathrm{d}x/\mathrm{d}t$, where x is the distance from a point on the organism to a 'growing tip'. The modellers here use the continuous mathematics of classicial analysis, employing such concepts as the relative elemental rate of increase of area of a growing organism, a, say, where

$$a = \operatorname{div} v = \frac{\partial \dot{x}}{\partial x} + \frac{\partial \dot{y}}{\partial y}$$

and x and y could be the co-ordinates of a point on the surface of, say, a leaf. This approach is obviously based upon the classical law

$$\operatorname{div} v = \frac{\partial p}{\partial t} \quad \text{where } p = \text{density}$$

and is a statement of the law of the conservation of mass which in biological considerations of growth phenomena is non-sense. Along similar lines, other workers study the hemisphere, core and ellipsoids in so far as these yield insight into relationships between ratios of meridional to circumferential rates of expansion.

The drawbacks of the above methods in their description of differentiated growth processes are legend but perhaps a major one is that none possess even the seeds of a rudimentary organic overall control process model which could begin to parallel that observed in the most lowly cellular organism when undergoing differentiated growth.

This chapter discusses an alternative approach to biological growth modelling which is believed capable of development to do just this.

2. BRIEF HISTORY OF DEVELOPMENTAL SYSTEMS

Developmental languages and systems has its origins in two fields of research, (see Fig. 1) automata, and computability on the one hand and natural language or symbol manipulation studies on the other—though nowadays these two fields are almost one and the same so far as their developed foundations and principles are concerned.

In what follows these two strands are briefly delineated to set the stage for their ultimate coalition in what is now termed Lindenmayer systems or L-systems for short.

2.1 Automata strand

In 1945, McCullough & Pitts published a paper called 'A logical calculus of the ideas immenent in nervous activity'. Whilst studying the behaviour of neurons they used Boolean logic as a main tool and added the concept of 'threshold logic'. That is, a particular neuron would fire if a certain Boolean function of its excitatory and inhibitory thresholds gave an integer value equal to or above the threshold value of that neuron. This 'logical neuron' model is a mathematical hybrid of Boolean logic and the additive group of integers.

This was then developed by various associates into a new model which is also a hybrid and combines string-recognizing finite automata with certain numerical manipulations usually based on integer rates and thresholds. The principal tool in this system was called an 'axiomatic enzyme'. This consisted of a 'black box' or finite automaton that is able to recognize strints of letters,

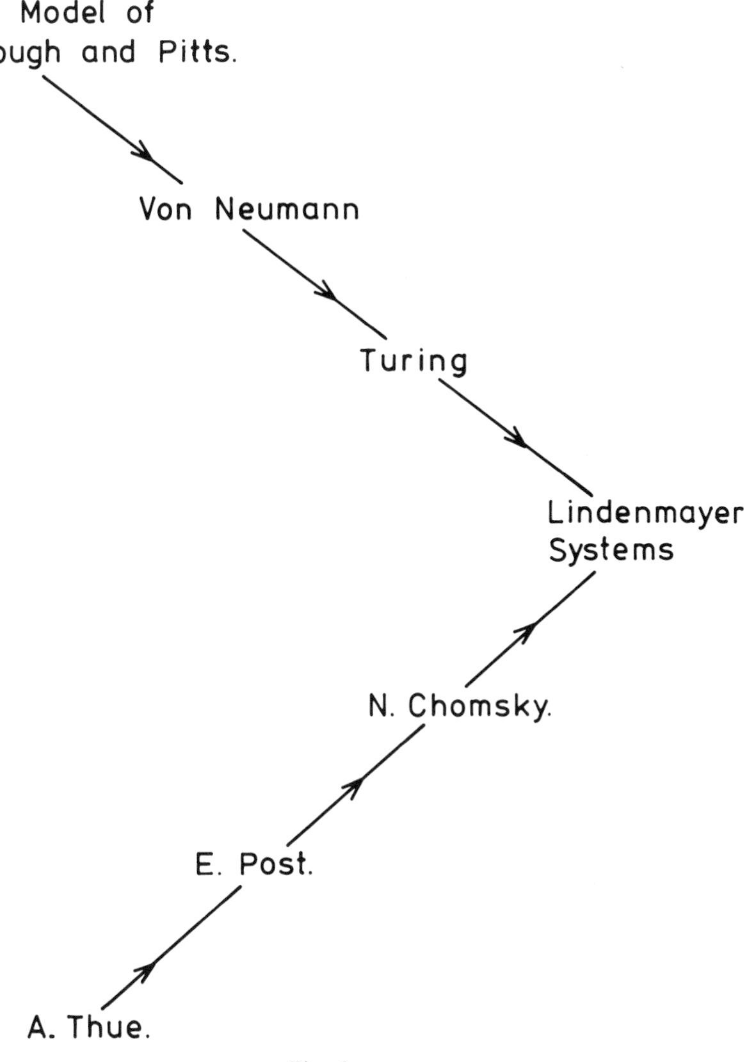

Fig. 1.

representing biochemicals and assess their numerical thresholds. If a number of threshold conditions were met, the algorithmic enzyme combines strings into a product operating at a fixed rate per unit time.

This string processing model was then used for the study of self-reproduction is automata. Von Newmann (1956) designed two types of self-reproducing systems based on automata. In his major work *The General and Logical Theory of Automata*, he remarks: 'Organisms can be viewed as made up of parts which to a certain extent are independent, elementary unitsaxiomatizing the behaviour of the elements means this: We assume that the elements have certain well-defined outside, functional characteristics; that is, they are to be treated as "black boxes". This being understood we may then investigate the larger organisms that can be built up from these elements and general theoretical regularities that may be detectable in complex syntheses of the organisms in question.'

The real cells of organisms are extremely complex structures. They contain over 10^{10} molecules, hundreds of thousands of proteins complex organized subcellular aggregates such as mitochondria and numerous copies of all biochemicals. Cells appear to operate in a probabilistic rather than simply deterministic manner. It has been pointed out that a typical cell manufactures several thousand new protein molecules including a variety of different enzymes, per minute. The mammalian genome probably contains 100,000 to 1000,000 genes encoded in billions of nucleotides. With these facts in mind, it may be anticipated that any mathematical model of a cell will be inadequate for a full study of cell processer. Nevertheless, the axiomatic or algorithmic enzyme is clearly a step in this direction. They provide a well-defined and plausible mathematical device for simulating differentiation, growth, malignant cell escape and other typically biological phenomena.

2.2 Symbol processing strand

People communicate verbally or in writing through the use of a natural language. Presumably having developed in step with the human animal, natural languages therefore are rich in expression and extensive in their coverage of man's experiences. The 'understanding' of a language is as much dependent upon circumstance, mood, intonation and delivery as upon any rules of syntax.

To study, or model, a natural language, or design a computer language, to model the language of a robot, assist the study of automata or examine the very fundamentals of mathematics, a natural language is altogether too imprecise. What is called for is a language defined in strict symbolic and manageable terms—via an artificial or formal language.

Emil Post's Canonical System of 1936, following upon A. Thue's 1914 work on rewriting systems, represented the beginnings of the development of formal languages. In Post's eyes, mathematics was a process of pure symbol manipulation. With this view in mind he developed a system for operating upon strings of symbols. A mathematical theorem, could then be reduced to a mere string of symbols. His basic idea was that given a set of symbols called an Alphabet and a set of starting strings termed axioms then by applying various rules, called productions (or rewriting rules), to the starting strings and to strings so derived, proofs of theorems could be shown to be in a sense just sequences of strings of symbols.

The following example of a Canonical System to produce palindromes is to be found in Minskey (1972) p. 228:

Alphabet A = {a, b, c}

axioms a
b
c
aa
bb
cc

Productions (i) $\$ \rightarrow a\$a$
(ii) $\$ \rightarrow b\$b$ } \$ = any string of the set.
(iii) $\$ \rightarrow c\$c$

Chomsky (1956) took Post's work and extended it to a study of natural language. In its development here a further set of symbols was introduced which allowed things to be said *about* symbols in the system's alphabet, i.e. a set of meta-symbols. The two sets of symbols constitute via the system's total alphabet are traditionally called the terminals (those that end or terminate the further generation of set elements) and the non-terminals (those that disappear from the scene as the elements of the set are derived).

A well-known example of Chomsky (1959) is the following

$$\text{Alphabet} = \{\underbrace{\text{S, Sp, Vp, N, V, Op, A}}_{V_N}, \underbrace{\text{man, dog, has, a, the}}_{V_T},$$

$$\underbrace{\text{S} \rightarrow \text{SpVp},\ \text{Sp} \rightarrow \text{AN},\ \text{Vp} \rightarrow \text{VOp},\ \text{Op} \rightarrow \text{AN},\ \text{A} \rightarrow \text{a/the},\ \text{N} \rightarrow \text{man/dog},\ \text{V} \rightarrow \text{has}}_{P}, \text{S}\}$$

where V_N = Non-terminals P = productions
V_T = Terminals S = axioms

Starting from S the following strings are derivable

$$S \rightarrow SpVp \rightarrow ANVp \rightarrow ANVOp \rightarrow ANVAN$$

substituting each non-terminal for a terminal yields

a man has a dog
a man has the dog
the man has a dog

etc.

These sets of rules which generate strings of symbols over an arbitrary alphabet are called grammars.

The above example is termed a phrase-structured grammar because the terminal strings are derived by successively substituting sub-strings (phrases) for other sub-strings (phrases) according to the predefined list of allowable substitutions via the productions.

Chomsky further sub-divided the phrase structured grammars by considering those which allowed only specific combinations of terminal and non-terminal symbols to give rise to further strings in the language. That is to say, the context in which a particular symbol occurred was important. Such grammars, he named 'context-sensitive' and called the languages derived from them, type 1 languages.

Type 2 languages, derived from grammars in which the left-hand side of a production consisted of one element only from V_N, he christened 'context-free'. Within this class of languages, he found those which derived from grammars that coincided with the rules of string construction obtained by traversing a directed graph. Here, on account of the parallel with Turing Machine and automata studies, Chomsky labelled this third class of grammars,his type 3 grammars, the finite-state grammars.

The twin rivers of thought met here and it remained for Lindenmayer to channel the stream into biological development. Before considering that phase, it is important to note two things about the formal languages developed by Chomsky. Firstly, in order to increase the generative capacity of the systems by providing for both literal and syntactic symbols the alphabet was divided into terminals and non-terminals.

Secondly the productions are applied sequentially to a string one at a time until a derivation is obtained which has only terminal symbols in the string.

Considerations can now be given to Lindenmayer's contribution to this field. In 1968 Aristid Lindenmayer published two papers showing how the growth development of an organism could be modelled based upon the concept that each cell in the organism could be regarded as a tiny automaton or 'black box' having unique response patterns to a given input signal pattern. This has been discussed more fully under the section 'automata strand'. The obvious (with hindsight) strong links between formal language theory as discussed under the section 'symbol processing strand' and Lindenmayer's work was not at first fully realized, but in 1971, Lindenmayer published a paper redefining his work in terms of the formal language approach and entitling it 'Developmental languages and systems'.

The major difference between Lindenmayer systems (L-systems) and formal language systems is that the former does not exhibit the two characteristics mentioned above. Lindenmayer systems are *parallel* rewriting systems *without terminals*. That is to say all strings generated by the system are in the language of the system, not just those consisting entirely of terminals. Also rewriting (or the applying of production rules) happens *simultaneously* (i.e. in parallel) to all symbols in a string during a scan.

The reasons for this are that in an organism development is proceeding everywhere at once and so all symbols in a string representing the string of cells making up the organism must 'develop' to some extent during the same time pulse.

The presence of terminal symbols implies dead cells which would remain permanently unchanged and unchanging in the organism.

Should a cell disappear for any reason this may be represented by use of a production involving λ, the empty string or the string containing no symbols.

The research undertaken initially by Lindenmayer, has been extended and rigorously formalized by such workers as Herman, Rosenberg and others (1975, 1976).

In developing his work described in the remainder of this paper Lindenmayer wrote the following admirable short critique of the current developmental models at that time. 'It appears to me that an important aspect of development has been grossly neglected, namely the study of integrating mechanisms which bring the multitude of individual acts of division, differentiation and cell enlargement under the control of the organism as a whole. Without an overall organization determining the proper times and places for these individual acts, the development of a mature organism from a single cell, the fertilized egg, could not take place. Undoubtedly, developmental instructions specifying such overall control occupy a considerable portion of the genetic make up of the fertilized egg, they may even constitute the major portion of it.' His work was to develop a theoretical framework within which intercellular relationships can be discussed, computed and compared. He concentrated on such aspects of development which involved the interaction of two or more cells and the timing and position of divisions. Once more than a handful of cells are present, the possible combinations of these interactions rapidly becomes unmanageable by the unaided intelligence, therefore these problems require either a sufficiently powerful mathematical theory or the application of computers.

As with all modelling, the basic assumptions are important. Lindenmayer lists twelve assumptions which underlie his work.

(i) Each cell is an independent unit with spatial and temporal boundaries.
(ii) Time progresses in discrete steps.
(iii) Each cell in any one time unit may be present in one of a number of states which may be interpreted as certain distinguishable physiological or morphological conditions like the presence or absence of a particular enzyme.
(iv) Each cell at each moment of time produces one of a certain set of outputs, which may be interpreted as the production of physical or chemical changes that may be propagated outside the cell.

(v) In each time unit, each cell can receive such impulses from its neighbours, the collection of which constitutes the input of that cell.
(vi) The next state of each cell is determined in each moment as a function of its present state and of its input.
(vii) The output of each cell is likewise a similar function.
(viii) An explicit rule states how outputs are converted into inputs.
(ix) A rule must be stated on neighbourhoods, i.e. the outputs of which cells are to be taken as inputs of which other cells.
(x) Cell divisions are to be taken into account by having a state sequence of two or more components as the value of the next state function under certain state input combinations.
(xi) Each cell preserves its identity from the time it arises till its death.
(xii) Cell death can be represented by introducing an empty state.

The correspondence between one dimensional cellular arrays with the above assumptions and the strings of symbols used in abstract language theory is very close. Just as the theoretical linguists are concerned with production rules or transformation rules by which certain types of words or sentences can be generated, so are we concerned with finding developmental instructions with which known kinds of organisms can be generated. Again quoting Lindenmayer, '. . . Thus we shall consider the generating constructs of linguists called grammars in relation to our systems. It is very helpful furthermore that an ordering relation exists already with respect to abstract grammars and languages (the Chomsky hierarchy of grammars) which we can utilize to provide us with an ordering of developmental generating systems and languages, giving us a measure of complexity of distinguishing primitive against advanced characters, whereas no such measure has been available before to biologists'.

3. DEVELOPMENTAL LANGUAGES

This section shows how developmental languages arose from formal languages. To clarify notation, several definitions must be given.

Definition An Alphabet V is a finite, non-empty, set of symbols. A word over V is a finite length string of symbols from V. V* is the set of all such words over V including the empty string Λ. A language L over V is any subset of V*.

As V* is infinite, a language L can also be infinite. Thus descriptions of L are not always possible. However, it may be able to describe an algorithm that decides membership or a procedure that generates members of L.

Definition A language L over V is recursive if there exists an effective procedure to decide membership in L. L is recursively enumerable if there exists an effective procedure to generate a list of members of L.

Recursive languages can be seen to be a subset of recursively enumerable languages and defining the generative process is giving a grammer to the language.

Definition A type O grammer G is a quadruple, $G = \langle V_N, V_T, P, S \rangle$ where V_N is a finite set of symbols called variables. V_T is a finite set of productions of the form $\alpha \to \beta$ where α is a word over V_N containing at least one non-terminal symbol and β is a word over V. S is an element of V_N called a start symbol.

Definition Given a grammar $G = \{V_N, V_T, P, S\}$ and strings ρ and σ in V^* we say that ρ directly derives σ in G if and only if the following conditions are satisfied.

(i) ρ can be written as $\gamma\rho\delta$
(ii) σ can be written as $\gamma\sigma\delta$
(iii) $\alpha \to \beta$ is a production in P.

This is denoted by $\rho \Rightarrow_G \sigma$. If $\{\alpha_1, \alpha_2, \alpha_3 \ldots \alpha_n\} \varepsilon V^*$ and $\alpha_1 \Rightarrow \alpha_2$, $\alpha_2 \Rightarrow_G \alpha_3 \ldots$. then the sequence $\alpha_1, \alpha_2, \alpha_3 \ldots \alpha_n$ is called a derivation and by convention $\alpha \Rightarrow^*_G \alpha$ for each string α.

Given a grammer G, the language L(G) generated by G is defined to be

$$L(S) = \{\omega | \omega \, \varepsilon \, V_T^* \text{ and } S \Rightarrow^*_G \omega\}$$

i.e. ω is in L(S) if it consists solely of terminals and can be derived from S.

If for every production $\alpha \to \beta$ in P.

Definition $|\alpha| \leqslant |\beta|$ then G is context sensitive α is a single variable and $|\beta| > 0$, G is context-free every production is of the form $A \to aB$ or $A \to a$ where A, B are variables and a is a terminal then G is said to be regular.

$|\alpha|$ denotes the length of the string α.

One can now form the Chomsky hierarchy of languages
regular languages
context-free languages
context-sensitive language
recursively enumerable languages
each language is a subset of the one below it.

The developmental languages studied by Lindenmayer differed in two significant ways to the formal languages of Chomsky. There was no distinction made between terminal and non-terminal symbols and the production rules were applied simultaneously to all symbols in each array. There were obvious biological justifications for these assumptions. Thus a Lindenmayer system can be defined as a triple $\langle \Sigma, P, \omega \rangle$ where Σ is the alphabet, P is the set of productions and ω is the starting configuration (called an axiom of G). A simple example will illustrate its properties.

$$\Sigma = \{0, 1, 2, \ldots, 9\}, \quad \omega = \{4\}$$

Productions are

$0 \to 10,\ 1 \to 32,\ 2 \to 3(4),\ 3 \to 3,\ 4 \to 56,\ 5 \to 37,$

$6 \to 58,\ 7 \to 3(9),\ 8 \to 50,\ 9 \to 39,\ ) \to\ ),\ (\to\ ($.

The tenth application of this scheme will produce the string

33(3333339) 33(333339) 33(33339) 33(33(9)3750) 33(3758) 33(56)33(4)3210

If the system is applied to a cell that could assume ten possible states during its development and the parentheses denote a branching point then the tenth sequence could be represented as

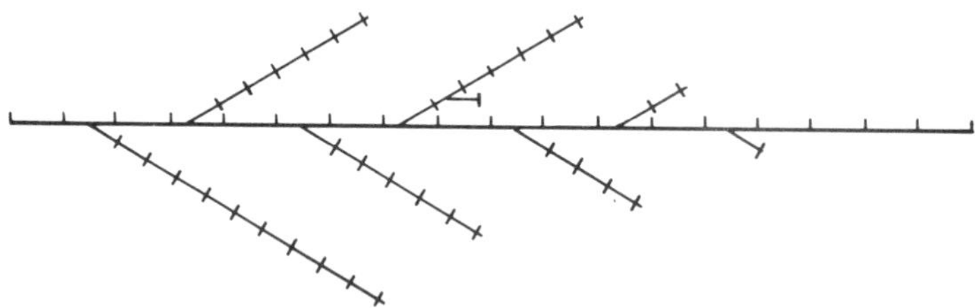

and further applications will model the development of a filamentous plant (i.e. an algae or fungus).

This model assumes no cellular interactions and is denoted as an OL scheme. If no production rules of the form $\alpha \to \Lambda$ exist then the scheme is known as propagating and if exactly one α exists such that $a \to_P \alpha$ for every a then the scheme is known as deterministic. Cellular interactions can take place on either side giving 1L and 2L schemes thus a whole family of developmental schemes can be created.

The previous example was a PDOL scheme and at first sight would seem to be a context-free language but it can be proved that it is in fact a context-sensitive language. Unfortunately formal context-sensitive languages are awkward to handle and do not reflect the biological organisms that are being modelled. The Chomsky theory is useful because it provides a wealth of theorems and proofs that can be adapted to the Lindenmayer systems. All further references will be to the new developmental systems.

4. AN EXAMPLE: THE FRENCH FLAG PROBLEM

This problem was initiated by Wolpert (1968). It is to create a mechanism by which a filament of initially identical cells turns into a French flag (one third red, one third white and one third green) and restores this pattern in spite of large external disturbances e.g. the cutting of the filament into two or more parts. This problem has been tackled by Herman and Liu (1973) and we now give an account of their solution. They use a PD2IL scheme, i.e. a propagating, deterministic, two-way interacting language generated by $G = \langle \Sigma;\ P,\ \langle g,\ 0,$

$0\rangle\rangle$ where Σ is the alphabet which consists of triples $M \times N \times N$ where M {e, i, j, k, r, w, b, p, l, n} and N = {0, 1, 2, 3, 4}. The cells are thus modelled by triples which represent three attributes. The first in the TYPE of cell, the second represents TIME and the third POLARITY. Various assumptions are now made

(a) there is an initial array of identical cells which is disturbed at one end;
(b) the region where the disturbance occurs becomes the red end;
(c) if the array is cut, then the arrangement of the colours will follow the original;
(d) if a middle portion of the filament is removed before the disturbance has reached there, then it would be impossible for cells to know how to form, thus no cuts must be made until the disturbance has reached the cutting point;
(e) when a break occurs the two end points are aware that they are now end cells;
(f) the propagating mechanism is modelled by the transmission of waves along the array. This mechanism is not intended to model any biological or physical processes but was devised as a means to achieving the end.

A brief description of the model can now be given. When one end is disturbed, three waves of different speeds (labelled i, j, k) are sent out. When the i wave meets the opposite end, it is reflected back and its intersection with the j and k waves determines the boundaries for the red, white and blue portions. When the array is cut, a new set of waves is sent out or a p-wave is sent to 'clear up' the array depending on which side the cut is. The attribute TYPE thus consists of the type of wave that is present in the cell at the particular time. If no wave is present, it will be in a neutral state. The speeds of the waves determine the thicknesses of the three colour bands and must be calculated accordingly. Production rules have to be formed for each possible cell combination and will be of the form

$$\langle i, o, m-1\rangle\langle k, n, m\rangle\langle x, y, m+1\rangle = \langle i, o, m\rangle \text{ for } n = 4$$
$$= \langle k, n+1, m\rangle$$
$$\text{for } o \leqslant n \leqslant b$$

where x, y stand for any permissible values of Type and Time.

The problem can then be stated formally as

for any filament $p = \langle a_1, 0, 0\rangle\langle a_2, 0, 0\rangle \ldots\ldots \langle a_x, 0, 0\rangle$
s.t $x \geqslant 3, a_1 = i, a_k = a$ for $2 \leqslant K \leqslant x$, t_0 s.t $t \geqslant t_0$
$\lambda^t(p) = \langle r, o, m_1\rangle\langle r, o, m_2\rangle \ldots\ldots \langle r, o, m_u\rangle\langle w, o, n_1\rangle \ldots$
$\langle w, o, n_v\rangle$
$\langle b, o, 1_1\rangle\langle b, o, 1_2\rangle \ldots \langle b, o, 1w\rangle$

where $|3u - x| \leqslant 2$, $|3v - x| \leqslant 2$ and $|3w - x| \leqslant 2$

Herman and Rosenburg use a special purpose simulation program called CELIA (Cellular, Linear, Iterative, Array Simulator) which is written in FORTRAN and provides a general framework for simulating linear arrays in which individual cells can develop, split into two or more cells or die. We have written a more modest program in BASIC for the ZX SPECTRUM and typical results are shown in Fig. 2. In Fig. 3, the array has been cut in two places and the development of a new flag is shown.

The French flag problem suggested to our non-specialist eyes, the problem of the growth of a worm which has the property that if it loses part of its anatomy, it can regenerate the missing part, i.e. if it loses its tail, it will simply grow tail cells at that particular point. This has also been programmed for the spectrum but we are still investigating the mathematical basis for this model and more work is needed here.

5. APPLICATION TO MODELLING COURSES

As discussed in the introduction, this type of modelling provides a refreshing alternative to the usual modelling of growth. It can also enable biologists and biochemists to model cell development and interaction at a microscopic rather than macroscopic level. Although it has not been mentioned here, there is obviously a close connection to Turing machines and the work of Church (1941). Thus, this could be an interesting application at the end of a computer science course.

Since the inception of the new maths in the 1960s, set theory, functions and boolean algebra have been introduced into all levels of the maths curriculum and a main criticism is that a lot of theory is taught but the practical results are trivial. Here is an application that uses these branches of mathematics extensively, is far from trivial and produces useful applications.

6. CONCLUSION

This chapter does not disseminate any original work on the part of the authors. It is written to bring to the attention of mathematical modellers, an aspect of modelling which is not widely known and yet has tremendous possibilities in the biological or life sciences. In our opinion, the work in this field has just ended its classical period where the forerunners, i.e. Lindenmayer, Herman and others have mapped out the theory and supplied elegant proof. One or two interesting problems have also been attempted. In this chapter, we have tried to give a flavour of this work and any interested modeller could fill the gaps by using the bibliography. But the interesting period is yet to come. We are sure

```
RUN
HOW MANY CELLS IN THE PROCESS?20
DO YOU WISH TO MAKE A CUT??N
ERR 5836 AT 60 : ARRAY INDEX OUT OF BOUNDS.      @ 042043
PRINT N
 21
RUN
HOW MANY CELLS IN THE PROCESS?16
DO YOU WISH TO MAKE A CUT??N
N01 N02 N03 N01 N02 N03 N01 N02 N03 N01 N02 N03 N01 N02 N03 N01
I01 N02 N03 N01 N02 N03 N01 N02 N03 N01 N02 N03 N01 N02 N03 N01
J01 I02 N03 N01 N02 N03 N01 N02 N03 N01 N02 N03 N01 N02 N03 N01
J11 T02 I03 N01 N02 N03 N01 N02 N03 N01 N02 N03 N01 N02 N03 N01
K01 J02 T03 I01 N02 N03 N01 N02 N03 N01 N02 N03 N01 N02 N03 N01
K11 J12 T03 T01 I02 N03 N01 N02 N03 N01 N02 N03 N01 N02 N03 N01
K21 B12 J03 T01 T02 I03 N01 N02 N03 N01 N02 N03 N01 N02 N03 N01
K31 B12 J13 T01 T02 T03 T01 N02 N03 N01 N02 N03 N01 N02 N03 N01
K41 B12 B13 J01 T02 T03 T01 I02 N03 N01 N02 N03 N01 N02 N03 N01
H01 K02 B13 J11 T02 T03 T01 T02 I03 N01 N02 N03 N01 N02 N03 N01
H01 K12 B13 B11 J02 T03 T01 T02 T03 I01 N02 N03 N01 N02 N03 N01
H01 K22 B13 B11 J12 T03 T01 T02 T03 T01 T02 N03 N01 N02 N03 N01
H01 K32 B13 B11 B12 J03 T01 T02 T03 T01 T02 I03 N01 N02 N03 N01
H01 K42 B13 B11 B12 J13 T01 T02 T03 T01 T02 T03 I01 N02 N03 N01
H01 H02 K03 B11 B12 B13 J01 T02 T03 T01 T02 T03 T01 I02 N03 N01
H01 H02 K13 B11 B12 B13 J11 T02 T03 T01 T02 T03 T01 T02 I03 N01
H01 H02 K23 B11 B12 B13 B11 J02 T03 T01 T02 T03 T01 T02 T03 L01
H01 H02 K33 B11 B12 B13 B11 J12 T03 T01 T02 T03 T01 T02 L03 T01
H01 H02 K43 B11 B12 B13 B11 B12 J03 T01 T02 T03 T01 L02 T03 T01
H01 H02 H03 K01 B12 B13 B11 B12 J13 T01 T02 T03 L01 T02 T03 T01
H01 H02 H03 K11 B12 B13 B11 B12 B13 J01 T02 L03 T01 T02 T03 T01
H01 H02 H03 K21 B12 B13 B11 B12 B13 J11 M02 T03 T01 T02 T03 T01
H01 H02 H03 K31 B12 B13 B11 B12 B13 L01 B02 T03 T01 T02 T03 T01
H01 H02 H03 K41 B12 B13 B11 B12 L13 B01 B02 T03 T01 T02 T03 T01
H01 H02 H03 H01 K02 B13 B11 L12 B13 B01 B02 T03 T01 T02 T03 T01
H01 H02 H03 H01 K12 B13 L11 B12 B13 B01 B02 T03 T01 T02 T03 T01
H01 H02 H03 H01 K22 L13 B11 B12 B13 B01 B02 T03 T01 T02 T03 T01
H01 H02 H03 H01 Q22 B13 B11 B12 B13 B01 B02 T03 T01 T02 T03 T01
H01 H02 H03 H01 H22 B13 B11 B12 B13 B01 B02 T03 T01 T02 T03 T01
H01 H02 H03 H01 H22 B13 B11 B12 B13 B01 B02 T03 T01 T02 T03 T01
H01 H02 H03 H01 H22 B13 B11 B12 B13 B01 B02 T03 T01 T02 T03 T01
H01 H02 H03 H01 H22 B13 B11 B12 B13 B01 B02 T03 T01 T02 T03 T01
H01 H02 H03 H01 H22 B13 B11 B12 B13 B01 B02 T03 T01 T02 T03 T01
H01 H02 H03 H01 H22 B13 B11 B12 B13 B01 B02 T03 T01 T02 T03 T01
H01 H02 H03 H01 H22 B13 B11 B12 B13 B01 B02 T03 T01 T02 T03 T01
H01 H02 H03 H01 H22 B13 B11 B12 B13 B01 B02 T03 T01 T02 T03 T01
H01 H02 H03 H01 H22 B13 B11 B12 B13 B01 B02 T03 T01 T02 T03 T01
H01 H02 H03 H01 H22 B13 B11 B12 B13 B01 B02 T03 T01 T02 T03 T01
H01 H02 H03 H01 H22 B13 B11 B12 B13 B01 B02 T03 T01 T02 T03 T01
H01 H02 H03 H01 H22 B13 B11 B12 B13 B01 B02 T03 T01 T02 T03 T01
H01 H02 H03 H01 H22 B13 B11 B12 B13 B01 B02 T03 T01 T02 T03 T01
CONTINUE??N
```

Fig. 2.

that the combination of the pure maths topics, the mathematical logic, the formal language theory and the development of simulation languages such as CELIA will result in a surge of interest and the solution of many problems in the near future.

```
RUN
HOW MANY CELLS IN THE PROCESS?16
DO YOU WISH TO MAKE A CUT??Y
INPUT THE POSITION OF THE CUT:_
                FIRST CELL??5
                 LAST CELL??16
AT WHICH TIME STEP DO YOU WISH TO CUT??7
N01 N02 N03 N01 N02 N03 N01 N02 N03 N01 N02 N03 N01 N02 N03 N01
I01 N02 N03 N01 N02 N03 N01 N02 N03 N01 N02 N03 N01 N02 N03 N01
J01 I02 N03 N01 N02 N03 N01 N02 N03 N01 N02 N03 N01 N02 N03 N01
J11 T02 I03 N01 N02 N03 N01 N02 N03 N01 N02 N03 N01 N02 N03 N01
K01 J02 T03 I01 N02 N03 N01 N02 N03 N01 N02 N03 N01 N02 N03 N01
K11 J12 T03 T01 I02 N03 N01 N02 N03 N01 N02 N03 N01 N02 N03 N01
K21 B12 J03 T01 T02 I03 N01 N02 N03 N01 N02 N03 N01 N02 N03 N01
                I02 I03 N01 N02 N03 N01 N02 N03 N01 N02 N03 I01
                J02 T03 I01 N02 N03 N01 N02 N03 N01 N02 N03 P01
                J12 T03 T01 I02 N03 N01 N02 N03 N01 N02 P03 P01
                K02 J03 T01 T02 I03 N01 N02 N03 N01 P02 P03 P01
                K12 J13 T01 T02 T03 I01 N02 N03 P01 P02 P03 P01
                K22 B13 J01 T02 T03 T01 I02 P03 P01 P02 P03 P01
                K32 B13 J11 T02 T03 T01 T02 I03 P01 P02 P03 P01
                K42 B13 B11 J02 T03 T01 T02 T03 I01 P02 P03 P01
                H02 K03 B11 J12 T03 T01 T02 T03 T01 I02 P03 P01
                H02 K13 B11 B12 J03 T01 T02 T03 T01 T02 I03 P01
                H02 K23 B11 B12 J13 T01 T02 T03 T01 T02 T03 L01
                H02 K33 B11 B12 B13 J01 T02 T03 T01 T02 L03 T01
                H02 K43 B11 B12 B13 J11 T02 T03 T01 L02 T03 T01
                H02 H03 K01 B12 B13 B11 J02 T03 L01 T02 T03 T01
                H02 H03 K11 B12 B13 B11 J12 M03 T01 T02 T03 T01
                H02 H03 K21 B12 B13 B11 L02 B03 T01 T02 T03 T01
                H02 H03 K31 B12 B13 L11 B02 B03 T01 T02 T03 T01
                H02 H03 K41 B12 L13 B11 B02 B03 T01 T02 T03 T01
                H02 H03 H01 Q02 B13 B11 B02 B03 T01 T02 T03 T01
                H02 H03 H01 H02 B13 B11 B02 B03 T01 T02 T03 T01
                H02 H03 H01 H02 B13 B11 B02 B03 T01 T02 T03 T01
                H02 H03 H01 H02 B13 B11 B02 B03 T01 T02 T03 T01
                H02 H03 H01 H02 B13 B11 B02 B03 T01 T02 T03 T01
                H02 H03 H01 H02 B13 B11 B02 B03 T01 T02 T03 T01
                H02 H03 H01 H02 B13 B11 B02 B03 T01 T02 T03 T01
                H02 H03 H01 H02 B13 B11 B02 B03 T01 T02 T03 T01
                H02 H03 H01 H02 B13 B11 B02 B03 T01 T02 T03 T01
                H02 H03 H01 H02 B13 B11 B02 B03 T01 T02 T03 T01
                H02 H03 H01 H02 B13 B11 B02 B03 T01 T02 T03 T01
                H02 H03 H01 H02 B13 B11 B02 B03 T01 T02 T03 T01
                H02 H03 H01 H02 B13 B11 B02 B03 T01 T02 T03 T01
                H02 H03 H01 H02 B13 B11 B02 B03 T01 T02 T03 T01
                H02 H03 H01 H02 B13 B11 B02 B03 T01 T02 T03 T01
                H02 H03 H01 H02 B13 B11 B02 B03 T01 T02 T03 T01
                H02 H03 H01 H02 B13 B11 B02 B03 T01 T02 T03 T01
CONTINUE??N
END   @   3000
```

Fig. 3.

REFERENCES

Chomsky, N., 1956, Three models for the description of Language, *IREE Trans. Inform. Theory*, **IT2**, 113–124.

Chomsky, N., 1959, On certain formal properties of grammars, *Information and Control*, **2,** 137–167.

Church, A., 1941, The calculi of lambda-conversion. *Annals of Mathematical Studies*, 6, Princeton.

Erickson, O. R. *et al*., 1976, Growth in two dimensions, descriptive and theoretical studies. Published in *Automata. Languages, Development.* Lindenmayer. A. and Rosenberg (eds). North-Holland.

Herman, G. T. and Liu, W. H., 1973, The Daughter of Celia, The French Flag, and the Firing Squad. *Simulation*, Aug., 1973, 33–41.

Herman and Rosenburg, 1975, *Developmental Systems and Languages.* North-Holland.

Lindenmayer, A., 1968a, Mathematical models for cellular interactions in development Part I, *J. Theoret. Biol.*, **18,** 280–299.

Lindenmayer, A., 1968b, Mathematical models for cellular interactions in development Part II, *J. Theoret. Biol.*, **18**, 300–315.

Lindenmayer and Rosenberg, 1976, *Automata, Languages, Development.* North-Holland.

Maynard-Smith, J., 1968, *Mathematical Ideas in Biology*, Cambridge University Press.

McCullough, W. & Pitts, W., 1945, A logical calculus of the ideas immenent in nervous activity. *Bull. Mathematical Biophysics*, **5**, 115–133.

Minsky, M. L., 1972, *Computation Finite and Infinite Machines.* Prentice-Hall.

Moscardini, A. O., 1981, *Modelling Plant Growth, Case Studies in Mathematical Modelling.* Stanley Thornes.

Post, E., 1936, Finite combinatory Processes Formulation I. *J. Symbolic Logic*, **1**, 103–105.

Thue, A., 1914, Probleme uber Veranderungen von Zeichenreihen nach gegebenen Regeln. *Skr. Vidensk. Selsk.*, **10**, 34.

Turing, A. M., Computing Machinery and Intelligence. *Mind*, **59** (n.s. 236), 433–460.

Von Neumann, J., 1956, Probabilistic logics and the synthesis of reliable organisms from unreliable components, *Automata Studies*, Princeton, 43–98.

Wolpert, L., 1968, The French Flag Problem. A Contribution to the Discussion on Pattern Development and Regulation, in *Towards a Theoretical Biology*, Vol. 2, E.U.P.

Simulation—Its Role in a Modelling Course

I. D. Huntley
Sheffield City Polytechnic

1. INTRODUCTION

Simulation seems to me to be one of the big forgotten areas in maths teaching. Not so long ago Operational Research, of which simulation is just one part, was firmly seen as a postgraduate activity—you first did your maths or your physics degree, and then (if you were bright enough) you went on to do an MSc in OR.

This is slowly changing, and nowadays all computing degrees (and a lot of other ones as well) contain an element of OR. However, COPRA is now pushing for an A-level option in Decision Maths (the same as OR to most of us), and the Spode Group is just starting to write a resource book of OR-type examples for middle school use. Things have come a long way in 10 years!

All that background perhaps explains, to an extent, why simulation is not included in current maths syllabi. With the recent (and, I hope, unfinished) growth in computing power available in secondary and tertiary education, however, the topic deserves another look.

2. OLD ONES ARE THE BEST ONES

Lecturers who have taught the ideas of Monte Carlo simulation for several years via the 'estimate pi using the circle and square method' are probably convinced that it is so boring that no students could possibly enjoy it! However, I have found that if students are allowed the facilities to discover this idea for themselves, the example can generate real enthusiasm and lively discussion.

The problem, then, is merely to estimate a value for pi. The class is sounded out, and all sorts of ideas put forward and written up on the board. Typical responses are:

use your calculator,
look it up in tables,
there was a magazine at school with it in,
22 over 7,
tie a bit of string round a cylinder,
pour a known weight of water into a measuring cylinder,
note the displacement when a sphere is placed in a bucket of water.

These are all good discussion points, and usually the student group will weed out the silly suggestions for you. Last time I ran this exercise, though, the previous week's television had been showing nothing but darts, and all I had to say was the magic word 'Jollees' for someone in the class to suggest the standard 'circle and square' method.

For those of you who have not met this before, you consider a circle within a square (Fig. 1).

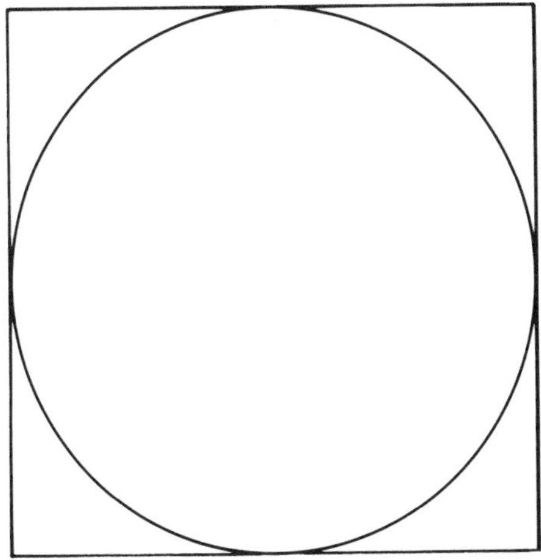

Fig. 1.

Clearly

$$\frac{4^*\ (\text{area circle})}{(\text{area square})}$$

gives a route by which pi may be calculated. The tie-up with darts comes when you consider a blind darts player throwing darts at the board at random—the number of darts inside the circle divided by the number in the square will give an estimate of pi/4! The students had made this quite awkward mental leap almost entirely by themselves—and had generated a method to solve the original problem to any required level of accuracy (given sufficient darts).

Of course, it was quickly realized that a computer would do the job far more easily, and I was soon asked if 'the machine' could generate random numbers. This led on to a couple of sessions where we discussed pseudorandom numbers, ERNIE, convergence, errors—all very respectable mathematical topics. In fact, we finished up by using Minitab (an exciting new stats package that the students were very eager to play with) to confirm that the random number generator on the mainframe was really very poor, and that for any serious work you probably ought to write your own. Not bad as a progression from the well-tried pi example—and with students who would have told you that they hated computing and anything to do with it!

3. THE RURITANIAN SPORTS AND SOCIAL CLUB

Having spent some time on the pi example, I then gave the students a simple example during which I showed them how to generate a random number from any given distribution (Fig. 2). This was quite valuable, since afterwards I found that they had already met the idea in statistics—I gave a non-specialist's treatment of a statistical technique (which helped the students), and learnt the 'proper way' to do it myself!

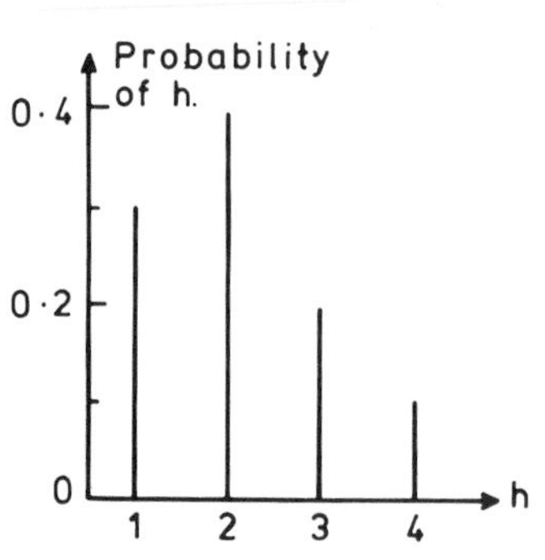

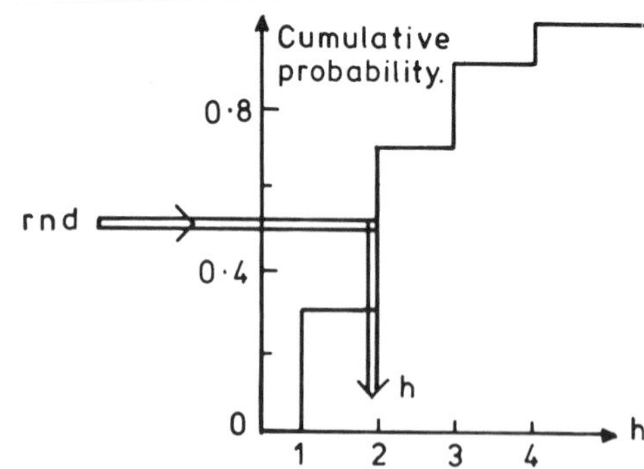

Fig. 2.

We were now in a position to look at the case study given in the Appendix. This is fully written up in Huntley & James (to appear) but let me show you the variety of solution methods which appeared from a class of first-year undergraduates. Incidentally, the example has also been used in maths classes and geography classes in secondary schools—we just changed 'Ruritania' to 'Glasgow Rangers'!

3.1 Method 1

The most obvious way of solving this problem—and the way that 3rd form children discovered for themselves—is merely to enumerate all the possibilities. Given a certain rainfall on Gala day we can work out from the question the likely profit. We then do the same on the Barbecue day, add the two profits together, and know the annual profit. In fact, the information given is very coarse, and only 30 outcomes are obvious—one of the 5 Gala types and 6 Barbecue types of rainfall. This leads to the following diagram (Figure 3 and Table 1) which is better presented as the graph in Figure 4.

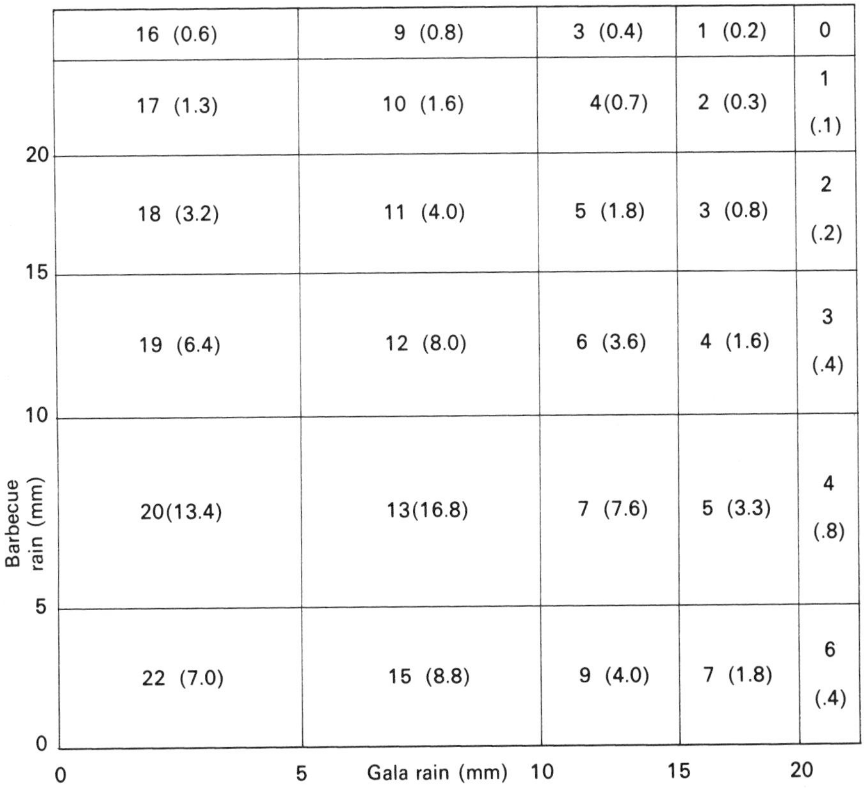

Barbecue rain (mm)	0–5	5–10	10–15	15–20	20
	16 (0.6)	9 (0.8)	3 (0.4)	1 (0.2)	0
	17 (1.3)	10 (1.6)	4(0.7)	2 (0.3)	1 (.1)
20 – 15	18 (3.2)	11 (4.0)	5 (1.8)	3 (0.8)	2 (.2)
15 – 10	19 (6.4)	12 (8.0)	6 (3.6)	4 (1.6)	3 (.4)
10 – 5	20(13.4)	13(16.8)	7 (7.6)	5 (3.3)	4 (.8)
5 – 0	22 (7.0)	15 (8.8)	9 (4.0)	7 (1.8)	6 (.4)

Gala rain (mm): 0, 5, 10, 15, 20

Fig. 3.

3.2 Method 2

Nearly all the undergraduates who came up with the above solution were clear that it wasn't very satisfactory, and wanted to use the Monte Carlo ideas to generate a better solution. The programs they came up with weren't very elegant—but they hadn't taken an age to develop, and they worked!

Table 1

Profit (k$)	% of times	Cumulative %
22	7.0	7.0
20	13.4	20.4
19	6.4	26.8
18	3.2	30.0
17	1.3	31.3
16	0.6	31.9
15	8.8	40.7
13	16.8	57.5
12	8.0	65.5
11	4.0	69.5
10	1.6	71.1
9	4.8	75.9
7	9.4	85.3
6	4.0	89.3
5	5.1	94.4
4	3.1	97.5
3	1.6	99.1
2	0.5	99.6
1	0.3	99.9
0	0.1	100

Here's an unedited example of the sort of thing I got in—written for an Apple.

```
5   DIM L(23)
10  FOR J=0 TO 23
12  L(J)=0
14  NEXT J
20  SEED=RND(-1)
30  INPUT "NUMBER OF REPLICATIONS?";NMAX
40  FOR N=1 TO NMAX
50  G=RND(1)
60  IF G < 0.32 THEN GG=16000:GOTO 110
70  IF G < 0.72 THEN GG=8000:GOTO 110
80  IF G < 0.90 THEN GG=3000:GOTO 110
90  IF G < 0.98 THEN GG=1000:GOTO 110
100 GG=0
110 B=RND(1)
120 IF B < 0.22 THEN BB=6500:GOTO 180
130 IF B < 0.64 THEN BB=4000:GOTO 180
140 IF B < 0.84 THEN BB=2500:GOTO 180
150 IF B < 0.94 THEN BB=1500:GOTO 180
160 IF B < 0.98 THEN BB=500:GOTO 180
170 BB=0
180 REC=GG+BB
190 REC=REC/1000
200 FOR J=0 TO 23
210 IF J < REC THEN L(J)=L(J)+1
220 NEXT J
230 NEXT N
240 FOR J=0 TO 23
250 L(J)=L(J)/NMAX
260 PRINT "PROBABILITY OF NETTING GREATER THAN $";J*1000;" IS ";L(J)
270 NEXT J
280 END
```

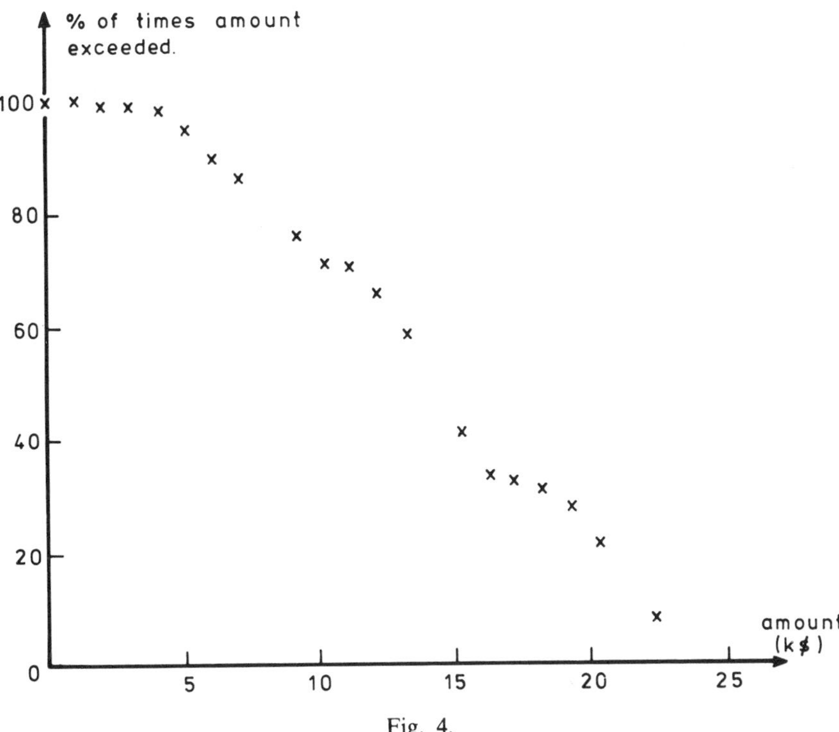

Fig. 4.

You'll see that the output here is really not very pretty at all—line after line of the same text—but the program certainly gave enough information for the student to answer the question posed. This seems to me a vital part of any modelling exercise.

In fact, only a couple of extra lines of coding are needed to produce a very nice graph. Not surprisingly, this is the same one as we had earlier in Fig. 4.

3.3 Method 3

My first-year undergraduates happened to be doing a parallel course in statistics and, much to my amazement, were quite willing to try out some of the ideas of pdf's and cdf's in this case. It's a really awful example to try this on, since the original data set was so crude, but they fiddled around with a simulation involving negative exponential distributions—the easy ones—and learnt quite a lot in the process. The point, though, is that the students were determining their study patterns for themselves, and so their motivation was considerably greater than it would have been otherwise.

4. CONCLUSION

While this chapter has been all about simulation—and I really am convinced that this subject is a very nice vehicle for the teaching of mathematics—it is also discussing a far more general point.

I hope that the examples I have shown have convinced you of the worth of active, student-driven learning. Students probably don't learn quite as many 'facts', but the increased motivation and interest generated easily compensates for any such lack.

REFERENCE

Huntley I. D., and James D. J. G. (Editors) More case studies in Mathematical Modelling (to appear).

APPENDIX: THE RURITANIAN SPORTS AND SOCIAL CLUB

COMMENT TO LECTURER

The case study which follows is strongly based on material from *Computer Models in the Social Sciences* by R. B. Coats and A. Parkin (Arnold, 1977). The study has been extended, however, to show how this sort of example can provide a useful modelling exercise at many different levels of maturity. As written, the example is probably best suited to numerate undergraduates, but with only minor changes (such as changing 'Ruritanian' to 'Glasgow Rangers'!) it has already been tried out with mathematics students in the third year and modern studies/geography students in the fourth year at secondary school. In fact, the ability to use the same modelling exercise across the ability spectrum seems to be one of the features of a good example: the approaches used and answers received will vary enormously, but—no matter what his ability—each student will be able to display his full potential.

BACKGROUND

The Ruritanian Sports and Social Club Committee was discussing its expenditure for the coming year. The debate was particularly pointed because, in the season just ended, two annual fundraising events (the Gala Day and the Barbecue Dance) had not proved very profitable.

'Look,' said the Membership Secretary, 'we can't possibly afford another year like the last—the members won't stand for it. We started the year with \$25 000; we authorized expenditure on new buildings and equipment to the tune of \$20 000; yet the Gala and Barbecue were such a washout that we only netted \$5 000, leaving us with a credit balance of a mere \$10 000. We must make quite sure we don't overspend this coming year, so that we can build up our reserves again.'

The Treasurer looked doubtful. 'What do you mean by *quite* sure? If you mean that there must be absolutely no risk of a loss, then there is only one course of action open to us—we should not spend anything on new facilities next year. Indeed, to be strictly accurate, we should not even authorize the preliminary expenditure on the Gala and Barbecue—after all, it's on the cards that we could make a loss on both events. Look what happened this year at the Gala.'

'That's a bit thick, isn't it?', interposed the Events Organizer. 'The only reason that we did so badly this year is that it poured with rain on both days. The chance of that happening two years running is so remote that we can ignore it.'

The Treasurer thought to himself, 'There's something wrong there. I don't see how this year's weather can affect next year's. The chances of bad weather next year must be the same, irrespective of what weather we had this year.' He opened his mouth to make this point but hesitated, wondering if the Events Organizer would understand him. Then he had a vision of an ensuing argument, a considerable detour from the matter in hand, and an even more protracted meeting. He shut his mouth again.

The Membership Secretary ended the hiatus with a bland inconsistency. 'We can't have *no* expenditure next year—there would be uproar from the members if we failed to provide any new equipment at all. Then there's the repairs to the pavilion roof, a new filtration system for the swimming pool, ...' He went on with a long list which made it evident that there were ample outstanding projects to soak up whatever funds were made available.

'Then what you must decide', declared the Treasurer, 'is this: exactly what risk of loss is the club prepared to accept?'

There was silence for a moment, as nobody knew how to answer this question. 'I should have thought', said the Chairman tentatively, 'that we want to be about 90% sure that we will more than break even.'

'And accept a 10% risk of loss?', queried the Events Organizer doubtfully. 'I suppose we could afford a small overdraft, but I don't think we want more than, say, a 1% risk of one of more than about $5000.'

'And I don't think we should accept any risk of a deficit as large as $20 000', added the Treasurer, 'since we would never find large enough overdraft facilities to cover it and that would mean we would have to fold up completely. Now let me get this straight: the feeling of the Committee is that the planned expenditure for next year should aim to give us 90% chance of balancing the books. We are prepared to accept a small risk, 1%, of a deficit of $5000 and no risk of a deficit greater than $20 000?' There were murmurs of assent. 'Of course', continued the Treasurer, 'the last constraint means we cannot contemplate more than $30 000 expenditure, for if we are avoiding any risk at all of a greater loss we must consider the worst case—with *no* receipts.'

The Membership Secretary looked glum, but no one was prepared to counter this argument. The Committee members fidgeted, many eyes on the clock. 'Give me until next meeting to think about it, and I shall try to find the figure which meets these criteria', concluded the Treasurer.

The Chairman closed the meeting.

Considering the problem in the comfort of his study, the Treasurer was beginning to realize that his problem was not an easy one. He was not worried much by what was meant by 'a 90% chance' of being all right. His concept was that there was a certain level of expenditure which (if the same policy were applied each year and other circumstances did not change significantly) would tend, in the long run, to produce receipts exceeding payments in 90% of the years. It did strike him that a very large number of years—perhaps more than his remaining years as Treasurer—might be needed before anyone could say whether his recommended figure had been right or wrong, but he found this more of an encouragement to continue than a philosophical obstacle.

No, his real problem was that the club had been running only five years and he did not have much past data to work on. He was fully aware that Ruritania enjoyed an extraordinarily stable economy, currency and population, and that Ruritanians were famous for their consistent habits. 'The only thing', he mused to himself, 'that can be effecting receipts from our annual events is the weather.' To test this hypothesis, he rang the Weather Bureau and asked for the rainfall figures during the time of the Gala (10 am to 5 pm) and the Barbecue (8 pm to 1 am) on the relevant dates in previous years (Table A.1).

Table A.1

Year	Gala		Barbecue	
	Profit ($)	rain (mm)	Profit ($)	rain (mm)
1	20 000	1	5 000	5
2	3 000	12	7 000	1
3	28 000	0	1 000	21
4	17 000	2	8 000	0
5	1 000	17	4 000	8

The Treasurer found the consistent correlation between rainfall and receipts an encouragement to his hypothesis. He noticed that the Gala seemed to be effected by poor weather more than the Barbecue and found this encouraging, for he knew that the Gala was entirely open air while the Barbecue was partly covered, and so he would not expect the Barbecue to suffer as badly when it rained. He decided to make the pragmatic assumption that the relationships were reliable and described what could be expected in the coming year.

It was now becoming clear that if he could predict the rainfall on the days of the Gala and Barbecue in the coming year, he would be able to make a prediction of the total receipts. The Gala was scheduled for 1 July and the Barbecue for 27 August, so he again contacted the Weather Bureau and asked them for the rainfall figures for the relevant periods. The Bureau's figures went back 50 years, and he was able to construct the histogram below from the data they supplied.

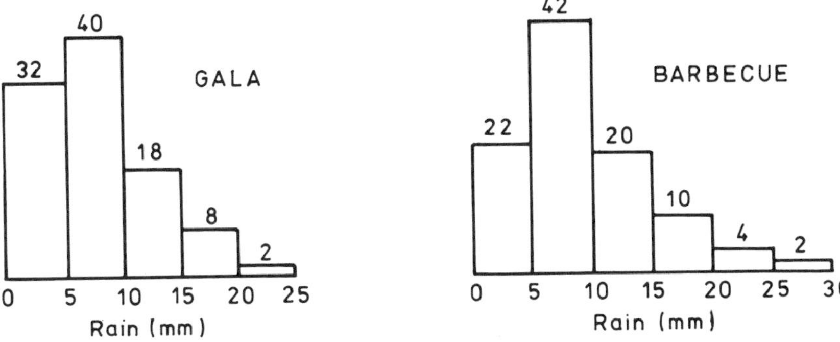

Fig. A.1 Percentage occurrence of rain

Armed with all this information, the Treasurer felt sure that he could work towards an answer to his problem. He thought that *at least* he would be able to make statements like 'There's only a 5% chance of a deficit if we authorize capital expenditure of $15 000 next year', which would go some way to meeting the Committee's requirements. What would you do in his place?

The Use of The Simulation and Case Study Techniques in the Teaching of Mathematical Modelling

R. R. Clements
Bristol University

SUMMARY

Traditional university mathematics courses concentrate on teaching mathematical skills and techniques. The modes of assessment used reinforce the tendency of students to develop a restricted range of mathematical skills chiefly oriented towards efficiently solving examination questions.

In contrast to this, the demands made on mathematicians working in industrial and commercial environments are much wider. The industrial mathematician must be able to recognize the underlying mathematical structure in a problem presented in real world terms or to develop a structure to describe the problem. He must be able to balance the mathematical and non-mathematical constraints placed upon him to arrive at a solution that is somewhere near optimal in the environment pertaining. He must, above all, be skilled at interpreting the results of his work and reporting them in an appropriate form that is acceptable and understandable to non-mathematical members of the organization.

The recognition of this dichotomy of requirement has led to initiatives in the mathematical education community including the development of mathematical modelling teaching. The simulations described in this chapter represent an associated but slightly divergent line of development. The simulations are based on problems that have arisen and been solved in industrial and commercial organizations. They are presented to students in a realistic way through a collection of printed source documents. The students work in modes that they might adopt in the industrial environment simulated, followed by 'post mortem' discussion.

The design, implementation, teaching and assessment of the course of simulations is described and its relationship to other forms of mathematical modelling teaching discussed.

1. THE NEED FOR EDUCATION IN THE USE OF MATHEMATICS IN AN INDUSTRIAL AND COMMERCIAL CONTEXT

There has been an increasing recognition, reflected in the literature on both sides of the Atlantic over the past fifteen or twenty years, that traditionally educated mathematics graduates do not possess all of the skills needed to make them effective practitioners of mathematics in industry and commerce. As a result of certain shortcomings in their education there is a period after entering employment in which they need to develop, in a more or less ad hoc way, these additional skills, and this requirement reduces their short-term and possibly also their long-term utility to their employers.

Klamkin (1971) discussed the skills used by mathematicians in industry. He pointed out that these are not restricted to the solution of well formulated problems in mathematical terms, but include the recognition of the mathematical properties of the original problem (almost certainly not a mathematical problem but a physical or organizational one), the formulation of something more precise to work on and then, subsequently, the explanation to others, less well versed than himself in mathematics, of the implications of the mathematical analysis for the original problem.

Handelman (1975) reported a discussion arranged by SIAM on the 'Mathematical Education of the Non-academic Mathematician'. The panellists in this discussion also identified the skills described by Klamkin as important for the non-academic mathematician and commented on their frequent absence in the university mathematics graduate. Many of the responses elicited by Gaskell and Klamkin (1974) in a survey of the views of employers on the needs of industrial mathematics further illustrated and reinforced these views.

In a comprehensive survey involving British graduate mathematicians McLone (1973) also identified these skills as being important for the effective use of mathematics and poorly developed by the University courses taken by his respondents. There is no reason to believe that his sample was untypical in this respect. Cornfield (1977) further reinforces these points in an article which includes a discussion of the non-mathematical constraints which affect the work of the industrial or commercial mathematician.

The need for a training in the use of mathematics thus identified has given rise to a number of initiatives to meet this need. The main weight of these has been in the area of mathematical modelling. The initiatives in this area have a foundation also in a philosophy of applied mathematics. To this extent initiatives in the teaching of mathematical modelling are doubly attractive to mathematical educators—they both fulfil a perceived educational need and represent a development in our understanding of the essence of applied mathematics. The philosophical basis of modelling is established in, for instance, Ford and Hall (1970). The history of the mathematical modelling movement and the development of the methodology of mathematical modelling has been reviewed recently by Clements (1982b).

In this paper we describe the development of a somewhat different but related educational exercise, a course of simulations with associated case study material in the use of mathematics in industrial and commercial contexts. This development is also associated with the training needs identified above.

2. THE EDUCATIONAL OBJECTIVES ADOPTED

Clements and Clements (1978) describes the background to the development of this course. The contrast is drawn between the demands that the solution of conventional University degree examination problems makes on students and the much wider range of mathematical and non-mathematical activities undertaken in solving problems arising in the world of industry. Students are led by their University experience to believe that mathematics is a subject in which well posed problems, with all their associated data (initial values, boundary values and the such like) given, are solved by the application of some given technique to yield a single 'correct' answer.

In contrast to this the industrial or commercial mathematician is usually presented with a problem expressed in the terms of the problem's owner. His task is firstly to discern the underlying mathematical structure of the problem or to create a usable and valid mathematical structure which represents (or 'models') the dominant features of the problem. Even here his mode of operation differs from the academic one. Commonly he must operate within institutional, non-mathematical constraints such as shortage of time, limitation of financial or manpower resource, use of particular types of computer equipment, all of which may result in his adopting a solution mode significantly different from that which would be chosen by his academic counterpart who, faced with the same problem, is governed by his own distinctive constraints and values such as elegance, compactness, and preference for the analytic over the numerical solution. The industrial mathematician must then interpret his solution into its implications for the original real problem, evaluate these implications and recommend actions based on a suitable explanation of his reasoning intuitively acceptable to his partners in the problem solving process who do not share his mathematical skills.

Another feature of this problem solving process which contrasts it strongly with usual degree course activities is its open ended nature. Typically there is a spectrum of possible models and methods of solution from which the modeller must choose any one which will give him an acceptable answer within the constraints pertaining. For his own satisfaction he will probably attempt to choose that one which does this most easily or efficiently but, so long as the constraints are met, this is not essential to the organization. Indeed effort expended on identification of the optimal mode of solution may itself violate an institutional constraint.

In the light of these perceptions, which are described more fully in Clements (1978), the following objectives were formulated:

1. To give students practice in synthesising mathematical models from engineers, and other non-mathematicians' descriptions of physical and industrial systems.
2. To give students practice in interpreting the results of mathematical models in physical terms and critically evaluating the implications.
3. To give students practice in evaluating the effects of the various sections of, and inputs to, models, and making appropriate simplications and approximations to aid efficient solution.
4. To alert students to the non-mathematical constraints under which mathematicians operate in the industrial and commercial environment.
5. To give students practice in critically examining the various possible approaches to, and models of, a system, and choosing an optimal or near optimal method of analysis within the constraints of the system.
6. To develop students' confidence in themselves as model builders and problem solvers.

A course was then designed to meet these objectives. A caveat which might be inserted here is this—whilst these objectives are important they are additional to more traditional objectives associated with the imparting of knowledge and technique so no University course can attempt to do more than begin to develop the skills described therein. The continued development of such skills will occur after entry to employment. It is our objective to facilitate that continued development by giving to students an awareness of the importance of these skills and a brief initial training in them.

3. THE DESIGN OF A COURSE TO MEET THESE OBJECTIVES

Simulation, the creation of an analogue of a real world situation, is increasingly used in many areas of education and training. Aircraft simulators for pilot training, business games for managers, and military manoeuvres are all examples of simulations. Simulations contribute to training in two basic modes. Firstly there is the adaptive learning that occurs during the progress of a simulation whereby the pupil improves his performance of the simulated task and secondly the post mortem learning that occurs when pupil and teacher together review, sometimes with the aid of visual, audio or other forms of record, the performance of the simulated task.

It was felt that simulation would be an appropriate mode of teaching the wide ranging skills involved in the creative use of mathematics in an industrial environment. Problems would be presented to students as a collection of source documents—correspondence, design drawings and data, technical reports, memoranda—such as an industrial mathematician might need to work from. These would set up an industrial context and outline a problem facing the organization. Data might be provided and constraints implicitly or explicitly given. The students would be required to approach the solution of the problem as they would in that industrial context. During the progress of the simulation the tutor adopts the role of a departmental head or section leader requiring a report within some time scale but, in the interim, offering advice and comments when asked and reviewing progress regularly. Subsequently, after the presentation of some form of final report, the tutor and students could review the progress of the simulation.

The problems to be solved could have been generated internally, within the University Department, but it was realized that, if sufficient suitable problems could be gleaned from industrial sources, not only would the realism of the exercise be enhanced but also the post mortem learning could be enriched by the availability for comparison of the solution adopted in the originating organisation. This course was therefore adopted. The identification of organizations likely to be able to assist with the creation of the simulations, the collection of material and its organisation into the final version of the teaching materials to be used is described fully in Clements and Clements (1978) and Clements (1978).

At the same time it was realized that a further aspect of industrial work could be illustrated in this course of simulations. In contrast to most undergraduate activity, much industrial work is corporate in nature. The mathematician must function smoothly as one part of a team tackling a problem. This calls for new skills again. The course was designed in such a way that students worked in groups of three on each case study making corporate reports at the conclusion. Although no formal instruction in group organisation and dynamics was given, the tutor has the function also of monitoring, as best he is able, the development of cooperative skills in each group and encouraging their development.

These two aspects of the simulations developed were in the nature of bonuses so far as the original objectives of the course were concerned. Two further objectives were added to the list given above to reflect the additional properties.

1. To give students practice in cooperative (as opposed to individual) working, and improve their group working techniques.
2. To demonstrate the range of problems that are modelled mathematically, and illustrate the scope of the industrial mathematicians' task.

4. STUDENT REACTION TO THE COURSE

The course of simulations was first used with students in the 1978/79 academic year and has been used each year subsequently. The course is integrated into the second year of a degree course in Engineering Mathematics. At the beginning of the course the format and purpose of the simulations is explained to the students in a short lecture, and the source material for the first study is distributed. Subsequently the students are asked to report progress to the tutor, role playing the supervising officer, at roughly weekly intervals. The students may seek advice at other times as well, just as they might in the industrial context, and the experience is that students who make use of this possibility progress most rapidly. Each simulation lasts from one to three weeks depending partly on the complexity of the problem and partly on their progress.

Observation of the students during the course and discussion of the course with students at the end of the year has provided some feedback on the student reactions to this course. As anticipated, the initial reaction tends to be one of bewilderment. The exercise is very different from any prior experience and, in the absence of firm expectations about the performance required from them, confusion is inevitable. However, if the tutor can draw from the students ideas and help them to structure and choose lines of approach to the problems, they quickly gain in confidence. One of the most gratifying aspects of teaching this course is to observe the rapidly increasing confidence which the students show in their own creative powers in mathematics. Once they have begun to succeed in answering some relatively unstructured problems successive problems hold little fear for them. The simulations also help to develop a wide variety of subsidiary skills in the area of mathematical modelling of course. The development of mathematical discretion, the ability to discriminate between important variables and effects and marginal or irrelevant ones, is brought out by many of the studies. In addition to these there are other non-mathematical skills, for instance report writing, which this course requires. It is possible to require a whole range of final report forms, written and oral during the course. The opportunity to ask for written reports for a variety of purposes (for example a briefing document for a section head, a technical memorandum to another department within the organization, a consultant's report to a client) enables practice to be given in choosing appropriate report structures and styles as well as in technical writing. These again are skills which are little practised in traditional mathematics degree courses and where guidance can be offered to students in the post mortem sessions on the simulations.

Finally it should be reported that students, after the course is ended, confirm their initial confusion but almost invariably claim to have found the course stimulating and, once the general format has been mastered, very enjoyable. They obtain great satisfaction from the feeling that they are tackling and succeeding in solving real problems not just academic exercises. This satisfaction, of course, provides further justification for the decision to seek real problems from industry and commerce rather than using problems generated within the academic institution.

5. THE ASSESSMENT OF STUDENT PERFORMANCE

It was a tenet of the original design and philosophy of this course that, if its objectives were worth attention and the mode of teaching was accepted as likely to achieve those objectives, any difficulty in designing an assessment procedure to accompany it should not be allowed to stand in the way of its implementation. In practice it was found that the natural enthusiasm of the students for this course obviated the need for the threat of assessment as a motivating device.

In the event an assessment procedure has been developed during the first three years of use of the course. In its initial experimental phase a pass/fail grade was given based on the subjective judgement of staff supervising the course. Subsequently a more complex procedure has been developed. The tutor in charge of each group makes, after each meeting, brief notes on progress in the simulation and grades the performance of the group as a whole on a nine-point scale. This grading takes into account any written work that he has asked for as well as the oral presentation made by the group. In addition the performance of each student relative to the group may be noted. Above average and below average grades are given respectively for obvious leadership roles and perceptive contributions and for evidence of lack of participation and lack of understanding of the proceedings. At the end of the course each group is awarded a composite grade and the above or below average marks of individuals summed to give an indication of individual placement relative to group average.

Groups who evidently function well as groups tend to get few above or below average individual marks as all members seem to make contributions appropriate to individual skills. Their group average grade is of course higher as a result. Groups who are more fragmented in their operation tend to get rather more individual grades. The satisfactory operation of the system is still dependent on the subjective judgement of the staff operating it but this judgement is operated at intervals over a period of one term or more and careful notes and records made. The worst faults possibly evident in uncontrolled subjective marking are thereby controlled. The scheme has, in all events, been found operable and acceptable to the staff of the department as a whole.

6. THE RELATIONSHIP OF THIS COURSE TO MATHEMATICAL MODELLING COURSES

The simulation technique obviously involves an element of mathematical modelling. Modelling is an important activity in the successful pursuit of an adequate resolution of the problems facing the students. The difference between this course and one on mathematical modelling pure and simple lies in emphasis and scope. Many modelling courses adopt a sequential development of modelling skills approach with elementary exercises building up to more comprehensive studies later in the course. Modelling courses also vary in the emphasis placed on reporting and the role assigned to staff in the course. In this course the staff role is very well defined and its adoption helps staff to determine exactly what their input to the model building process should be. The format of the exercises also demands that students place nearly as much emphasis on the devising of a suitable reporting format as on the model building itself. This of course is both an asset and a drawback. Students probably learn less about a formal methodology of model building in the simulation course than they do in courses with a more strictly modelling emphasis.

It could be argued, and the author would have considerable sympathy with this suggestion, that these simulations would best be used as a more advanced course with students who had already received a basic course in mathematical modelling. Doubtless in an ideal world this would be the optimum solution. Constraints placed upon the degree course in which this exercise is located militated against this arrangement. In the event students learn a considerable amount about the practical techniques of modelling in a relatively unstructured way. This is probably made more possible by the high standard of entry to the course (average entry is slightly better than one A and two B grades at A level).

The only other directly comparable course which the author is aware of was developed independently and in parallel at Oklahoma State University and is described in Agnew and Keener (1980). The format of their course is almost identical to ours, the main significant difference being that they were able to involve the industrial donors of the problems directly in the running of the course. Most significantly they have drawn almost identical conclusions concerning the response of the students, the skills they have learned and even that their students, like ours, gained a great deal of satisfaction and enjoyment from the course.

7. ACKNOWLEDGEMENT

The initial development of this course during the period February 1976–July 1977, was supported by the Nuffield Foundation through their Small Grants

Scheme for Developments in Undergraduate Teaching. The author acknowledges his indebtedness to the Foundation for this support.

REFERENCES

Agnew, J. L. and Keener, M. S., 1980, *Amer Math Monthly*, **87**, 55.
Clements, L. S. and Clements, R. R., 1978, *Int. J. Math. Educ. Sci. Technol.*, **9**, 97.
Clements, R. R., 1978 *Bull. Inst. Maths Applics.*, **14**, 295.
Clements, R. R., 1982a, *Int. J. Math. Educ. Sci. Technol.*, **13**, 111.
Clements, R. R., 1982b, *Teaching Maths Applics.*, **1**, 125.
Cornfield, G. C., 1977, *Bull. Inst. Maths Applics.*, **13**, 13.
Ford, B. and Hall, G. G., 1970, *Int. J. Math. Educ. Sci. Technol.*, **1**, 77.
Gaskell, R. E. and Klamkin, M. S., 1974, *Amer. Math Monthly*, **81**, 699.
Handelman, G. H., 1975, *SIAM Rev.*, **17**, 541.
Klamkin, M. S., 1971, *Amer. Math Monthly*, **78**, 53.
McLone, R. R., 1973, The Training of Mathematicians, SSRC Research Report, Social Science Research Council.

Simulation Modelling: How to Assess Model Credibility

S. S. Stone
The Open University

SUMMARY

Complex simulation modelling used to be the exclusive province of the expert mathematician. The introduction of the System Dynamics (SD) modelling technique means that complex models can be created by people with little mathematical expertise. For both modellers and model users there is then a problem of credibility assessment.

A series of techniques for analysis of SD models have been developed at the Open University to aid the modeller in constructing a model. They also provide information, not normally available, which improves credibility assessment.

This chapter shows how one modeller used the path analysis technique of this series to develop his model. Also reported are some of the results from analysis on two further models which show that: (a) fundamental flaws in models remain undetected; (b) the basic structure of the model is opaque; (c) the computer program and some simulation results do not provide adequate information on which to base a credibility assessment.

1. INTRODUCTION

Complex simulation modelling used to be the exclusive province of expert mathematicians. However this created a lack of communication between modeller and user which Fromm (1974), Greenberger (1976), and Sage (1977) claimed as the major cause of failure to make use of a model.

Attempts have been made to provide modelling methods which could help bridge this gap. One such method is System Dynamics which because of its ease in actual use is sometimes proposed as a method for bridging the intellectual and organizational 'distance' between modeller and policymaker. These models are nonlinear, dynamic and complex.

The problem for the novice modeller and non-mathematical user is how to assess the credibility of a complex model. At present it appears that the most one can expect in terms of information accompanying an SD model is: (i) a listing of the model equations, plus a brief verbal explanation of them; (ii) a flow diagram of the model; (iii) computer output of a few variables relating to specific simulation runs.

I maintain that this is not enough information about the model to make a good evaluation of its credibility. These models, being complex and highly nonlinear, may produce 'credible' or 'plausible' behaviour which results from all kinds of spurious factors and interactions (de Vries, 1979). A few runs relating to one or two parameter sets do not provide an adequate picture of the model and of how it reacts to parametric variation. Often these models are used for some kind of prediction purpose which creates difficulty in assessing credibility from simulated model output. This problem is compounded if there is any trace of what Forrester (1971) calls 'counterintuitiveness' in the simulated behaviour; such behaviour implies a need to establish whether it is a true reflection of the reference system, or simply a reflection of errors in the modelling.

To enable the various participants in the construction and use of the model to assess its credibility, methods of analysing model structure and behaviour which generate the requisite information are required. These methods must be holistic in both spirit and practice, i.e. they must deal with the properties of the whole model, not with linearizations and other simplifications, or with limited subsets of simulation runs (Stone, 1982b). The most commonly used contemporary analytical techniques in this area tend to violate one or other of these requirements. They use mathematical techniques and concepts which are not easily accessible to non-mathematicians, and there is thus a danger that the communication gap between modeller and user which System Dynamics has, to some extent, reduced will re-emerge.

At the Open University a suite of analytical techniques have been developed to attempt to overcome some of these problems (Stone, 1982a), based on mathematical concepts taught to 11–16-year-olds in English schools. The techniques can be used in analysing four different aspects of a model: the internal structure and behaviour of the model in general; the internal structure and behaviour of the model for a specific set of quantitative parameter values (normally those for the modeller's 'standard' run); a complete sensitivity analysis of a run; and the boundary properties of the model. A suite of computer programs carry out the numerical computation and presentation of the results in a concise manner which can be used in further analysis.

This chapter examines the use of path analysis techniques to find the internal structure of a model. It is shown how this is used to help a modeller develop his model and to provide additional information for assessment of the credibility of two further models.

2. DEVELOPMENT OF A SYSTEM DYNAMICS MODEL

The model in this case study is that of a Laboratory Recirculating System for Intensive Fish Culture constructed by G. Mantle (1983). It has three interconnected submodels—food conversion, ammonia conversion and that for oxygen use and replenishment. It is the relationship between these submodels which form the major part of the model development. The analysis used in this demonstration uses concepts and language from graph theory. The System Dynamics flow diagram of the original model is given in Fig. 1. A digraph of the model is created by the variables being treated as vertices and the functional relationship B = f(A) being represented by an arc from vertex A to vertex B. The sign associated with an arc obeys the rule: When variable A increases *and*:

(1) variable B always increases then the sign (+) is used.
(2) variable B always decreases then the sign (−) is used.
(3) neither conditions (1) or (2) hold then the sign (?) is used.

The digraph is shown in Fig. 2a (flow diagram symbols are used for vertices to keep continuity between the two diagrams). Both flow diagram and digraph have different symbolization for arcs. The different symbols refer to different submodels in the diagram. In the flow diagram these submodels refer to level and information respectively (Forrester, 1968), while in the digraph they are time-sliced, (where $B_{t+1} = f(A_t)$), and event-sequenced (where $B_t = f(A_t)$). These latter two submodels are used in the analysis.

If Forrester's principles are adhered to the level and time-sliced submodels are identical. Often this identity is not possible for theoretical or modelling reasons.

If CLIP, SWITCH, functions are used in the model it is possible to identify other submodels present within the model, operative when different CLIP and SWITCH function conditions are in force. To distinguish these submodels from the time-sliced and event-sequenced submodels, they will be referred to as modes of the model.

In the original Fish Culture model there was one CLIP function which activates the feeding equations once every twenty-four hours. The feeding mode involves the complete model and the nonfeeding mode isolates vertices AI, ASI, F, WIR, A, by removal of arcs to and from them.

The next stage in the analysis is to find the basic time-sliced structure of the model. This requires that the event-sequenced paths between two time-sliced arcs are replaced by a single arc signed with the combined sign of all the arcs. This arc is then combined with the time-sliced arc it leads to. One then has a

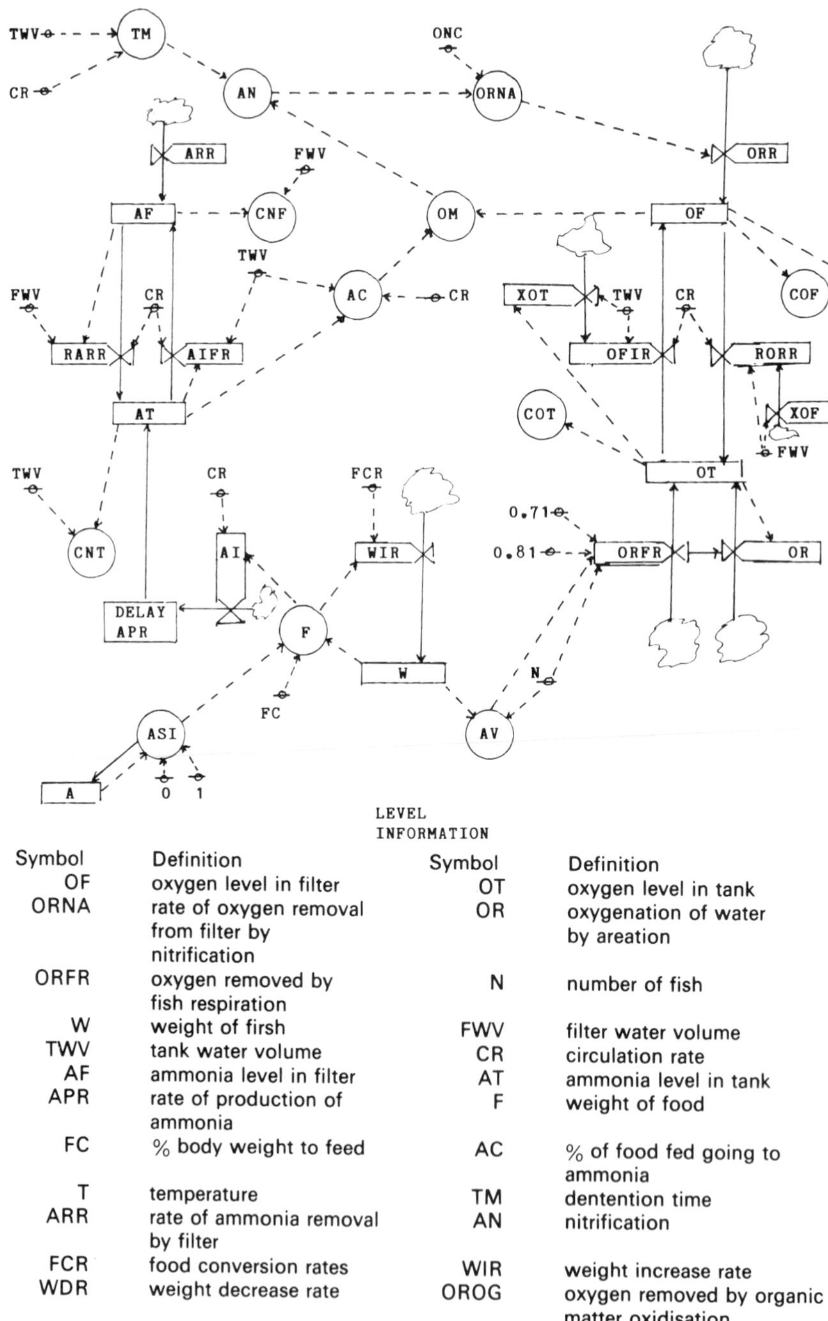

Symbol	Definition	Symbol	Definition
OF	oxygen level in filter	OT	oxygen level in tank
ORNA	rate of oxygen removal from filter by nitrification	OR	oxygenation of water by areation
ORFR	oxygen removed by fish respiration	N	number of fish
W	weight of firsh	FWV	filter water volume
TWV	tank water volume	CR	circulation rate
AF	ammonia level in filter	AT	ammonia level in tank
APR	rate of production of ammonia	F	weight of food
FC	% body weight to feed	AC	% of food fed going to ammonia
T	temperature	TM	dentention time
ARR	rate of ammonia removal by filter	AN	nitrification
FCR	food conversion rates	WIR	weight increase rate
WDR	weight decrease rate	OROG	oxygen removed by organic matter oxidisation

Fig. 1. System dynamics flow diagram of the model of a recirculating system

complete time-sliced digraph of the model. The complete time-sliced digraphs for both modes of the model are given in Figs. 2b, 2c.

3. FIRST STAGE IN THE DEVELOPMENT OF THE ORIGINAL MODEL

Figs. 2b, 2c were used to consider the theoretical implications in the model. The first point to be considered was to the behaviour of weight (W) during a run. In Fig. 2b, there is only one arc to W (weight) which is from W and this is positive. In Fig. 2c there are no arcs to W, thus weight cannot decrease.

If the modeller made the supposition that the model would not be used to investigate behaviour in adverse conditions, this might be credible. However, as the model was to be used eventually to investigate adverse conditions thus changes to the model are required. A set of SWITCH and CLIP functions were

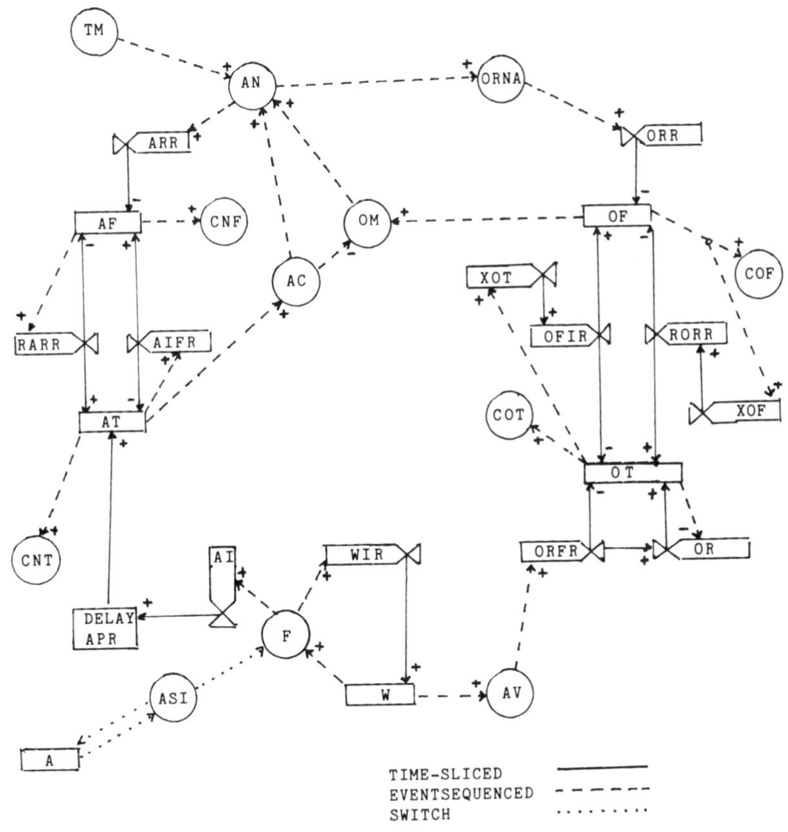

Fig. 2a. Digraph of recirculating model

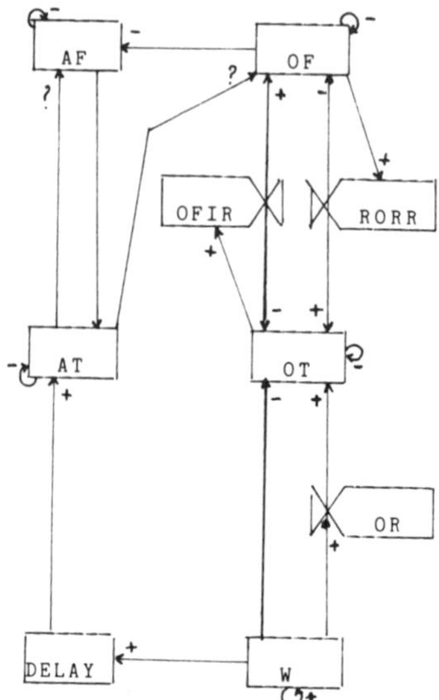

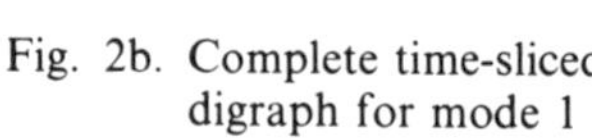
Fig. 2b. Complete time-sliced digraph for mode 1

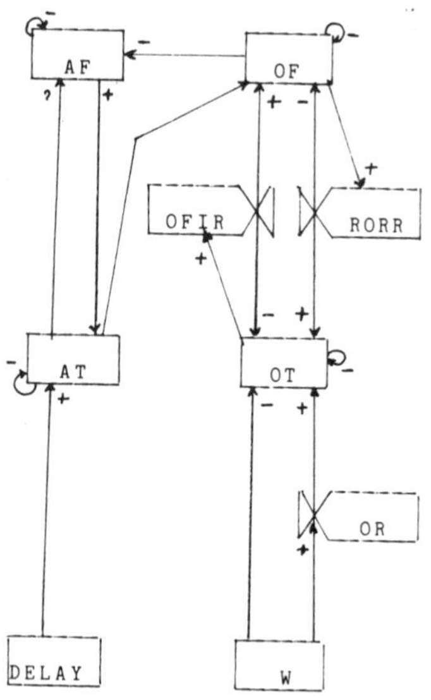

Fig. 2c. Complete time-sliced digraph for mode 2

used to allow new weight loss and death functions to operate instead of weight increase under given conditions. For weight loss, the condition was AT $\geqslant 900$. For death, the condition was AT $\geqslant 18000$.

The lack of a time-sliced arc between AF and OF led to questioning the validity of the arc AT to OF. It was then realized that the connection should be from AF and not AT to OF. It was also realized that no account had been taken of the occurrence of ammonia nitrification inhibition when oxygen levels in the filter fell below a given level. The equations were altered appropriately for the paths to be from AF not AT. There was also a CLIP function introduced which inhibited nitrification when OF < 630.

The lack of an arc between W and OF was considered theoretically incorrect. The appropriate equations were added to the model.

The digraph of the modified complete model is given in Fig. 3a. There are now eight modes of behaviour—four occurring once every twenty-four hours and four holding at other times—while the fish live. Similar differences occur between the two sets of modes as in the original model, therefore it was decided to concentrate on four modes that occurred once every twenty four hours, the other modes being found by the removal of the arc to W. The modes are

(a) OF $\geqslant 630$, AT < 900—mode 1 (fish fed, nitrification occurs)
(b) OF < 630, AT < 900—mode 2 (fish fed, no nitrification)

(c) OF < 630, AT ⩾ 900—mode 3 (fish not fed, no nitrification)
(d) OF ⩾ 630, AT ⩾ 900—mode 4 (fish not fed, nitrification occurs)

The complete time-sliced digraphs for these modes are given in Figs. 3b, 3c, 3d, 3e.

4. RESULTS OF TWO FURTHER CASE STUDIES

The following two case studies rely for their model credibility on the equations of the model together with background theory and computer output for a few variables. For reasons of space I can only make selective comments from the results of our analysis.

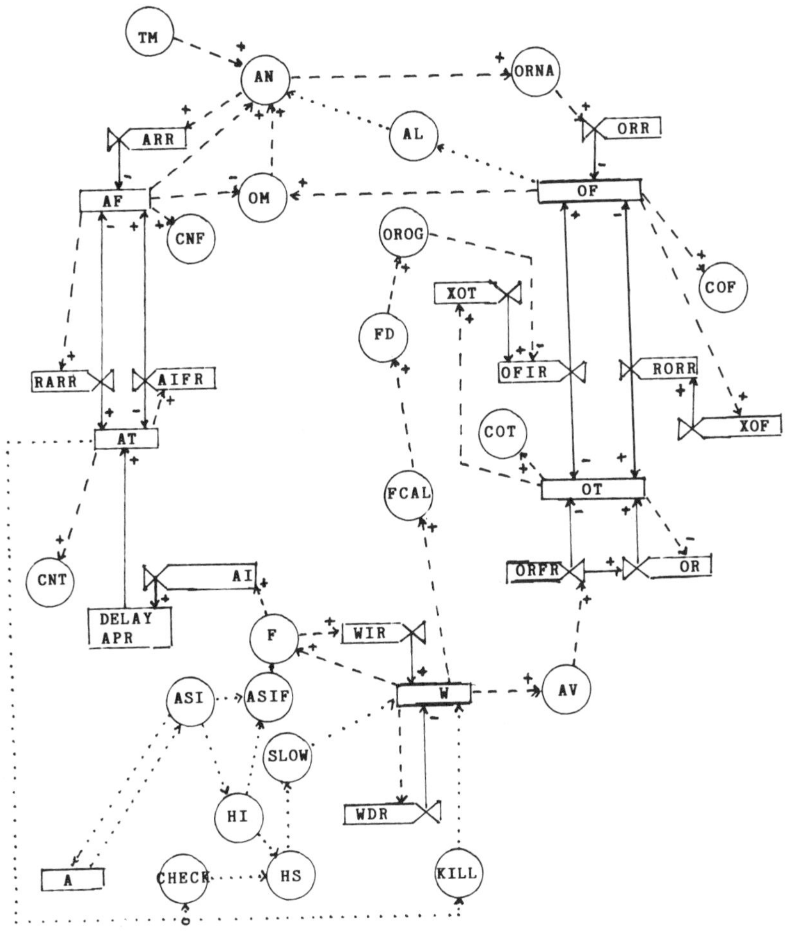

Fig. 3a. Digraph of modified model

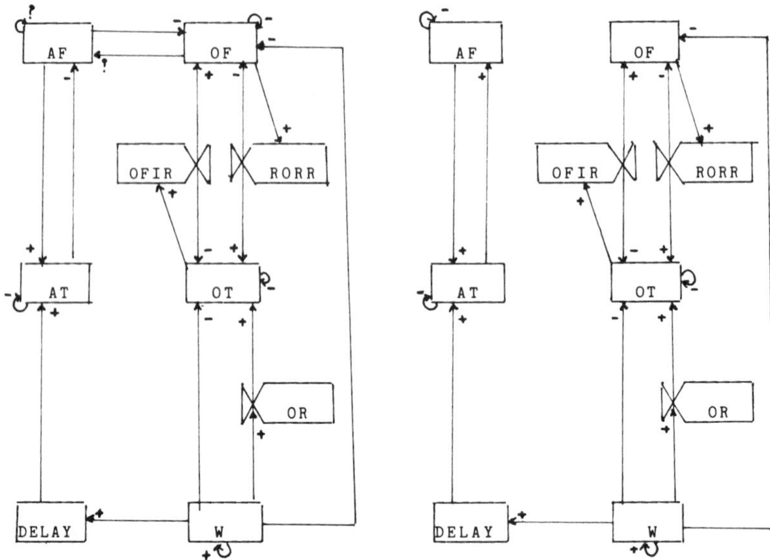

Fig. 3b. Complete time-sliced digraph for mode 1

Fig. 3c. Complete time-sliced digraph for mode 2

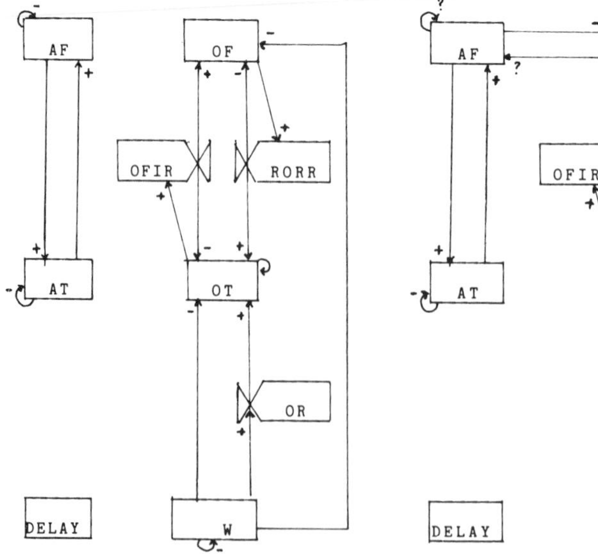

Fig. 3d. Complete time-sliced digraph for mode 3

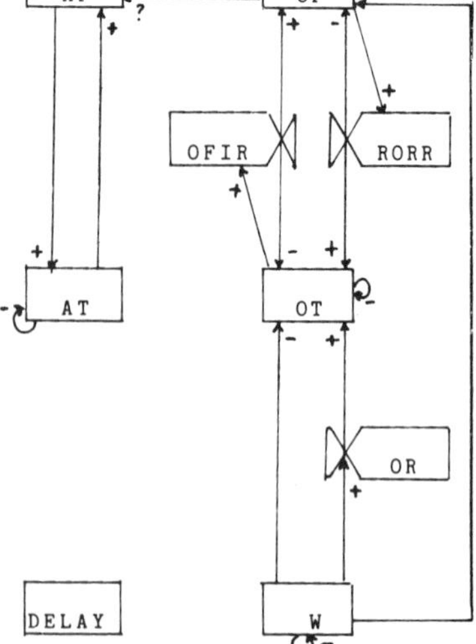

Fig. 3e. Complete time-sliced digraph for mode 4

4.1 World Petroleum Tanker model

The World Petroleum Tanker Industry model was used by Hammad (1977) to describe the structure and interrelationships of the world petroleum marine tanker capacity. The model is supposed to be based on those characteristics of the real system considered necessary to give 'behaviour characteristics of interest'.

The primary objective of the model is to 'improve and deepen the understanding of the complexities of the petroleum tanker industry system rather than just come up with a number of final results'. Hammad considers that the model's validity is based on how its dynamic behaviour compares with the real system it represents.

There are two state variables:

SR—Super tanker capacity

RR—Regular tanker capacity

In the results three other variables are also considered important:

TOTCAP—total tanker capacity

TDR—Total demand for tankers

PDM—World petroleum demand

The model has four equations involving switch type functions, but two of these are used to initialise the model for various experiments. They therefore changed the model for the complete run not for parts of the run. The remaining two conditions occur during a run thus providing four possible modes of behaviour for the model which are:
These modes of behaviour are:

(i) SR $\geqslant$DSR, RR $\geqslant$DRR known as mode 1
(ii) SR $\geqslant$DSR, RR $<$DRR known as mode 2
(iii) SR $<$DSR, RR $<$DRR known as mode 3
(iv) SR $<$DSR, RR $\geqslant$DRR known as mode 4
DSR—DEMAND LEVEL FOR SUPER TANKERS
DRR—DEMAND LEVEL FOR REGULAR TANKERS

One can conclude from our analysis that the model is in essence two noninteracting nearly identical models running concurrently, one for the supertankers and one for the regular tankers.

Of the results Hammond provides only two, SR and RR, are part of the feedback submodels, one in each submodel. TOTCAP and TDR are calculated from the results of these submodels and other exogenous parameters. PDM is directly calculated from an exogenous time dependent parameter.

Thus in the model Hammad does not allow for any market substitution of one type of tanker by another apart from that predetermined by the exogenous parameters. Also an increase in total effective capacity, TOTCAP, does occur without an increase in tanker fleets when experiments involving closure of Suez Canal and/or an oil pipeline are performed.

There is also an implicit assumption that the closing of the canal will increase the demand for tankers whereas the closing of the pipeline will not.

The most surprising result from the analysis was a division by zero. One would have expected that such a step would either cause an error message from the computer or spurious results. In this case neither happened. The actual division by zero occurred between two of the calculation time points.

The equations in question are:

SCANS.K = CLIP (SAN.K,O,SR.K,DSR.K) 9

SAN.K = SR.K*(1/(SR.K-DSR.K)) 10

SCANS—SUPER ORDERS CANCELATIONS

SAN—SUPER TANKER EQUILIBRIUM

SR—SUPER TANKER CAPACITY

DSR—DEMAND LEVEL FOR SUPER TANKERS

According to Hammad equation 9 selects SAN if SR > DSR and zero otherwise. However the definition of CLIP gives that when SR.K ⩾ DSR.K then SCANS.K = SAN.K. Therefore when SR is equal to DSR, SAN is selected and not zero so that SCANS becomes infinite.

SAN is also supposed to change 'in accordance with the changes in the supertanker capacity and the demand for supertankers. The cancellations will increase as the demand decreases in relation to the capacity. If SR = DSR, the effect will be zero' Hammad (1977).

However, when DSR < SR and DSR → SR, SCANS → ∞, which contradicts the statement above. Thus the model is not verified.

4.2 The magazine publishing firm model

Hall (1973) attempts to investigate the problems of large publishing firms involved in magazine publication. There had been several reported cases of magazines failing when they had record circulation and advertising revenue. Hall developed a System Dynamics model to investigate this seemingly paradoxical behaviour. He developed the model in two forms:

(1) a generalized form which is only usable when information of a particular firm has been incorporated.
(2) the model for the *Saturday Evening Post.*

Hall hoped that the model would provide a vehicle for studying the impact of management decisions on the system. He points out that the general model was only tested on a single firm and therefore one can only consider the model as a single-firm model. He also pointed out that his model was of a firm publishing a single magazine and not a multi-magazine publishing firm as in the case of the *Saturday Evening Post.*

The model has six state variables:

TREV—total revenue
TEXP—total expense
SNRTE—Subscription rate
ADRTE—Advertising rate
CNPEX—Circulation promotion expense
TOTRDS—Total readership

The analysis of this model is of its structure and behaviour in both its general form and its free running form. The experiments were all based on this free running model and represented different management intervention, therefore the credibility of this model is important to all Hall's work.

The switch type functions were examined and it was found that there were three modes in which the model operated:

(a) Mode 1 when no pulses occurred
(b) Mode 2 at the start of a year except in 1945
(c) Mode 3 at the start of 1945 (in the time period of the model).

There is only one strongly connected submodel which is the same for all modes. This submodel forms the feedback core of the model, and comprises total readership (TOTRDS), the trial subscription rate (TSSR), the regular subscription rate (RSSR), the trial (TSE) and regular (RSE) subscription expiration and expiration rates (TSER and RSER), the magazine volume page rate (MVPR), the advertising rate (ARPTR) and 'rate' per thousand readers (ADPSR) and total reported readers (TOTRDR). Only one state variable is part of the feedback core of the model namely TOTRDS—total readership. Three of the state variables are constants which can only be changed by exogenous managerial decision namely SNRTE, ADRTE, CNPEX. The other two state variables TREV, TEXP—total revenue and expense—are calculated from the feedback submodel. Fig. 4a gives the time-sliced digraph for mode 1.

The feedback core of the model is dependent only on the subscription rate, advertising rate and circulation promotion expense which implies that it is not dependent on the bulk of the expenditure on the magazine, nor on the total revenue received. Mode 3 allows a change in value in SNRTE, ADRTE, CNPEX which will cause a change in the constant values used in the feedback submodel. Mode 2 provides a similar yearly change to the other two state variables TEXP and TREV.

Using another part of the analysis, inspection, it was found that when the model was run as free running model TSSR, TSE, TSER were constant. This reduces the size of the feedback core of the model still further, Fig. 4b. These variables are constant because of the state variables SNRTE, ADRTE, CNPEX being constant and because a regression constant—B2—is assumed zero. There are eight parameters in the free-running model equal to zero, which leads me to question whether the results from this particular run could ever be considered sufficient to assess the credibility of the model in general.

5. CONCLUSION

The results presented here form part of the complete analysis and come in the main from the use of path analysis technique. Specific model structure can be a very reduced form of the general model structure (Hall's model). By presenting the model in a different manner, as with path analysis, errors and misunderstandings can be observed which are not otherwise apparent.

Fundamental error such as division by zero may not necessarily be found by the computer, nor be detected by inspection of the results. Thus the credibility of the model cannot be assured by the simulation output.

Both Hall's and Hallam's model were shown not to be as complex as might be expected. Only part of each model involved feedback loops where one might have expected the whole of the model to be involved.

Though this chapter has presented only a small part of the information available from using these techniques, it is evident that relying on the computer

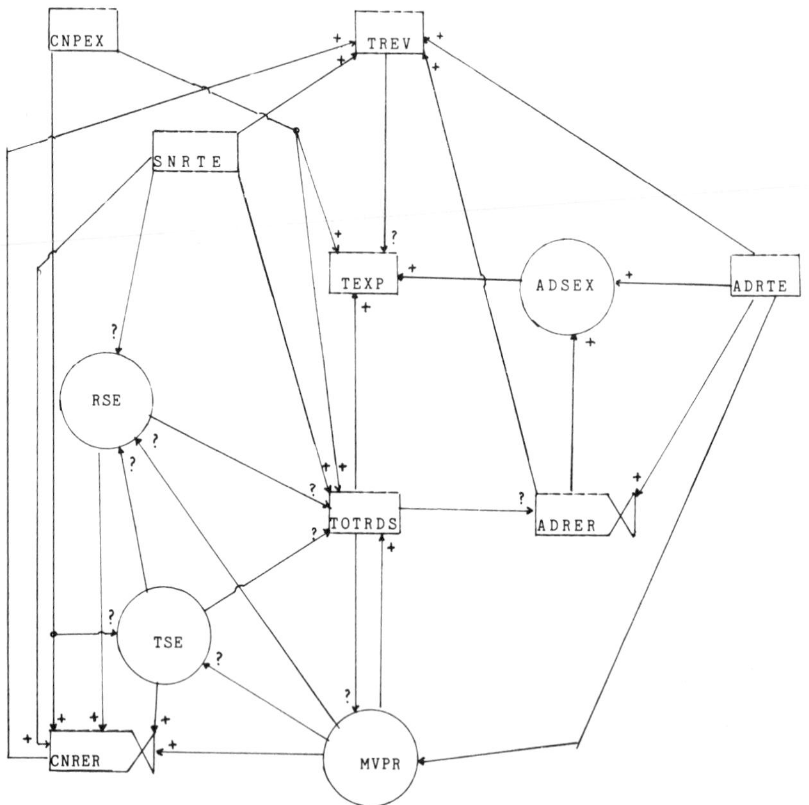

Fig. 4a. Complete time-sliced digraph of magazine model, mode 1

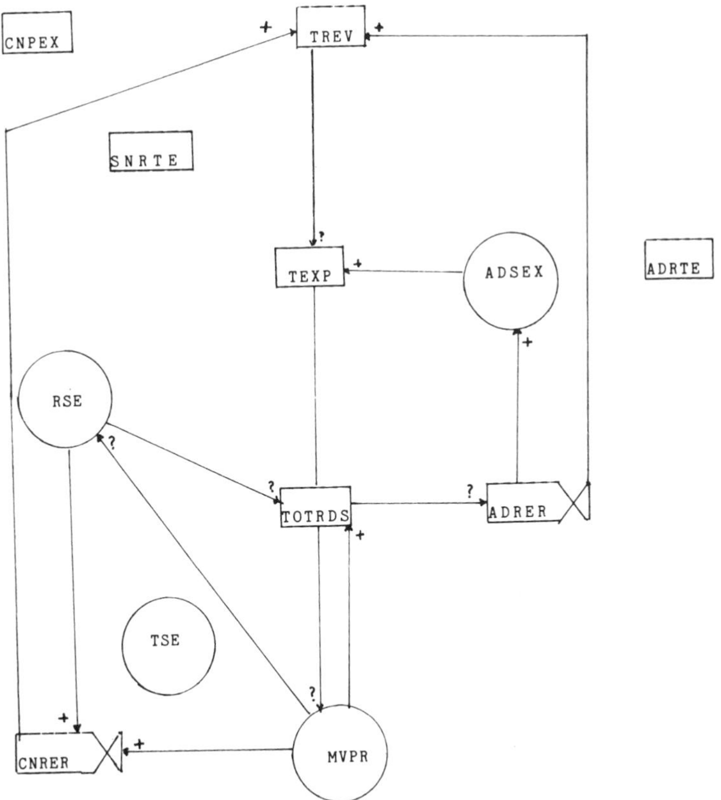

Fig. 4b. Complete time-sliced digraph of magazine model, mode 1

program and some simulation runs is completely inadequate for making a positive assurance of credibility. With models as complex as those described here a more careful analysis is required.

The techniques I have developed provide another view of how the model behaves thus adding to the amount of information available for assessment of credibility. As the techniques are based on simple concepts and the results presented in a concise manner, it is hope that this will make them accessible to the less expert mathematician.

REFERENCES

Forrester J. W., 1968, *Principles of Systems*. Wright-Allen Press Inc.

Forrester J. W., 1971, Counterintuitive Behaviour of Social Systems. *Simulation, February*

Fromm G., Hamilton W. L. and Hamilton D. E., 1974, Federally Supported Mathematical Models: Survey and Analysis. Report for RANN.

Greenberger M., Crenson M. A., and Crissey B. L., 1976, *Models in the Policy Process.* Russell Sage Foundation, New York.

Hall R. J., 1973, A Systems Model of a Magazine Publishing Firm. Ph.D. thesis, University of Washington.

Hammad A., 1977, The Development of a System Dynamics model for the World Petroleum Tanker Industry. Ph.D. thesis, George Washington University.

Mantle G., 1983, An Investigation into the Behaviour and Performance of a Recirculating System for Intensive Fish Culture. Ph.D. thesis, Open University.

Sage A. P., 1977, *Systems Engineering: Methodology and Applications.* IEEE Press.

Stone S. S., 1982a, Simple Analytical Methods for Elucidating Structure and Behaviour of System Dynamics models of Social Systems. Ph.D. thesis, Open University.

Stone S. S. and Naughton J. J., 1982b, Techniques for Analysing System Dynamics models: A Critical Survey. International AMSE Conference: Simulation and Modelling, Paris-Sud.

de Vries G., Naughton J. J., 1979, Economische Theorieen en Macro-economische. *Kennis en Methode*, Jaargang 111, no. 3 (English translation from J. Naughton, Open University).

On the Use of Simulation Software in Higher Educational Courses

A. O. Moscardini,[1] M. Cross[2] and D. E. Prior[1]

1. Sunderland Polytechnic,

2. Thames Polytechnic.

1. OVERVIEW OF SIMULATION

Simulation today is applied to a large number of areas of endeavour, including politics, management, ecology, engineering, medicine and pharmacology (see for example, the papers in *Simulation*). One reason for this diversity is the cheap and widespread use of computer technology combined with the tremendous upsurge in mathematical modelling over the last twenty years. Simulation and modelling are obviously intimately connected and to understand simulation, one must first understand something of mathematical modelling.

The art of mathematical modelling is not a recent phenomenon (Aris 1970). It can be traced back to the planetary work of the early Greeks, the work of Copernicus in the middle ages and, more recently, the theories of Einstein on Photoelectricity and Relativity. There are many definitions of a model but, for this purpose, it could be described as a 'simplified conception of reality designed to satisfy a particular purpose'. Scale models provide an obvious example here. Model aeroplanes, for example, can now be built with amazing accuracy and attention to detail. If one analyses this type of model, then it consists of a set of component parts and static relationships between them: the aileron is in the appropriate position on the wings, the joystick is in the pilot's cabin etc. The model components and interrelationships are copies of the real situation, a feature which may be termed the structure of the model. When dealing with mathematical models, the components parts will be the variables and their relationships are represented by the governing equations. If these variables are given differing values or the ailerons and joystick are manoeuvred, then the relationship between the components changes with respect to time. Such changes may be referred to as the processes of the model. The distinction between structure and processes defines mathematical modelling and simulation, respectively.

Simulation may be described as a dynamic or operating model, i.e. changes over time in the model correspond isomorphically with changes over time in the situation being modelled. A simulation of a situation involves abstracting not only the static, structural relationships but the dynamic process relationships as well. Once the variables that have been selected are given values and the relationships between the variables are specified, the model is allowed to operate. It may operate through the interaction of people who play roles within the model or it may operate on a computer. This chapter will concentrate on computer simulation which has become the most widely used method for analysing and predicting system behaviour.

Model building and simulation, in effect, are techniques for bringing order and predictability out of an apparent chaos of multiple variables. A unique advantage of simulation is the ability to push ahead in time, allowing a user to pretest possible alternatives under variable conditions representing future periods. If the model is realistic this glimpse into 'possible futures' permits a user to evaluate and choose the most favourable of many alternatives. Simulation is not a panacea. In many situations, other methods of problem solving are quicker and more economical. In the areas of personnel and organization planning, the inability to quantify facts may render simulation impracticable. Simulation may also be costly in time and money so its pay-off potential must be significant. The key to simulation is to apply it in areas where it provides substantial return for its high investment cost.

2. SIMULATION IN PRACTICE

2.1 Simulation methodology

In recent years the methodology of model construction has been researched in depth Aris (1978), Ziegler (1976). By comparison, in Education simulation is often treated as consisting of simply running the model on a computer. There is, of course, a great deal more to it. Fig. 1 shows a methodology for simulation which is essentially comparable to that for model building. This diagram also illustrates the relationship between simulation and model building.

A major decision in simulation is the selection of a suitable host language and as a general principle, familarity and available hardware generally provide the main guidelines. Most practitioners are familiar with BASIC or FORTRAN and these languages are available on every computer installation. Useful simulation packages have been developed based upon the above languages and their educational advantages are discussed later. However, recently, a large class of simulation languages have been developed to enable

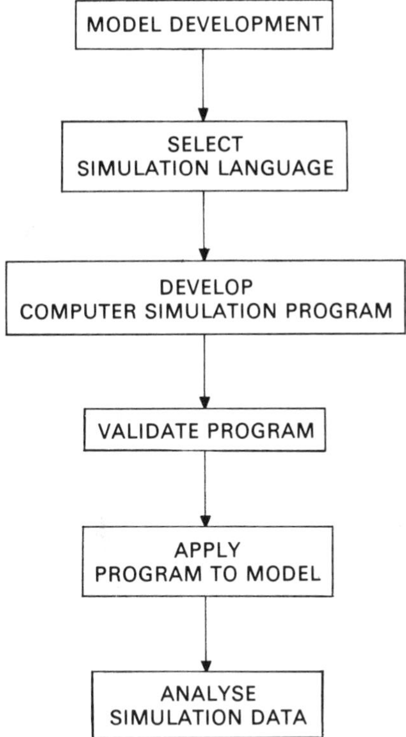

Fig. 1. Flowchart indicating the constituent phases of simulation

the analyst to concentrate on the model development and operation rather than the programming difficulties. Typical languages in this class are GPSS (General Purpose System Simulator) Schriber. (1974), ACSL (Advanced Continuous Simulation Language) Mitchell (1976), CSMP (Continuous System Modelling Program) Green (1980), DYNAMO Shaffer (1980), and GASP-IV Pritzker (1974).

Even though simulation languages were developed specifically to ease the programming difficulties, such a wide variety of options are available, that the user must be rather familiar with the language in order to obtain the maximum advantage from it. For example, parameters may have to be set early in the program if complicated output is required. Often companies will call in outside consultants at this stage to help develop these skills, however, students under training do not have this facility!

In one sense, the program forms a model of the original model. Thus one has not only to ascertain the validity of the model but also to check that the program accurately reflects the model. The problem becomes acute if the results obtained, suggest an error is present. For a complicated situation, it is difficult to know whether the error is in the model equations or in the program. This emphasizes the importance of the model implementation stage. The simulation program should, therefore, be developed in a well structured manner with each section capable of validation. Most simulation systems have sophisticated debugging and TRACE commands that can help to discover programming errors. More subtle errors can be detected by observing what happens when the parameters take extreme or simple values and checking the results. It must not be forgotten that even though each individual section is working, there could be an error in the way that these subsections interact. Such a possibility may be checked by a complete dry run using simple values and known data. Care must be taken here, however, to avoid pitfalls such as a one-time-only correlation. A simulation model is not static but must react dynamically and it is important that validation of the simulation program is recognized as a problem in its own right. For example, a significant number of continuous systems turn out to be numerically 'stiff'. The simulation program may still produce results, but in such cases they are worthless, unless the embedded numerical software is designed to cope with stiffness.

When the program is eventually applied to the model, it is generally accepted that often small changes and additions will have to be made. Flexibility during the development stage is a necessity. The original objectives must constantly be kept in mind. Also, each additional factor introduced may result in greatly increased complexity which may have a direct influence on storage and running costs and so from the economy in programming is of crucial importance in real life simulation.

A large selection of statistical and graphical options are provided with most simulation systems. Professionals involved in simulation utilize such facilities to optimize the impact of the presentation of the results as a valuable aid to understanding an acceptance.

2.2 Problems of simulation

Computer simulation has many advantages and great advances have been made in the development of facilities and capabilities; nevertheless use of simulation is not problem free Hermann (1967), Ignall (1978), Naylor (1978), Wigan (1972). A simulation is only as good as the model upon which it is based and the development of a good model is often a substantial problem itself.

A second problem with simulation is in the area of experimental design. Considerable effort is often expended in an attempt to obtain information; unfortunately, this is frequently done in an inefficient manner based on 'poor' input data. It is therefore necessary to consider computer simulation as an equivalent to a laboratory experiment. Before any simulation is undertaken, areas of pay off are established and the feasibility of the completion of the project with estimates and budgets provided. Organization of the data, the use of statistically designed experiments are all incorporated where appropriate. The latter becomes particularly important when trying to state on a rigorous basis, the comparison of one system with another.

Factors such as kind of memory, speed of computation and errors due to rounding must all be taken into account in simulation. Simplification is often necessary due to the speed of computation or limitations of memory. The method of filling and retrieving information may also be significant problems. Which data to accumulate and at what point in time are often difficult decisions to make also.

3. TEACHING OF COMPUTER SIMULATION

3.1 Educational objectives

The aim of simulation courses we teach are to expose the student to as many aspects of computer simulation as is possible given the background of the staff and the computer hardware available. This may be resolved into the following objectives:

(a) to teach efficient and effective programming
(b) to demonstrate a variety of simulation languages
(c) to introduce a wide range of useful mathematical techniques
(d) to inculcate good presentation habits.

3.1.1 *Efficient and effective programming*

In order to be effective at programming simulations then expertise in a number of areas is required. For a start an overview of computer hardware is required with special emphasis given to those factors which impede speed of computation. It is likely that large-scale simulation generates the need to process substantial amounts of data and so a working knowledge of data processing techniques are also necessary. In fact some exposure to database technology would not come amiss either. In addition it is likely that much of the software will be written as user oriented and so some attention must be given to the development of user friendly environments. 'Friendliness' is usually a processor overhead and there is, therefore, a balance to be sought here with the overall economics!

With regard to programming procedures, it is vital that structured techniques are acquired and utilized from day one. Also it is important to make students aware of a reasonable library of debugging techniques such as TRACE statements, etc. Above all, clear and detailed documentation should be demanded of simulation programmers from the start. Professional computer scientists will often argue that all programming should be taught through PASCAL, but most simulation is still performed in FORTRAN and it would, therefore, seem most practicable to start with the latter language—warts and all.

3.1.2 *Simulation languages*

Over the last decade or so a large number of special purpose simulation languages have been developed and many are now widely used in industry and in research. Often the languages are based upon FORTRAN (or BASIC) and consist of a set of special commands. Until recently such languages were often tied to one processor operating system due to the implementation of the simulation commands. However, portability has been a concern of the simulation fraternity for some time and all of the major simulation languages are now generally fairly portable.

Good simulation languages are not necessarily good vehicles through which to teach the art of simulation. As such, a number of highly interactive simulation languages have been designed (usually in BASIC) which provide a suitable learning framework for students. One such language is IPSODE Moscardini (1980), Cross (1982), and this will be considered in more detail below. Of course, students must eventually become familiar with the languages they will use in professional practice. One of these languages, DYNAMO is reviewed in more detail below.

3.1.3 *Useful mathematical techniques*

When teaching mathematical topics as part of a simulation course it is appropriate to take a pragmatic approach. The amount and level of mathematics taught obviously depends upon the sophistication of the course. At more advanced levels, it is tempting in such courses to include all the latest techniques and explore each in some detail. In such cases, the course is in danger of becoming one in numerical mathematics rather than computer simulation. In the case of both discrete and continuous simulation, it may be argued that the associated mathematical techniques should be taught in a hands-on fashion so that students acquire a working capability rather than 'formal' training.

Finally, it is important that students grasp the concepts associated with system feedback and this may be handled by either some simple control theory (or through a systems dynamics methodology).

3.1.4 *Presentation*

With the recent boom in microcomputing and their associated graphics, it would be most useful for students to have some exposure to these techniques. Presentation of results may be improved tremendously by using good graphics software and it is important for students to be aware of the impact of presentation in simulation as a vehicle for communicating with clients.

3.2 *Teaching method and assessment*

As with the teaching of mathematical modelling, a course in simulation demands special organization and skills. We believe most of the teaching should be done at the terminal in what could be called 'workshop sessions'. In these sessions the lecturer must adopt a more informal approach. Except for the mathematical techniques and initial sessions in programming, none of the topics to be covered lend themselves readily to formal lectures. The lecturer must have a substantial amount of prepared material which demonstrates all the aspects of the course. This will include working packages, complete simulations which are very like, if not the same as, the industrial simulations and well documented programs. The class should attempt to work as a systems team with the lecturer playing the role of the senior consultant. Gradually, as the skills of the students develop more responsibility for each project development will be transferred to them.

More formal lectures will cover the teaching of the techniques and the philosophy behind the different languages. Careful timetabling with both types of teaching should be attempted. Workshop sessions will provide the motivation for techniques that should have been previously covered in the formal 'techniques' lectures. Lecturers must thus work as a team where each is not only aware of what the others are teaching but also the relevance of his own area in the overall framework of the course. It is no use if one lecturer is demonstrating the virtues of efficient programming if his colleagues are using badly structured and documented programs of their own. Handling a workshop session successfully requires skills different to those for conventional lectures. Material will rapidly become outdated and the lecturer accepts that interested students will often produce new and fresh ideas or methods of procedure. Such ideas will inevitably produce what may be perceived as challenges to the lecturer's competence. Such assaults are best met with fair-mindedness. A necessary skill needed by lecturers is that of generating and maintaining the full participation of everyone in the group. This will contain extroverts and introverts, leaders and the led, the talkers and the silent ones. A lecturer has to control the more voluble students (without damping their enthusiasm) and at the same time encourage the more timid members. A crucial factor here is the size of the group—obviously the smaller the better.

The skills highlighted above that have been mentioned are not easily mastered by academic staff and although the best way of developing them is through experience, it is a long process. Shared experiences can speed up this process and for this reason, more summer schools, conferences and groups such as the National Workshop on the teaching of Mathematical Modelling should be encouraged and well attended.

Since the main teaching vehicle is the workshop session, the principal course assessment must be based on what is being taught (or rather learnt) during these sessions. Some of the skills that are being developed by the students are:

- analysis of real life situations
- choice of a suitable model
- choice of relevant language
- efficient programming and application
- good presentation.

Formal examinations are not an appropriate way of assessing the above skills and some form of continuous assessment is necessary. The presentation of individual projects is another way of assessment but it is difficult to provide enough for large classes. Oral examinations are used extensively on the continent and perhaps the British polytechnics and universities should utilize this method of assessment more widely, too.

4. USING SIMULATION LANGUAGES

The central role of specialist languages (or packages) in the implementation of models for computer simulation has already been highlighted. However, each language is generally based upon a particular philosophical approach to modelling. As such, implementation of a model via a simulation language inevitably places constraints on the formulation. These constraints obviously influence the degree to which the computer implementation can reflect the mathematical model (and, of course, the actual situation). In this section we compare how two simulation languages implement the same problem.

4.1 The example

Consider the problem of the supplier responding to the demands of customers. Here, the input to the system is CRATE orders/week whilst the output is simply CORATE customer deliveries/week, where each order only demands one item (and all of these are identical). Processing delays lead to a pool of unfilled orders, UFORDS. Also, the management generally increase pressure to downdate the paper work in relation to the quantity of orders awaiting processing. In addition, there is a desired inventory level (DESINV, items), the actual inventory level (INV, items) and the rate at which orders are placed to replenish stocks (PORATE, items/week). The latter, of course, are supplied by an external source (with an assumed infinite supply), but which induces a delay, D3 (weeks) into the system after which time the stock is replenished at IDRATE items per week.

In the implementations that follow the dependent variables are considered to be the level of unfilled orders (UFORDS) and the actual inventory level (INV).

4.2 Using DYNAMO

The DYNAMO modelling framework emerged in the 1950s to analyse and predict the behaviour of industrial/commercial systems. In the 1960s the Systems Dynamics approach of Forrester was encapsulated in the DYNAMO series of computer software packages. Versions are now available to run on most mini and mainframe systems. DYNAMO is a continuous system simulation language which utilizes an interconnected system of levels and rates to yield a set of difference equations. A special form of diagrammatic representation has been developed to illustrate the relationships between variables. A simple example is shown in Fig. 2, where a population level (POPLEV) is illustrated by a rectangular box whilst the rate of change of that level (PRATE) is indicated by the valve symbol ⋈▭. Every rate has an associated constant (K_1). The diagram also shows that the population level is influenced only by the rate, by the connecting dotted line. Mathematically, the resulting equation is

$$\frac{dP}{dt} = K_1 P.$$

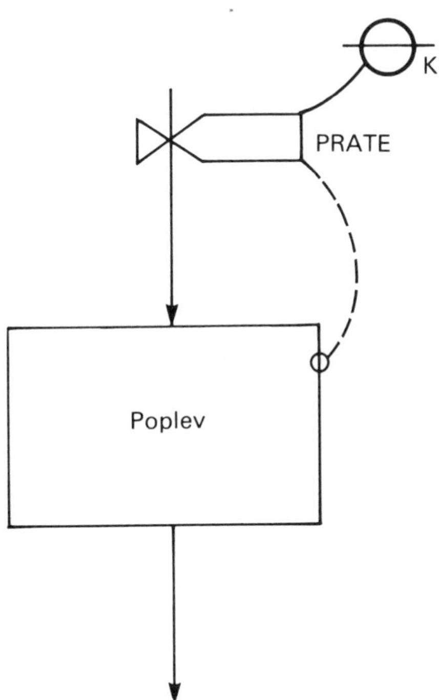

Fig. 2. A simple process represented by System Dynamics notation

Using, the System Dynamics approach this equation is written as the pair of first order difference equations

POPLEV = POPLEV + DT*PRATE

PRATE = POPLEV/K1

Of course, it is assumed that the rate constant does not change throughout the whole time interval DT. In the DYNAMO system time is divided into past, present and future, represented by J, K and L respectively. Thus, implementation of the above ideas in DYNAMO is as follows:

L POPLEV.K = POPLE.J + DT*PRATE.JK

R PRATE.KL = POPLEV.K/K1

Note that rates are assumed to be averaged over time, hence the double subscript attached to the rate variables.

Two important functions in the DYNAMO system are SMOOTH and DELAY. SMOOTH (Rate, K) serves to smooth a variable prediction over the last K cycles of the model run, so that

$$x_t = x_{t-1} + (x_t - x_{t-1})/K$$

Hence, large K implies that past history is a dominant feature and vice versa. It is also possible to introduce exponential smoothing fairly simply.

Since real life systems do not react immediately to changes in the input conditions, it is important to include a delay facility in any modelling system. DYNAMO essentially deals with system delays through cascades of the simple first order delay.

A DYNAMO representation of the example described in section 4.1 is shown in Fig. 3. The DYNAMO implementation of this model is shown in Fig. 4. The cycle was selected as one week and the time step 0.1 weeks used. Fig. 5 shows a tabular output of the simulation whilst Fig. 6 illustrates the corresponding graphical output.

In the authors' experience a simulation course based upon DYNAMO can prove to be extremely beneficial in stimulating students to investigate and identify the structure of the system under consideration. However, the concepts involved in the system dynamics approach require time to be assimilated. As such, DYNAMO is not a computer package to be used as a superficial aid in a course on simulation.

4.3 *Using IPSODE*

The IPSODE package (Interactive Programming with Systems of Ordinary Differential Equations) was specifically designed for students to implement models easily for computer simulation. The package is highly interactive and attempts to protect the user as much as possible by

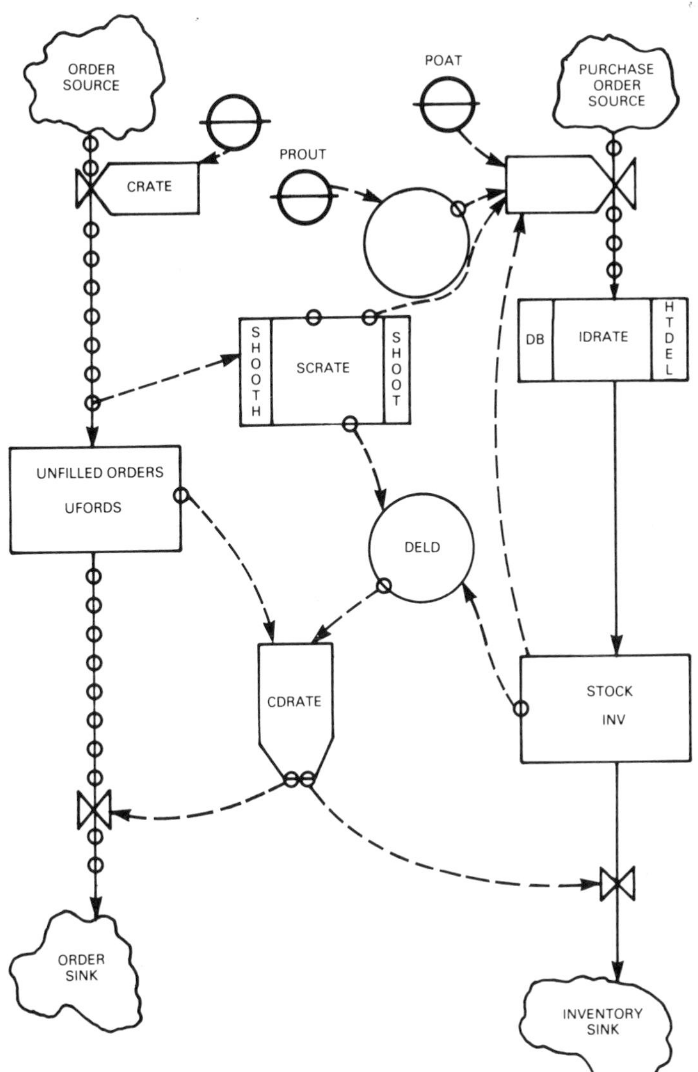

Fig. 3. System Dynamics representation of the customer–retailer system

(i) not requiring any previous computing experience,
(ii) not requiring any familiarity with numerical techniques,
(iii) making the system secure and incorruptible by user-generated errors,
(iv) trapping errors and providing a comprehensive help-facility to steer the user to the 'correct' way or provide information on systems options, etc.,
(v) providing a wide variety of output options.

```
P-  1 RUN-4          MINI-DYNAMO  * VERSION 1.00


L UFORDS.K=UFORDS.J+DT*(CRATE.JK-CDRATE.JK)
L INV.K=INV.J+DT*(IDRATE.JK-CDRATE.JK)
A SCRATE.K=SMOOTH(CRATE.JK,SMOUT)
A DELD.K=TABLE(RATIO,INV.K/SCRATE.K,0,12,4)
T RATIO=4,2,1,0.75
A DESINV.K=PROVT*SCRATE.K
R CRATE.KL=ICRATE+STEP(HT,T1)
R CDRATE.KL=UFORDS.K/DELD.K
R PORATE.KL=SCRATE.K+(DESINV.K-INV.K)/POAT
R IDRATE.KL=DELAY3(PORATE.JK,MTDEL)
N UFORDS=DELD*CRATE
N INV=DESINV
C SMOOT=8
C PROVT=8
C HT=100
C T1=5
C POAT=4
C MTDEL=6
C ICRATE=50
PRINT UFORDS,INV,CRATE,CDRATE,PORATE,IDRATE
PLOT INV=I(0,1500)/UFORDS=U,CRATE=O,CDRATE=D,PORATE=P,IDRATE=S(0,450)
SPEC DT=0.1,LENGTH=100,PRTPER=5,PLTPER=1
     COMPILED -
```

Fig. 4. DYNAMO implementation of the customer–retailer problem

P- 2 RUN-4   MINI-DYNAMO * VERSION 1.00

TIME E 00	UFORDS E 00	INV E 00	CRATE E 00	CDRATE E 00	PORATE E 00	IDRATE E 00
0.00	50.00	400.0	50.00	50.00	50.00	50.00
5.00	50.00	400.0	150.00	50.00	50.00	50.00
10.00	320.87	208.9	150.00	109.90	237.83	82.41
15.00	433.94	265.0	150.00	149.11	298.48	201.03
20.00	331.04	629.2	150.00	180.55	247.24	270.79
25.00	193.29	1069.4	150.00	173.16	158.42	246.77
30.00	141.26	1333.7	150.00	152.24	103.64	176.39
35.00	140.22	1314.7	150.00	148.59	114.43	122.33
40.00	152.10	1159.1	150.00	144.55	156.54	119.13
45.00	173.62	1092.9	150.00	148.33	174.81	147.85
50.00	168.98	1134.8	150.00	153.04	165.24	167.48
55.00	151.69	1206.3	150.00	152.18	147.88	164.78
60.00	147.42	1244.1	150.00	150.23	138.69	151.36
65.00	148.02	1224.9	150.00	149.60	143.62	142.21
70.00	151.34	1189.3	150.00	148.73	152.60	143.87
75.00	155.17	1178.7	150.00	149.89	155.27	150.51
80.00	153.25	1189.9	150.00	150.73	152.50	153.92
85.00	149.97	1204.6	150.00	150.26	148.85	152.64
90.00	149.39	1209.9	150.00	150.01	147.51	149.69
95.00	149.66	1203.9	150.00	149.90	149.02	148.16
100.00	150.56	1196.7	150.00	149.75	150.81	148.93

Fig. 5. Tabular output from DYNAMO

IPSODE has been used for several years at Sunderland Polytechnic and has proved to be useful in helping novice modellers become familiar with some of the main tasks involved in computer simulation.

The user implements the model in IPSODE by responding to a series of questions which establish the variables, $Y(\mathrm{I})$, number of equations and their form together with expressions for the parameters, $P(\mathrm{I})$. The latter may be

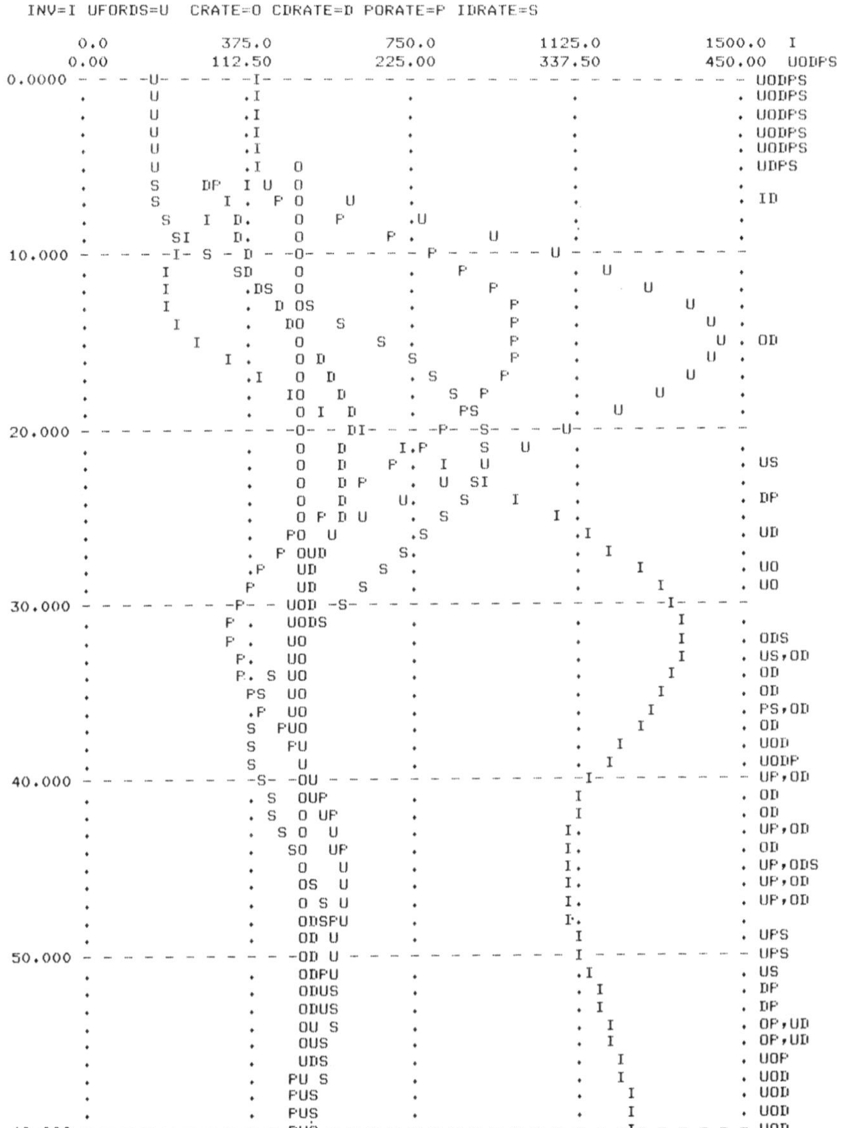

Fig. 6. Graphical output from DYNAMO

constant or vary either discretely or continuously. The independent variable is always the time, *T*. When the user has finished implementing his model, the information is stored in a file, MODEL. This file is then appended to a program called SECOND where the actual parameter values are specified. Such an arrangement is useful because it allows many simulation runs to be made on the same model.

The problem specified in section 4.1 was implemented and run using IPSODE. The input parameter CRATE changes from 50 to 150 after 5 time steps and this was represented by a discretely varying parameter, $P(5)$. The smooth parameter, SCRATE, is input as a continuously varying parameter, $P(1)$. The parameter, CORATE, is input as the continuously varying parameter, $P(2)$. The rate at which stock is replenished, IDRATE, is given by $P(4)$ and involves the use of a third order delay (in DYNAMO terms) using the rate of orders to replenish stocks, $P(3)$.

The computer implementation of the model via IPSODE to be used in the simulation is illustrated in Figs. 7 and 8. There are a number of ways to display

```
If at any stage you wish to change any data already input,press CONTROL+S keys
Do you want help with the input of your problem?N

How many dependent variables are there?2
Are there any control variables?N
Give descriptive names for other dependent variables as Y(I)
Variable Y( 1 ) is called?UFORDS
Variable Y( 2 ) is called?INV
How many equations are there?2
Except for your declared variables,all other constants and
coefficients must be input as parameters P(I).
Do any parameters change their value during the calculation?Y
How many of these parameters are there?5
How many constant parameters are there?3
Would you like an example of varying parameters?Y
Take an equation of the form :- DX/DT=CX+F
                    Suppose    C=1   0<=T<1
                               C=2   1<=T<2
                               C=3   2<=T
then C is DISCRETE varying parameter
                    Suppose    F=X-2   0<=T<2
                               F=3-X   2<=T
then F is a varying parameter that can be represented by
CONTINUOUS functions.
How many varying parameters vary continuously?4
Parameter P( 1 ) is represented by a continuous function
Does this function change during the calculation?Y
You are allowed only one conditional change.This is in the form:-
IF condition THEN P( 1 )=expression1 ELSE P( 1 )=express2
Please input the condition as a combination of Y(I),P(I) and T
Condition=?T<5
Now input the expressions.
Expression1=?50
Expression2=?150-100*EXP(-ABS(P(6)*(T-0.5)/(T-P(6)-0.5)))
IF T<5 THEN P(1)=50
ELSE P(1)=150-100*EXP(-ABS(P(6)*(T-0.5)/(T-P(6)-0.5)))
Is this correct?Y
Parameter P( 2 ) is represented by a continuous function
Does this function change during the calculation?Y
You are allowed only one conditional change.This is in the form:-
IF condition THEN P( 2 )=expression1 ELSE P( 2 )=express2
Please input the condition as a combination of Y(I),P(I) and T
Condition=?Y(2)<8*P(1)
Now input the expressions.
Expression1=?4*Y(1)/(12-Y(2)/P(1))
Expression2=?16*Y(1)/(24-Y(2)/P(1))
IF Y(2)<8*P(1) THEN P(2)=4*Y(1)/(12-Y(2)/P(1))
ELSE P(2)=16*Y(1)/(24-Y(2)/P(1))
Is this correct?Y
Parameter P( 3 ) is represented by a continuous function
Does this function change during the calculation?N
Input the function as a combination of Y(I),P(I) and T
P(  3 )=?
```

Fig. 7. IPSODE implementation of the customer–retailer problem, Part I

```
RUN
  Input of varying parameters.
  How many values does P( 5 ) take?2
        T , P( 5 )
  Pair  1 ?0,50.0
  Pair  2 ?5.0,150

  Input of constant parameters.
  P( 6 )=?8
  P( 7 )=?4
  P( 8 )=?6
  The initial value of UFORDS is?50
  Do you wish to restrict UFORDS?N
  The initial value of INV is?400
  Do you wish to restrict INV?N
Do any of the varying parameters have a non-zero initial value?Y
The initial value ofP( 1 ) is?50
The initial value ofP( 2 ) is?50
The initial value ofP( 3 ) is?50
The initial value ofP( 4 ) is?50
The initial value ofP( 5 ) is?50
  Which interval (A,B) are you considering?
  A=?0
  B=?10
  Do you want the automatic stepsize?Y
  Give a title to your differential equations problem
  ?STOCK
  Do you require your output
             1) Stored in a file
             2) As a print-out
             3) In graphical form
  Please type 1,2 or 3?2
  Are there any varying parameters you wish to observe?Y
  How many are there?4
  Please input 4 pairs of data X,X$ where:-
          X is the number of the parameter
          X$ is a name for the parameter
?1,SCRATE
?2,CDRATE
?3,PORATE
?4,IDRATE
```

Fig. 8. IPSODE implementation of the customer–retailer problem, Part II

the computer predictions and Fig. 9 illustrates the use of its graphical output.

It is probably clear by now that the model formulation described in section 4.1 is really oriented towards the use of DYNAMO. However, IPSODE can cope with the DYNAMO formulation quite well and the results from both packages are comparable. In fact, the formulation could also be implemented using CSMP and ACSL with an IPSODE-type approach. Although comparable, there is a difference in the predictions by the two implementations. These discrepancies can occur for a number of reasons including:

(i) different numerical techniques employed with inherently distinct accuracy levels,
(ii) the fact that the implementations are not precisely equivalent.

It would be interesting to see what the simulation might predict if the model was formulated with, for example, output variables CDRATE (orders delivered to customers) and PORATE (purchase ordering rate).

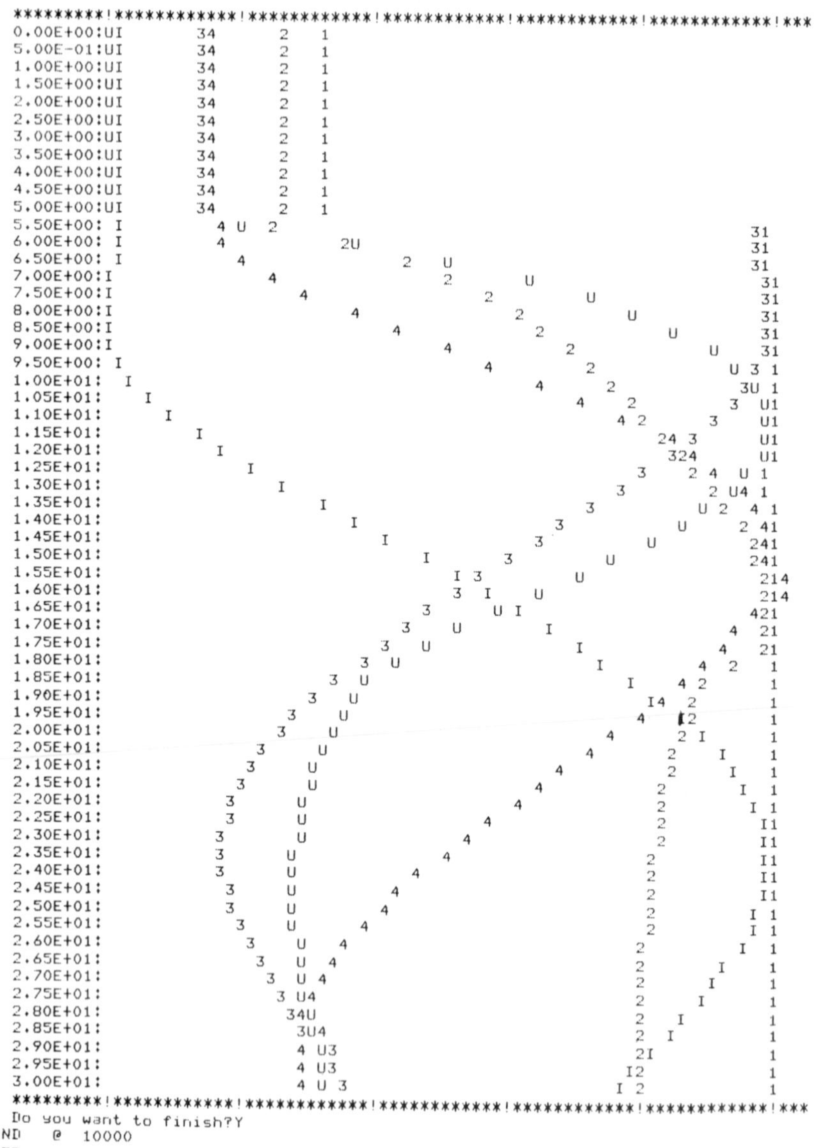

Fig. 9. Graphical output from IPSODE

5. CONCLUSIONS

It is certainly the case that once a model has been suitably formulated, then implementation via an appropriate 'user friendly' simulation language can be straightforward. Students also learn to use the facilities of such languages quite quickly so that they can experience most aspects of simulation fairly readily. As such, there is adequate software currently available upon which to base or utilize in courses on computer simulation. However, although software availability means that rather more complex (and possibly realistic) situations can be handled, it also means that the teacher has a greater responsibility to alert his students to all the attendant pitfalls and difficulties associated with the *computing* in computer simulation.

REFERENCES

Aris R., 1978, *Mathematical Modelling Techniques*, Pitman.

Atos I. and Mussgrave A., 1970, *Criticism and the Growth of Knowledge*, Cambridge University Press.

Cross M. *et al.*, 1982, 'Interactive computer simulation tools for use in teaching mathematical modelling', *Int. J. Maths. Ed. Sci. and Tech.*, **13**, 763.

Green W. and Speckhart F., 1980, 'CSMP', *Simulation*, **34**, 131.

Hermann C. F., 1967, Validation problems in games and simulations with special reference to models of international politics, *Behavioural Science*, **12**, 216.

Ignall B. J. *et al.*, 1978, Using simulation to develop and validate analytical models: some case studies, *Operational Research*, **26**, 237.

Mitchell E. and Ganthier, J., 1976, Advanced Continuous Simulation Language (ACSL), *Simulation*, **28**, 72.

Moscardini A. O. *et al.*, 1980, Interactive computer simulation without programming, *Ads Eng. Soft.*, **2**, 117.

Naylor T. H. and Finger J. M., 1967, Verification of computer simulation models, *Management Scienee*, **14**, 92.

Pritzker A. A. B., 1974, *The GASP-IV Simulation Language*, Wiley.

Schriber, T. J., 1974, *Simulation using GPSS*, Wiley.

Shaffer, W., 1980, DYNAMO, *Simulation*, **34**, 134.

Wigan, M. R., 1972, The fitting, calibration and validation of simulation models, *Simulation*, **18**, 188.

Zeigler B. P., 1976, *Theory of Modelling and Simulation*, Wiley.

Mathematical Modelling in the School Curriculum

David Burghes
School of Education, University of Exeter

1. INTRODUCTION

Although the major experiments in the introduction of mathematical modelling courses have been in higher education, the same feeling as experienced in H. E. of wanting mathematics to be more practical, more relevant and more useful has been prevalent in schools over the past decade or more.

This urge to make mathematics at school more relevant has been particularly fuelled by

(i) the introduction of modern mathematics which has led to the belief that mathematics, as taught, is fairly irrelevant;
(ii) the pressure from children, who realize that for them school mathematics is the end of their mathematical careers, and consequently are totally unmotivated by the normal mathematical topics.

So the demand to make mathematics at school more relevant is certainly there, the pressure coming from both pupils and teachers. Responding to this pressure is not easy. Note also that this is a demand for relevance, not necessarily for mathematical modelling.

If we stick with the present C. S. E. syllabus, just how do we make, say, matrices relevant to the average pupil at school. The usual sort of 'shopping basket' question is just not convincing for this audience. There have though been a number of projects aimed at making mathematics relevant, and we will consider two, the 'Maths Applicable' project and the work of the Spode Group.

2. MATHEMATICS APPLICABLE [1]

This was an expensive project, financed by the School's Council with the aim of producing A/O level in mathematics for the non-specialist mathematician—i.e. for the physicist, chemist, biologist who requires mathematics post O-level but not a full A-level. So the emphasis was very much on making the pupil feel able to apply his mathematics.

In hindsight, it is quite clear that this course, aiming as it was at an A/O level, was never likely to succeed. In the event, the course turned out to be really quite a challenge, and should have been a full A-level. It originally had a project element, but this only lasted one year, the argument coming from the validating body (the University of London) that projects were too difficult to mark!

Personally, I have never been convinced that the material was presented in the right way. I have two major criticisms

(i) much of the mathematics uses 'whimsical models', that is mathematical models which have little relevance, but are just there in order to dress up the mathematics to look relevant. For example

In a certain area of Suffolk it is expected that the quantity of blackberries (w tonnes) ready for picking at a time x days after 1st August will be given by

$$w = \frac{x}{50000}(96000 - 5200x + 120x^2 - x^3)$$

Find the rate of increase of w, in tonnes per day, when $x = 20$, 40 and 60.
Also find the value of w when $x = 60$.
For what values of x could the model apply?

(ii) much emphasis was put on designing mathematical models for use in hypothetical situations—projective modelling as it has been called. I would have preferred to see mathematical models which are used in practice.

Despite these criticisms, I think that the project deserves praise as it did really try to change the way in which we teach mathematics, and it has at least produced a valuable resource for others to use, and incorporate into existing courses.

3. THE SPODE GROUP

Almost three years ago, the Spode Group first met in order to produce a series of case studies, aimed at showing the practical uses of school mathematics. So the aim of the Group was far less ambitious than that of maths applicable, but of course the costs were minimal (the work being achieved mostly at weekends).

The Group has met regularly over the past three years, producing three books of real problems [2], two more in press [3], and three more in various stages of preparation [4]. Throughout these books the emphasis has been on finding examples that are as close as possible to the real world. The case studies in the first three books have all been written in a standard form

1. Problem Statement
2. Possible Solution
3. Teaching Hints

and we have allowed photocopying of the problem statements so that these can be handed out to groups. In this way we aim to encourage group work with the teacher playing the role of problem originator, and giving little or no mathematical help.

An example of this material is given below.

Where to Cut

A basic problem faced by some industries is how to best cut up lengths of tube which are supplied with fixed dimensions, in order to meet the required demand, with as little wastage as possible.

A simple example of this basic problem is to determine how many lengths of tube 10 m long are required to meet the following order and how should they be cut.

60 – 3 m lengths

49 – 4 m lengths

12 – 7 m lengths

If the basic lengths of tube are supplied in 12 m lengths, what is the answer?

Another problem faced by industry is that sheets of metal are supplied with fixed dimensions 10 m × 10 m and the following order is to be satisfied

60 – 3 m × 1 m

49 – 4 m × 2 m

12 – 7 m × 5 m

How should the sheets be cut? What happens if the sheets are 12 m × 5 m?

Of course, the teaching approach that we are encouraging here is somewhat different to the usual exposition/exercise approach, and it needs confidence as well as humility to use this approach well.

4. COCKCROFT AND BEYOND

For those of you not familiar with the findings of the committee of inquiry into secondary mathematics teaching, chaired by Dr. Cockcroft, I will bring to your attention three points which seem of particular importance

(i) many adults cannot cope with simple sums in everyday life.

(ii) too much maths in secondary schools is not only too difficult for the majority of children, but entirely irrelevant to their future jobs and adult lives
(iii) in general, employers had few complaints about the mathematical skills of their young employees

Also the range of mathematics actually needed by all pupils for future life is small indeed:

percentages
giving change
weighing and measuring
understanding inflation
understanding 24-hour clock
reading timetables

Last (but certainly not least), six components of good mathematics teaching are:

(a) practise of routines
(b) exposition
(c) discussion
(d) problem solving
(e) practical work
(f) investigations

We all know that teaching modes (a) and (b) abound in our educational system but that there is very little of (c)–(f). Mathematical modelling does provide a useful vehicle to launch into these other teaching modes.

For example, take the following problem, based on a problem in 'Solving Real Problems with C. S. E. Mathematics' by the Spode Group

Design a £1 book of stamps to cater for the present postage rates

	1st class	2nd class
Not over 60 g	16p	$12\frac{1}{2}$p
Not over 100 g	23p	17p
Not over 150 g	29p	21p

This problem certainly should provoke discussion—for example,

(i) What will maximize P. O. profit (left over $\frac{1}{2}$p, 1p etc.)?
(ii) What do the public want in a book of stamps?
(iii) Can stamp machines easily give change? etc.

There is a real problem to be solved, and practical work could include a survey of what class of post people use, and even an investigation of how long it takes letters (1st and 2nd) to reach prescribed destinations from your local area. But time spent on this sort of work will necessarily lead to fewer mathematical topics being covered for the average pupil. I, for one, would welcome this. So much of the mathematics we teach to 'average' pupils really is not relevant, and the syllabus has been designed on the 'top-down' principle—University entrance dictates A-level, which in turn leads to O-level and C. S. E. has become a watered down O-level.

Another recommendation from Cockcroft was for syllabuses to be designed on the 'bottom-up' principle. Decide what we really want to teach to the bulk of school pupils, and move on upwards to O-level and A-level.

5. FINAL REMARKS

This chapter is not a plea for mathematical modelling to become an integral part of all school mathematics teaching. It is though a plea that we at least look at what we are presenting closely and consider some of the alternatives, which could include modelling.

I would stress that I do not think it is that easy to teach modelling to average pupils. We urgently need mathematical applications resulting from problems which

(i) have intrinsic interest level and/or relevant to the pupil's everyday life;
(ii) are easily understood;
(iii) stimulate discussion, communication and critical judgement;
(iv) show the role of mathematics in society today and so teach the skills of formulation, communication, judgement.

Success in criterion (i) is often difficult to achieve. In fact, a rather whimsical problem, like the one below might well achieve greater pupil participation and enthusiasm.

Travelling-Spy Problem [5]
A spy has to travel regularly between the cities given in the table below. Find a path, passing through all the cities and back to the initial city, which has minimum distance

	Berlin	London	Madrid	Moscow	Paris	Rome
Berlin	—	7	15	11	7	10
London	7	—	11	18	3	12
Madrid	15	11	—	27	8	13
Moscow	11	18	27	—	18	20
Paris	7	3	8	18	—	9
Rome	10	12	13	20	9	—

Not only that, it leads on to the topic of algorithms and discrete mathematics—topics which must surely figure in any new syllabus construction.

So my final thoughts are not necessarily those of great enthusiasm for modelling everywhere in the school syllabus, but a more cautious yet innovative approach. What approaches provide teachers with material which pupils can feel real enthusiasm for and so provide them with a rewarding mathematical experience at school? As yet, I don't think we know the answers!

REFERENCES

[1] Mathematics Applicable Series, Heinemann, 1900.

[2] *Solving Real Problems with Mathematics*: Vol. 1, The Spode Group, Cranfield Press (1981).
Solving Real Problems with Mathematics: Vol. 2, The Spode Group, Cranfield Press (1982).
Solving Real Problems with C.S.E. Mathematics, The Spode Group, Cranfield Press (1982).

[3] *Enterprising Mathematics*, The Spode Group (to be published by Oxford University Press).
Motivating A-level Mathematics, The Spode Group (to be published by Oxford University Press).

[4] *Case Studies in Mechanics*, The Spode Group (in preparation).
Numeracy for Arts Students, The Spode Group (in preparation).
Discrete Mathematics for Schools, The Spode Group (in preparation).

[5] Open University, Course TM 361 (1982).

Mathematical Modelling = Mathematical Motivation?

M. S. Townend
Liverpool Polytechnic

SUMMARY

In the first part of this chapter the value of presenting all students with mathematical examples relevant to their own disciplines is discussed. Both mathematical models and mathematical modelling are seen as ways of achieving this end.

The second part of the chapter deals with a particular modelling course given to sports science students. A comprehensive list of the sports problems used is given together with an anecdotal report on the progress of the course. Many of the examples could easily be incorporated as motivational material on a wide range of existing sixth-form and undergraduate courses.

1. WHY MODELLING?

How often have students asked each of us the question 'Why do I need to study all this mathematics?' or 'Why do I need to study such and such a topic?'. The answer is probably, 'More times than we care to admit', given the fact that we are teaching from syllabuses which are 'relevant'.

With service classes particularly, the answers to these questions must be convincing if the students' interest in both the subject and its wealth of applications is to be stimulated.

Although there are notable exceptions, many of the mathematics texts used on undergraduate courses present topics in a stereotyped way even though the 'target audience' may be any group from amongst mathematicians, mechanical engineers, biologists, pharmacists, sports scientists etc. For example most undergraduate mathematics syllabuses (either service or specialist) contain the first order differential equation

$$\frac{dy}{dx} = \lambda y,$$

and the exponential function

$$e = \lim_{n \to \infty} \left(1 + \frac{1}{n}\right)^n.$$

A biology undergraduate is unlikely to be persuaded that these are important topics to discuss unless their introduction is related to biology. Telling him how a physicist or economist might interpret the equations would probably do more harm than good—he possibly chose biology because he did not like physics!

Within the last few years there has been an increase in the incidence of mathphobia students who can be identified by their cry of 'I won't be able to do this because it involves some mathematics. I have never been able to do mathematics'. Such students have to be coaxed very gently in order to give them some confidence and then to persuade them to use even the most elementary mathematics. This is a difficult and time consuming task but it can, and should, be done.

Given the often poor mathematical background of freshmen I believe that there is frequently too much material in mathematics courses (service and specialist) at universities and polytechnics. In my opinion the volume of material should be reduced so that the application of the material which is left can be expanded upon. There is very little point in teaching students a vast range of sophisticated mathematical techniques if at the end of the course they do not have the confidence, ability or inclination to apply them.

Mathematical models can be used successfully to introduce many of the topics currently found on undergraduate syllabuses and once a certain level of mathematical expertise has been reached, mathematical modelling can be used to develop the applications skill. Students who have seen the power of mathematics used to solve a problem drawn from their own discipline and, more importantly, have developed and solved a modelling exercise of their own in the same area are likely to be better motivated towards mathematics.

Apart from the development of motivation and problem solving skills a most important benefit is the observed improvement in the students' skill in written and oral communication through the writing of reports and discussions. Employers rate this skill very highly as is evidenced by the statement of McLone (1973) that 'mathematicians and their employers rank skills in problem formulation and communication at least as highly as ability in solving mathematical problems'.

As mathematical educators, we would be guilty of neglect if we do not modify our courses to provide students with the skills demanded by employers.

The inclusion of a mathematical modelling component in our courses is precisely the modification required to achieve the above aims.

The second part of this paper presents an anecdotal report on a mathematical modelling course given to a group of sports science undergraduates.

2. EXPERIENCES WITH A MATHEMATICAL MODELLING COURSE FOR SPORTS SCIENCE STUDENTS

During the last academic year, the opportunity arose to introduce a mathematical modelling course to a group of final year Sports Science students. The staff–student contact time was one hour per week for eighteen weeks plus an allocation of the students' own time which is normally used for homeworks and any other assignments on traditional courses.

Informal conversation with the students revealed that while they were intrigued by the title of Mathematical Modelling they had little idea what was in store. Prior to the commencement of the course the students were given a brief handout (about six pages) which described the philosophy, methods and aims of a modelling course. Although modelling should be an unfettered process I thought it prudent that these introductory notes should give the students some idea of a suitable path to follow. Consequently the Open University seven-stage system described by Berry and O'Shea (1982) was included. My first lecture cast the students in the role of observers while I demonstrated the modelling process with reference to an analysis of the world weightlifting records as presented by Burghes and O'Carroll (1980) in order to establish who is the best weightlifter. The intention was to illustrate the rationale behind the assumptions that are made, the formulation of the problem in mathematical terms, its solution, a critical analysis of the results possibly leading to a revision of the model and finally the presentation of a report. These students are usually very self-confident people and little difficulty was experienced in getting them to participate in the discussions. This is completely different from other undergraduates with whom I have used modelling exercises where the major difficulty has been in initiating their participation!

After this introductory example it was important for the students to become *actively* involved in their own modelling exercise as soon as possible. At such an early stage, I thought it reasonable for them to model some activity related to their major discipline, sport, in order to gain some confidence in the application of a modelling strategy. The students were invited to form themselves into several small groups (each of about three members) and then select a problem from amongst an extensive list of observations or brief statements on sports performances, see Appendix I.

Their brief was to develop a mathematical model of their chosen system and produce an explanation for the observed phenomenon. Their first reaction was

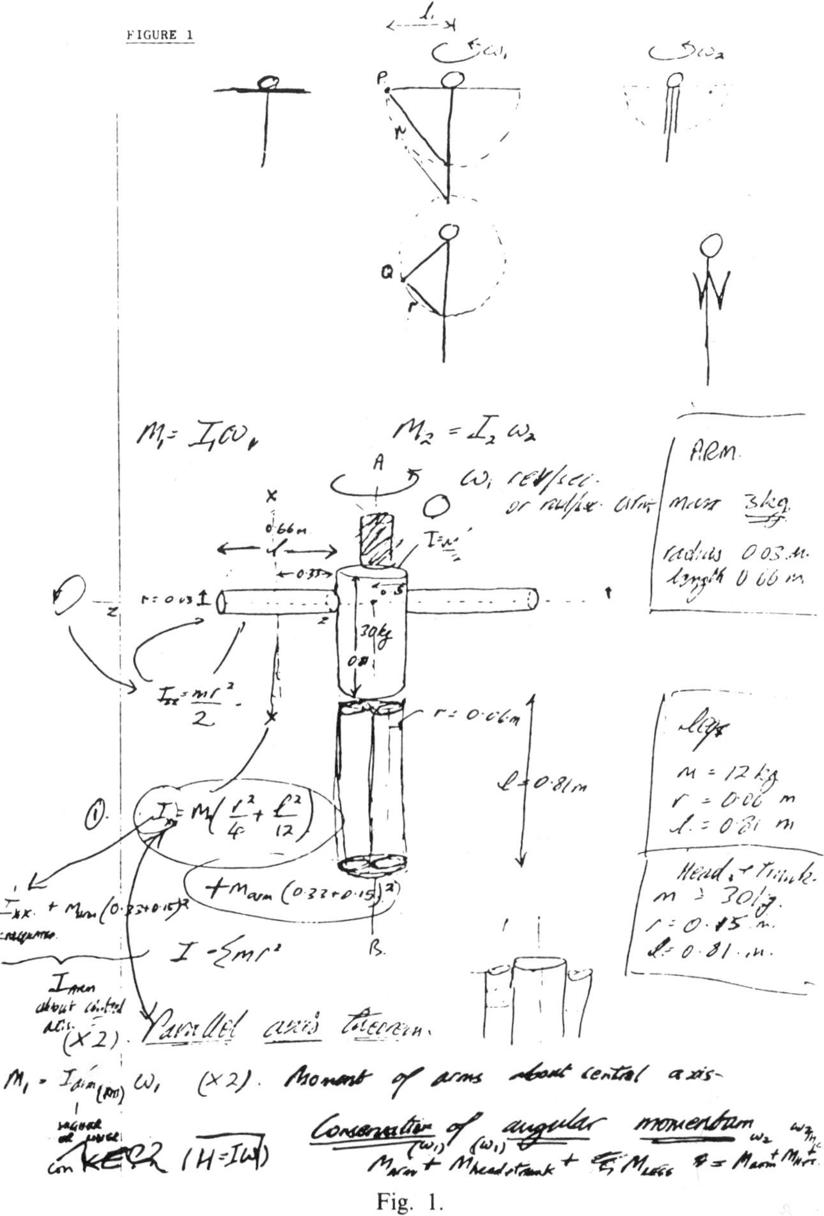

Fig. 1.

a stunned silence and many quizzical looks which seemed to say 'What exactly do you want us to do?'. After some discussion within their groups, the various spokespersons approached me and asked for data. They were quite startled when told that they must specify precisely the data they required! Despite this tentative start all groups had put pen to paper and begun their initial modelling by the end of the class. I encouraged the students to make rough notes of any ideas which occurred to them about the sort of model they might use, the data they might need and any physical or mathematical principles they might expect to be useful in the problem solution, Fig. 1 shows one group's work sheet. At the time I found it difficult to convince the students that such untidy scribblings of their ideas constituted probably the most important aspect of their modelling assignment. It was only after much persuasion that the students reluctantly allowed me to reproduce the rough working shown in Fig. 1. Instead they wanted to submit a neatly written transcript which would, of course, have completely destroyed the displayed spontaneity of their problem formulation. As they left the room at the end of their initial modelling session one of the students commented that 'this is real mathematics isn't it, we have to decide what data we need and what we shall do with it. It is not just crankturning'. At last the light was beginning to dawn!

The groups had obviously been working on their models during their free time during the following week because at the next class some requested more data and some asked for advice with regard to their mathematical solution. At the close of this class a couple of the groups had obtained an answer to their problem formulation. As mentioned earlier, the introductory notes contained a statement of the Open University's seven-stage approach together with other comments about the need to be critical of the answer and prepared to revise the original model in the light of this criticism. Despite these, the critical assessment and subsequent revision of the model proved to be the most difficult task to persuade the students to perform. Perhaps it was the fault of the instructor—may be this stage was not emphasized strongly enough in the notes and lectures or, possibly, the students were loathe to reject an answer which they had worked hard to obtain.

An anecdote about one group's modelling of the shot putting event serves to illustrate this point. As might be expected the group produced a standard projectile approach and concluded by, correctly, identifying the release velocity as the most important parameter. Their critical review stated that air resistance effects had been ignored which is perfectly reasonable for an initial model. When asked why they had not considered its inclusion as a refinement, they replied that although they subsequently tried to include the effect, the mathematical solution of the problem had proved too difficult for them. Consequently that section of the work was not complete and therefore had not been included in their report. They could not believe that inconclusive work was acceptable in their report—to them an answer was paramount. This occurred despite stressing at the outset that the object of the course was to develop their modelling, and not their mathematical expertise.

After one month, each group submitted its work in the form of a written report. The groups were given a free hand in how the production was apportioned between their members. Several teaching colleagues were sceptical about this, and suggested that it was open to abuse since some students would do all the work and some would merely be 'passengers'. In the event their scepticism was unfounded. Each member of the class appeared to have done their share of the work. At the end of the course I asked the students for their comments about the above criticism. While they admitted that it could happen they were adamant that they were not 'carrying any passengers' in their groups!

For a first attempt the standard of the reports was fair and produced an average mark of 53 % when marked on a system similar to that employed by the Open University and described by Berry and O'Shea (1982), although they lacked the usual structure of abstract, introduction, report and conclusion. The reports were returned to the students in order to give them some feedback which could be incorporated in their second and final report submitted on a topic of their own choice at the end of the course. The marks obtained for these showed an improvement (new average mark 60 %) mainly due to an improved criticism and revision section.

After submitting their first reports, one of the students came to see me. He is very keen on powerlifting (a form of weightlifting competition) and showed me a copy of the results of a recent World Powerlifting Championship competition. Could we, he asked, use these data as the basis of a lecture, similar to their introductory, passive, lecture on weightlifting models. As soon as a graph of log (lift) versus log (body weight) was drawn, it was apparent that the results did not conform to the pattern established by the Olympic weightlifting results, see Fig. 2. The difference was apparent to a greater or lesser extent regardless of whether the total lift or a single lift component was used. Clearly there were two different mechanisms in operation. Neither the group nor myself had expected this and initially we were unable to explain it (the critical appraisal stage of modelling) although after some discussion our 'expert' was able to provide a reasonable explanation. From my point of view as the instructor this impromptu exercise was one of the highlights of the course. It reinforced dramatically the need for the modeller to be able to communicate his results to a client (McLone, 1973) and showed the advantage of having access to specialist knowledge at the critical assessment stage of the modelling process.

While the students were working on sports models to develop their modelling ability (modelling in an area of familiarity) two of them made the comment

> We can see how mathematics can help us analyse sports problems, but could we try some situations where our fore-knowledge and intuition is not so good?

Although such a comment had not been anticipated, non-sports models were naturally planned for inclusion later in the course. The problems chosen for these more general modelling exercises included, *inter alia*, the warehouse

Weightlifting and Powerlifting

Weightlifting as seen in the Olympic games involves two types of lift, the snatch and the jerk.

An elementary modelling approach using dimensional analysis leads to a proposed relation between the lift (L) and the body weight of the athlete (W) of the form

$$L = kW^{2/3}$$

Powerlifting is a different type of competition and involves three lifts, the squat, the bench press and the dead lift.

While reasonable adherence to the above model is shown by recent Olympic weightlifting results, the results of a recent World Powerlifting Championship showed considerable divergence, regardless of which lift is used for the value of L.

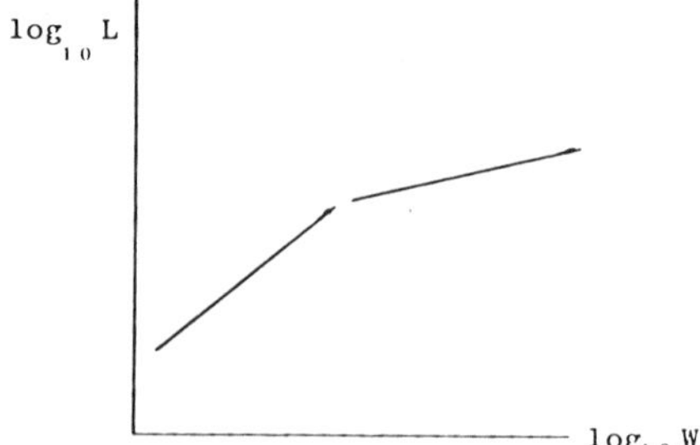

FIGURE 2

location problem, changes in the Retail Price Index and the newsvendor's dilemma (Croasdale, 1982). Each of these might almost be considered to be a classic in the modelling literature. On completion the students all expressed surprise that realistic problems such as these could be given a plausible answer using only (at most) mathematical and statistical techniques met by first-year undergraduates.

The remainder of the course involved the student groups in modelling other problems such as vehicle separation in a tunnel (Burghes, 1981), and selection of the best strategy for an average darts player to adopt for scoring a treble twenty. Concurrently, the student groups were each working on their second modelling report. As reported earlier there was an improvement in the quality of their second reports showing the benefit of the earlier feedback from their first reports.

The marking of the reports can be a very time-consuming activity. They certainly take much longer to mark than a traditional mathematics homework. In fact it is probably more akin to the marking of essays by our arts colleagues. As modelling courses become more widespread the demands made on staff time by their preparation and assessment *vis-à-vis* traditional mathematics courses will have to be taken into consideration by those of our colleagues who are responsible for timetabling.

3. CONCLUSIONS

The students responded well to what was a very different sort of course for them. After some initial hesitancy they challenged statements, queried results and generally became more discerning as the course progressed. Of equal importance was the fact that there was an improvement in the standard of their written work; that alone would make the course worthwhile.

At the end of the course the students were requested to complete a short questionnaire (see Appendix II) in order to assess the value of the course as a whole and the suitability of the examples used. The results consisted mainly of '2's on the arbitrary scale used suggesting that most students thought that the blend of the course was about right and that they had benefited from it. The group commented that in retrospect they could identify areas of all three years of their course in which knowledge of a modelling methodology would have been beneficial. This suggests that the modelling course would have an even greater impact if offered during the first year.

Like any other modelling course, this was one iterate in a hopefully convergent sequence. I look forward to incorporating the lessons learned in next year's course.

Since the completion of the course some of the sports examples have been included as motivational examples with other groups of students. Most people have some interest in sport, even if only in an armchair capacity, and as expected the examples proved to be popular with the vast majority of students, regardless of their particular discipline.

APPENDIX I

This appendix contains a selection of some of the sports problem observations or questions from which the students selected their first modelling exercise.

(i) Why does the rate of spin about the vertical axis of an iceskater increase as the arms are brought in towards the body?
(ii) 'I am a shot putter. Why does my coach emphasize the importance of release velocity so strongly?'
(iii) How fast does a racing cyclist accelerate?
(iv) Should an average darts player, trying to score a treble 20, actually aim at the treble 20?
(v) Why does a sprinter raise his knees high while running whereas a marathon runner does not?
(vi) Does a sprinter exert a constant propulsive force throughout the 100 m sprint?
(vii) 'I am a sprinter. There are three different types of sprint start which use starting blocks. Which would be best for me?'
(viii) In basketball, why are free shots released at such an acute angle to the horizontal?
(ix) Why do élite downhill skiers add small weights to their skis?

Cases (i), (iii), (viii) and (ix) are described fully by Townend (1980), (1983a), (1983b) and (1981) respectively.

APPENDIX II

As mentioned in the text, the following brief questionnaire was answered by the students at the completion of their course. For each question, the students were asked to circle one entry on the arbitrary scale.

1. Do you consider that this course has increased your ability to formulate and solve problems?
   Greatly increased = 1 2 3 4 = decreased
2. Did you find the project useful?
   Very = 1 2 3 4 = Not at all
3. The skills learnt will be very useful elsewhere
   Very useful = 1 2 3 4 = Useless
4. Circumstances dictated that you studied this course in your final year. Do you consider that you would have benefited more from the course if it had been available at an earlier stage of your undergraduate career
   Definitely = 1 2 3 4 = Don't know

5. Were you 'caught out' at any stage of the modelling course because you lacked the necessary mathematics to solve the problem you had formulated.
   No—never = 1 2 3 4 = Yes—often
6. Did the wide range of examples used help you to appreciate the use of ability in mathematical modelling?
   Yes = 1 2 3 4 = No

REFERENCES

Berry J., O'Shea T., 1982, *Int. J. Math. Educ. Sci. Technol.*, **13**, 715.
Burghes D. N., 1981, The Humber Tunnel Authority, in *Case Studies in Mathematical Modelling*, Ed. D. J. G. James and J. J. McDonald, Stanley Thornes, p. 101.
Burghes D. N., O'Carroll M., 1980, *Bull. I.M.A.*, **16**, 155.
Croasdale R., 1981, The News-Vendor Dilemma, in *Case Studies in Mathematical Modelling*, Ed. D. J. G. James and J. J. McDonald, Stanley Thornes, p. 170.
McLone R. R., 1973, The Training of Mathematicians, S. S. R. C. Report.
Townend M. S., 1980, *J.M.M.T.*, **3**, No. 1.
Townend M. S., 1983a, Teaching Mathematics and its Applications, 2, No. 1.
Townend M. S., 1983b, (Mathematics in Sports, Ellis Harwood).
Townend M. S., 1981, *Proc. of Sports Biomechanics Study Group* (Sports Science Society), **5**.

The Mathematics of a Modern Series of Texts—A Case Study

J. W. Searle
West Denton High School, Newcastle upon Tyne

SUMMARY

The view that children form of mathematics, whilst at school, would seem to be influenced both by their teacher's attitudes and approaches to mathematics teaching, and also the resource materials they use. A resource material in common use in a secondary school mathematics course is a series of text books. Several such courses were published and revised in the 1960s and 1970s reflecting the modernization of school mathematics, the influence of learning theories, and also the introduction of comprehensive schools and the CSE examination.

The 'Modern Mathematics For Schools' course of the Scottish Mathematics Group is evaluated here with particular reference to the nature and structure of the mathematics that it embodies and contemporary learning theories. These materials have about a 20% share of the U.K. market, and a discussion of a small sample survey of how teachers find them is included.

It is argued that in the main course there is too great an emphasis on the drilling of structural principles and their terminology and notation at the expense of relating mathematics to 'real world' problems. Also, abstractions are found to be made too rapidly for most children to appreciate them, and a need is indicated for more concrete experiences relevant to the child, and for more open, exploratory work in which children can use their own language and notation. The course is found to be a sound preparation for examinations for the more able pupil, but as such, it distorts the creative aspects of the nature of mathematics. In particular a misleading view as to the nature of mathematical modelling is given. The impression that most pupils gain from this course may well be one of mathematics as meaningless names and symbols, remote from their own experience, which is 'done' for examinations.

Teachers seem to find this a rather dull course lacking in stimulating ideas. Although it is suited to the more able pupil together with a well qualified teacher, most pupils may well be mystified by its modern approaches. The success of the aim of the course to make mathematics enjoyable and relevant for all children is questioned. They are unlikely to gain much mathematical understanding of their world from it, or to gain much appreciation of mathematics itself. It is argued that if such an aim is to be achieved, what is needed is a modelling approach, probably involving computers, in which 'real' identifiable problems can be investigated. It is argued further, however, that the convenience, permanence and authority of a series of text books such as this course together with the pressure of time and examinations, and expectations of the traditional secondary classroom, all act against the adoption of such an approach to mathematics teaching.

1. INTRODUCTION

The materials known collectively as 'Modern Mathematics for Schools' were prepared and developed by the Scottish Mathematics Group. They are the project materials of one of many projects that arose during the 1960s in response to a generally felt need to modernize school mathematics. The S.M.G. reviewed the secondary school syllabus in Scotland and prepared new materials with a view to making the school mathematics course interesting, relevant and enjoyable for all pupils and introducing some modern ideas whilst forming a sound foundation for those pupils who would study mathematics at higher levels. The S.M.G. organized their main course under separate section headings, principally of algebra, geometry and arithmetic, to allow teachers flexibility in their development of the course, but it is claimed that throughout the course, ideas, methods and materials are integrated within each branch of mathematics and across the branches so that it is mathematics as a whole that is presented. The S.M.G. also claim that their course encourages pupils to find out facts and discover results for themselves and to observe and study the themes and patterns that pervade mathematics today. In addition it is claimed that while the course emphasizes fundamental principles and reasonable attention is made to the matter of structure the width of the course should be sufficient to provide a useful experience of mathematics for those pupils who do not pursue the subject beyond school level. The aims of this course raise several questions such as just what is meant by 'mathematics as a whole', and 'discovery learning' and what constitutes a 'useful experience of mathematics' for the vast majority of children in secondary schools.

Questions such as these are discussed here, as part of an evaluation study of the S.M.G. materials, undertaken to investigate to what extent the aims of the course are being achieved. There is a need for evaluation of project materials, for if a project is to develop further, areas where its materials might benefit from modification in content or in use need to be discussed and brought to light. It can be noted here that evaluation of the projects initiated in the 1960s was not built in at the start, and the main criteria of a successful project seemed to be the number of children using its materials or achieving success in examinations both of which are of limited value as regards any further development (Griffith and Howson, 1974; Ormell, 1980).

The S.M.G. materials have about a 20% share of the U.K. market (90% Scotland, 12% England and Wales—publisher's estimate) and although this course is virtually compulsory in Scotland, owing to the centralized nature of the education system there, teachers in England and Wales particularly need information from evaluation studies in order to choose resource materials from the vast range available. This need is further enhanced by the continuing controversial debate over what mathematics is appropriate for schools (Mathematical Association, 1963; Hammersley, 1968; Thom, 1971; Kline, 1973), especially when schools themselves have been undergoing rapid changes with the coming of comprehensive schools and mixed ability classes, and the CSE examination, all within the past two decades. The S.M.G. materials like all such projects initiated in the 1960s were originally designed for pupils capable of attaining O' level standards, and the course materials were necessarily adapted in order to try and cater for a much wider ability range.

The evaluation study offered here has two main aspects. The nature and structure of mathematics embodied within the course are discussed in the light of contemporary learning theories, and also the results of a small survey of teachers who actually use some of these materials is discussed. It was considered particularly important to try and ascertain how teachers react to the materials for apparently text books determine to a great extent what is taught in schools and when (Wooton, 1965; Austin and Howson, 1979). Also the view children get of mathematics would seem to be influenced by their teachers' attitudes and approaches to the subject.

2. HISTORICAL BACKGROUND

Firstly, it will be useful in the light of the debate as to what constitutes appropriate mathematics for most children, to discuss the origins of modern mathematics in secondary schools and of the S.M.G. The idea of 'reform' in mathematics education is not new. Factors such as the needs of society, the utility of mathematics and its applications, how best to integrate the subject branches and the place of rigour in teaching mathematics, were all present in the debate that 'reformed' mathematics education early in this century. However, there were teachers then who saw no reason for teaching mathematics other than 'mental discipline', and found convenience and security in working for examinations from set textbooks. To just what extent reform did take place is debatable, and it has been noted that the principles underlying the early reforms as published by the Mathematical Association in 1911, would still seem very much valid today (Howson, 1973). The reform to 'modern mathematics' in the 1960s was influenced by much the same factors as above, except that in addition, advances and developments in mathematics itself made at university level were putting extra pressures on the schools to be 'up to date' (Griffith and Howson, 1974). It is not that the concepts of 'modern mathematics' are in any way modern, for their origin probably lies with the unity and cohesiveness Galois gave to abstract algebra in the 1830s; but concepts such as set theory, relations and functions, and ideas of structure in the number system and vector spaces, together with statistics probability and computing were developing rapidly in universities in the post-war years, but not in schools to any real extent.

The first major milestone of the new reforms in England was the Oxford Conference of 1957, when mathematicians from schools, universities and industry met virtually for the first time. In the United States the School Mathematics Study Group was set up to lead a massive reform in school mathematics and internationally, things came to a focus at the OEEC conference (now OECD) at Royaumont in 1959, and the conference report *New Thinking in School Mathematics* (OEEC, 1961) and others that followed greatly influenced the changes in mathematical education in Scotland, and the S.M.G. course materials. Although many aspects of mathematical education were discussed at the Royaumont conference, it was syllabus content that dominated the debate which itself apparently was dominated by professional mathematicians interested only in furthering the supply of potential research mathematicians (Ormell, 1969). At the time Goodstein warned that changes in school syllabi should be to add interest to the school course and to deepen understanding and not simply to advance the starting point of university courses (Rollett, 1963). The conference itself raised the problem of whether mathematics as a liberal education, mathematics as a basis for living and work and mathematics as a foundation for university study could all be achieved with one curriculum suitable for all children. However, when the conference came to publish its *Synopses* (OEEC, 1961) as a basis for classroom texts and experimentation it was a modern view of mathematics itself aimed at the more able child that dominated. Applications and the development of formal abstract principles through concrete and intuitive experiences and discovery learning were only mentioned briefly.

The large influence of the OEEC publications on the original S.M.G. materials is apparent from Robertson (1969). In 1964, a committee that was to become the S.M.G. issued a draft syllabus to all schools. The general aim of this syllabus, also gives the flavour of its content and was as follows.

> To provide a course that would interest the pupils in mathematics and would train them to use the language in which popular, technical and professional

texts were then being written; to give an early introduction to the concept of a set, the structure of a number system, the use of vectors and the idea of a group; to relate mathematics to the solution of up to date problems by means of linear programming, the use of matrices, iterative processes and calculating machines—in short to include the kind of mathematics that reflected modern developments and would lead to useful links with later work at school, college or university, and in the long term to increase the number of persons qualified in mathematics.

Between 1964 and 1968 the S.M.G. published the seven books of their main course to Ordinary Certificate level, each book following piloting in schools, feedback and discussion before the final version was published. Thus this project in Scotland was typical of what came to be called the research, development and diffusion model of curriculum dissemination (Havelock, 1971; see Fig. 1). This mass dissemination literally imposed the new curriculum on teachers, with the attendant dangers of teachers not really understanding or accepting new materials, using them superficially or not as intended, or, in short, 'garbling the message' (Griffith and Howson, 1974). In Scotland the last traditional syllabus Ordinary Certificate examinations were sat in 1972, and the books started to find favour in England and elsewhere and the materials, with revisions to meet the changes in education that occurred in the 1970s, now cover a comprehensive range. These include Books 1 to 7, which constitute both an 'O' level and CSE course, and Books 8 and 9 at Additional Mathematics level, plus specialist topic books for use in the sixth form and beyond. Books of worksheets, which are very simplified versions of the early main course, are available for the less able enabling flexibility in mixed ability classes, and booklets of multiple choice questions match each of the main course books. These offer the teacher a quick and convenient method of assessment, but the extent to which multiple choice questions really test a child's understanding and appreciation of mathematics is highly questionable. It is of particular note, in the context of 'real world mathematics', that as late as 1979, the S.M.G. published an 'interest pack', one of whose purposes was to enhance mathematics classrooms with permanent visual demonstrations of mathematical ideas and how they relate to the man made and natural world, perhaps recognizing the need, missing from their materials, to demonstrate mathematics in use.

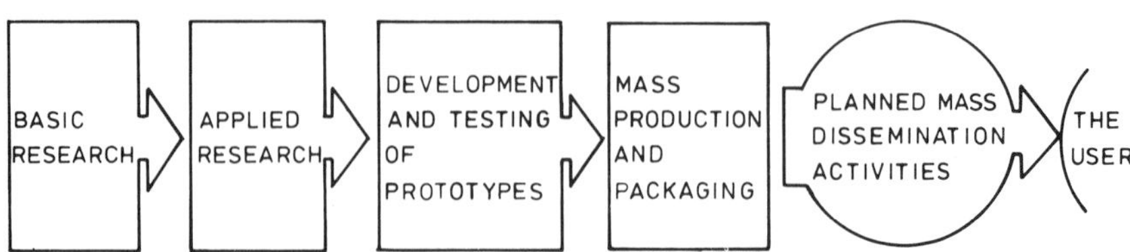

Fig. 1. The research, development and diffusion Model (after Havelock, 1971).

3. EVALUATION STUDY

The evaluation that follows will argue that the above is an important need, and largely it is indeed missing from the S.M.G. main course.

The general approach to learning throughout the S.M.G. course is said to be through discovery and investigatory methods, reflecting new theories in the psychology of learning, particularly those based on the work of Piaget. The theories argue that through these methods children are more motivated and interested in their studies and gain a better understanding and retention of what they learn. However, there has been controversial debate on the matter, and seemingly much disagreement as to just what discovery learning is (Ausubel, 1964; Friedlander, 1965; Shulman and Keislar, 1966). Most psychologists are seemingly agreed, however, that concrete and intuitive experiences, involving physical apparatus, are a necessary prerequisite to understanding a more formal treatment of mathematics (Dienes, 1971). Book 1 of the course suggests plenty of such experience in introducing sets and the geometry, but this is really the only explicit such suggestion in the whole course. It seems then that the S.M.G. assume children are ready to leave physical experiences behind at the Book 1 stage, or age about 11 or 12, whereas, for instance, some research has suggested 80% of 14-year-olds are still in Piaget's concrete operations stage, normally associated with primary schools (Shayer, 1970).

This question of passing from concrete experiences to more formal and abstract mathematics also relates to the nature of mathematics, which could be said to be concerned with what mathematics is about, where it comes from and how and why it was and is created. Some writers (Kline, 1968; Baron, 1972; Wilder, 1973) argue that origins of mathematics lie essentially in experiences in the 'real world', in the need to solve physical or commercial or economic problems and the formal tidying up of mathematics into logical structures gradually develops later. If children are to appreciate what mathematics is about and be interested in it and value it, then they must be given opportunity to experience the creative process, and not have 'cut and dried' mathematics imposed on them. In Book 1 children may appreciate that symbolization and abstraction are useful ways of representing their experiences with sets but they too may wonder what use they are. They are simply told sets are basic, and throughout the course pupils are repeatedly told various ideas and topics are important, but they rarely get the chance to find out how. Most ideas are introduced through a "real world" situation, but the concepts are rapidly abstracted and then built up through drill exercises and any problems that could be described as at all 'real' are usually met only at the end of a section. Much of the algebra is seen to have no 'real' origins at all, but is concerned with introducing the terminology of modern mathematics and building skills based on structural principles and using these to solve equations for their own sake. In many ways the algebra starts and goes nowhere and pupils meeting equations in science, for instance, may fail to make any link at all. The algebra exercises in Book 1 are largely of an abstract drill type, involving such terminology as open, true and false sentences, variable on a set, set of replacements and solution sets. Such exercises will probably be meaningless to most pupils but they are expected to 'get it right', thus suppressing the creative aspects of the nature of mathematics. Investigatory topics which could encourage creativity appear only intermittently throughout the course, and in a course of this length time may well prohibit pursuing them and there are none in the early algebra anyway.

Another example from Book 1 shows the rapid abstraction of ideas. Coordinates are introduced using maps and a game, but the notation $A \cap B$ is soon met describing the intersection of two straight lines defined as infinite sets of points. Sets may be a unifying theme of mathematics, but to Book 1 pupils this will mainly be seen as an unnecessarily awkward way of saying where two lines meet.

The geometry is introduced intuitively in Book 1, with concrete experiences suggested, but it becomes deductive in Book 2 and the idea of an axiom, based on 'hole filling and tiling' is introduced. The pupil is told that the kind of thinking that involves going from an assumption to a deduction is important in mathematics, but 12- to 13-year-olds will wonder what is important about deducing obvious results that are already known. They will probably find the mapping approach involved awkward to use, and not appreciate the significance in turning around a rectangle and discovering it fits its own outline. This would appear to be too formal too soon, but then is some concept of 'proof' a 'useful' aim of mathematics teaching anyway. The same question can be asked of structure, and it is notable that in the early 1960s some eminent psychologists argued strongly for the inclusion of structure in mathematical education (Bruner, 1960; Dienes, 1963). Structural ideas are present throughout the course, from the introduction of the commutative, associative, and distributive laws in Book 1, until the ideas are explicitly drawn together in Book 9. During the course, and particularly the first part, such as in the relations and mapping chapter in Book 3, children may become mystified as to what it is all about. For example, children are required to identify arbitrary and abstract relations as being or not being mappings, or to write similar relations as sets of ordered pairs.

Matrices and vectors are introduced in Book 5 through 'real origins', but these are soon left behind in favour of abstract structural principles, and any applications, such as matrices in transformation geometry, are largely within mathematics itself. It may appear to most pupils as playing rather pointless games, and lacking any useful, real application. Children are supposedly to build a sound understanding of the basic structural principles which underlie the extensive graded drill exercises in algebra of Books 3, 4, 5 and 6 but it could well result in half-understood, misconceived ideas, and amount to no more than rote learning of skills for examination purposes. It has been argued (Griffith, 1978) that the obtaining of useful results is what matters, before worrying about and analysing just how they were got. Griffiths argues for opportunity for children to take part in 'untidy' and sometimes wrong mathematics, based on guesswork in which children can invent their own language and notation. The very heavily guided discovery learning of the S.M.G. course does not allow this. It does not give children opportunity to set up and modify problems, make them their own and care about them. This is the sort of thing that modelling of real problems could offer, but in the S.M.G. course the problems have 'right' answers obtainable by 'right' techniques, which is usually the one just learned. But does this not make children and teachers feel secure; knowing clearly what to do, as opposed to the open-endedness of a real modelling situation? This balance between security and insecurity has been noted (Quadling, 1978) as a large problem in contemporary mathematics teaching. The appropriate language to use is equally a large problem, and Griffith again notes that introducing the formal language of mathematics too early may be mathematically unproductive, as it is far removed from children's normal language. The use of set language and notation may well cloud ideas for children rather than clarify them. For instance, is there any need, as in Book 3 to define the y axis as the set of points $\{x,y\colon x = 0\}$? Extensive use of this type of set notation is made in what is no more than drawing straight lines and shading regions. In Book 5, in introducing functions there is some very precise and formal language leading up to the set of ordered pairs $\{(x, f(x)\colon f(x) = x^2\}$ and its one–one correspondence with the set of points in the plane $\{(x,y)\colon y = x^2\}$. What is wrong with calling it the graph of $y = x^2$? Is this what Quadling means by the unacceptable face of modern mathematics (Quadling, 1975)?

Is there then any opportunity for modelling in the S.M.G. main course? The idea is introduced late in the algebra of Book 3 as a reason for solving equations and inequations. The 'applications' involve problems like finding the length and breadth of a rectangle given its perimeter and the ratio of its sides. Such a problem has a unique formulation and a unique solution and does not represent the modelling process in practice (Sida, 1975). There is no choice of model, no assumptions need to be made and the result needs no interpretation and success should be guaranteed. There was opportunity to discuss assumptions in Book 3 in the treatment of probability, where interpretations of theoretical results do need to be made but the idea of a probability model is not reached until the sixth form. In the main course modelling is only identified with the unique algebraic representation of some situation and it is only in the advanced sixth form books on probability and mechanics that modelling in the usually accepted sense (Hall, 1978) is discussed. Apart from the utilitarian aspects of the social arithmetic the main course may come across to most children as rather meaningless and something you learn to pass an examination in.

4. THE TEACHER SURVEY

It proved difficult to identify schools who used the S.M.G. materials, but eventually 35 teachers from 9 comprehensive schools in N.E. England returned questionnaires. There were further difficulties in that the degree of familiarity of teachers with the whole course varied, and different opinions would be held on different parts of the course and teachers would view it differently for children of differing ability. The latter problem was acknowledged by asking teachers to respond for a CSE or an 'O' level ability pupil, but other than that it was only possible to seek an overview. The main part of the questionnaire comprised an attitude scale covering several aspects of the course. Statistical analyses were carried out on the results, but owing to the small sample, these were of dubious validity, but the patterns of response to the items in the scale are of interest and are discussed below.

Teachers were also given opportunity to express their own opinions in open questions, and their answers showed a wide range of opinion. Some of the teachers comments are given in Table 1. Heads of Department indicated that they had chosen the S.M.G. course for its middle of the road balance of traditional and modern mathematics and its suitability for syllabus C 'O' levels.

Table 1. Teachers' comments about the S.M.G. course

1. 'I feel that some of the geometry is badly approached (particularly in Book 2). It seems to take a long time to achieve very little.'
2. 'Book 1 and 2 exercises—questions tend to be phrased in a manner which is too difficult for the children to understand, and so discourages them.'
3. 'I don't believe maths books of this sort are "teach yourself" and should not attempt to be.'
4. 'From other projects I have seen the S.M.G. strike the best balance.'
5. 'The books are useful in as much as they cover the majority of the 'O' level course. However, I do not feel that they make maths as exciting as they seem to want to.'
6. 'The majority will never appreciate it.'

Re—the importance of 'modern mathematics'

7. 'An early introduction for the good mathematician is good and important. However, this does not apply to 90+ % of the population.'
8. 'Use of relations and sets are important in that they appear to help children grasp many aspects of maths (especially the non-specialist). Specific reference to algebraic laws seems unnecessary (i.e. use of words such as associative).'

Re—whether 'mathematics as a whole' is presented in the course.

9. 'The objective is achieved but at the expense of puzzling those of lower ability.'
10. 'Achieved to quite a large extent, although more emphasis on applicable maths would be welcome.'

5. IMPLICATIONS AND DISCUSSION

The teacher survey found the S.M.G. course to be enjoyable, well presented and balanced and a good preparation for examinations. However, teachers found that the course does not relate well to real world experiences of children, does not stimulate creativity nor arouse much curiosity and interest in mathematics. Teachers had divided opinions over whether the course does offer learning by discovery and whether it does demonstrate mathematics in real problem solving, and links with other subjects, and this may well indicate different interpretations as to just what these are. Other items indicated that teachers find difficulty in assessing what children make of the course and what use the mathematics will be to them outside of examinations.

Overall, this course provides well for the future mathematician, but this is probably at the expense of mystifying most other children. If a 'useful' experience of mathematics is an encounter with modern structural principles, terminology and notation and ideas of axiomatic deduction, then the course does offer this, but it seems few children will really understand it and appreciate it. The course does largely fail to show the usefulness of mathematics in explaining and solving problems of the real world, and many children will seemingly be left wondering why they are required to study mathematics, with examinations seen as the principle, if not only, motivation. If the requirements of the examinations were changed, then perhaps the course materials could be adapted to a form more suitable for most children.

If a modelling approach were to be adopted, involving problems that could be developed at various ability levels, then children might not only be more motivated, but there would also be potential for natural links to other curriculum areas and for the use of computers. By working with problems that interest them, some children may eventually come to discern some structural principles, albeit with a teachers help, and this could then be developed for the few 'future mathematicians'. An approach through problems and developing the mathematics necessary to solve them, and from there to appreciating developments in mathematics itself is arguably presenting 'mathematics as a whole' more genuinely than is achieved by the S.M.G. course. This approach should lead children to what Ormell (1980) has called the integrated model of mathematics as opposed to the bifurcated model in which mathematics exists isolated from the real world. The needs of an industrial and technological society, and the needs of professional mathematicians, as opposed to the needs of children as regards mathematical education is still a controversial debate. The question of why teach mathematics to all secondary children and just what and how to teach it is still very much to the fore (see, for instance, Green, 1977; Harper, 1976; Lomas, 1979; McNelis, 1977; Mathematical Association, 1977; Mathews, 1976; Reynolds, 1975; Turner, 1978). However, the so-called modern mathematics 'experiments' now seem to be accepted as permanent, as about 80% of O' level candidates now take a 'modern syllabus'.

Teachers need to examine closely what it is they are trying to do for their pupils, and to query whether succumbing to society's pressures to get them through the present examination system is in their own, and society's, best interests. In many ways the S.M.G. course will be found dull, not to say boring, by most children, but then there is security and convenience for teachers and pupils in a sequential series of books, which can be easily used in traditional, expository, whole class teaching methods. The books possess a permanance and authority and pupils and teachers know just what is required of them, and to move to more open investigatory problems may bring insecurity to both, and the nature of the work may find disapproval from those of traditional views and values as regards teaching.

To adopt the real world modelling approach will be challenging and demanding on teachers in overcoming the many barriers. However, if mathematics is to be relevant and interesting to most children they must be overcome, and it may well be that courses like that of the S.M.G. are just inappropriate. New materials are needed, and they are emerging, and reform and change are again in the air. If the recommendations of Cockcroft (1981) are taken up, much of what was said in that report would seem to accord with what has been argued and proposed here.

REFERENCES

Austin, J. L. and Howson, A. G., 1979, *Educational Studies in Mathematics*, **10**, 161.

Ausubel, D. P., 1964, *The Arithmetic Teacher*, **5**, 290.

Baron, M., 1972, The Nature of Mathematics—Another View, in *The Process of Learning Mathematics*, Ed., L. Chapman, Pergamon.

Bruner, J. S., 1960, *The Mathematics Teacher*, **53**, 610.

Cockcroft, W., 1981, *Mathematics Counts*, Report of the Committee of Enquiry into the Teaching of Mathematics in Schools, H.M.S.O.

Dienes, Z. P., 1963, *The Arithmetic Teacher*, **10**, 115.

Dienes, Z. P., 1971, *Building up Mathematics*, Hutchinson Educational.

Friendlander, B. Z., 1965, *Harvard Educational Review*, **35**, 18.

Green, J., 1977, *Mathematics in Schools*, **6**, 26.

Griffith, H. B., 1978, The Structure of Pure Mathematics, in *Mathematical Education*, Ed., G. T. Wain, Von Nostrand Rheinhold.

Griffith, H. B. and Howson, A. G., 1974, *Mathematics, Society and Curricula*, Cambridge University Press.

Hall, G. G., 1978, Applied Mathematics, in *Mathematical Education*, Ed., G. T. Wain, Von Nostrand Rheinhold.

Hammersley, J. M., 1968, *Bulletin of the I.M.A.*, **4**, 66.

Harper, E., 1976, *Mathematics in Schools*, **5**, 9.

Havelock, R. G., 1971, *British Journal of Educational Technology*, **2**, 84.

Howson, A. G., 1973, *Educational Studies in Mathematics*, **5**, 157.

Kline, M., 1968, The Nature of Mathematics, in *Mathematics and the Modern World*, Ed., M. Kline, Freeman.

Kline, M., 1973, Why Johnny can't Add, St. Martin's Press.

Lomas, C. E., 1979, *Mathematics in Schools*, **8**, 22.

McNelis, S. and Dunn, J. A., 1977, *Int. J. Math. Educ. Sci. Technol.*, **8**, 175.
Mathematical Association, 1963, *Mathematical Gazette*, 47.
Mathematical Association, 1977, *Mathematics in Schools*, **6**, 2.
Mathews, G., 1976, *Int. J. Math. Educ. Sci. Technol.*, **7**, 253.
Organization for European Economic Co-operation, 1961, *New Thinking in School Mathematics.*
Organization for European Economic Co-operation, 1961, *Synopses for Modern Secondary School Mathematics.*
Ormell, C. P., 1969, Ideology and the Reform of School Mathematics, in Proceedings of the Annual Conference of the Philosophy of Education Society of Great Britain, p. 37.
Ormell, C. P., 1980, Mathematics, in *Values and Evaluation in Education*, Ed., R. Straugh and J. Wrigley, Harper and Row.
Quadling, D., 1975, *Mathematical Gazette*, **59**, 61.
Quadling, D., 1978, Issues at Secondary Level, in *Mathematical Education* Ed., G. T. Wain, Von Nostrand Rheinhold.
Reynolds, P., 1975, *Mathematics in Schools*, **4**, 1.
Robertson, A. G., 1969, Mathematics, in *Scottish Education Looks Ahead*, Ed., J. Nisbet, p. 74.
Rollett, A. P., 1963, *Mathematical Gazette*, **47**, 299.
Shayer, M., 1970, *Education in Chemistry*, **7**, 182.
Sida, D. W., 1975, *Int. J. Math. Educ. Sci. Technol.*, **6**, 205.
Shulman, L. S. and Keislar, E. R., 1966, (Eds.), *Learning By Discovery. A Critical Appraisal*, Rand McNally.
Thom, R., 1971, *American Scientist*, **59**, 695.
Turner, A. D., 1978, *Mathematics in Schools*, **8**, 14.
Wooton, W., 1965., *S.M.S.G. The Making of a Curriculum*, Yale.

The Claremont Colleges' Mathematics Clinic

Professor E. Cumberbatch
Claremont Graduate School, California

SUMMARY

In response to the decline in enrolments in mathematics, and the shift in student interest to applied mathematics, the Clinic was started in 1973 as a way students can gain 'hands-on' experience in solving a 'real-world' problem. Briefly, a team of undergraduate and graduate students and faculty formulate, analyse and solve a problem submitted by an industrial or government sponsor. To date there have been 48 such projects from a diverse set of clients.

This chapter will describe the academic setting, the educational goals and benefits, the general organization, and the relationship with the sponsor.

1. INTRODUCTION

Conventional courses in applied mathematics too often seem to be techniques in search of applications, and examples of the latter are often forced. Real world applications mostly come in the reverse order with technique following problem definition and modelling. Hence the interest in modelling courses and the need for conferences such as this one.

Still, the teaching of modelling is like the teaching of swimming; the concepts don't seem to apply once wet water surrounds you. Getting somewhere, preferably solidly attached, is all that matters. Indeed the latter can be adopted as a 'model' precept; any model, however unsophisticated, helps understanding and contributes to progress. So getting wet in the real world of applications has the benefits of anything self-taught. Insight and confidence are generated, and hard work flows from motivated heads and hands.

The Mathematics Clinic at the Claremont Colleges allows undergraduate and graduate students to experience the 'hands-on' application of mathematics by involving them in the formulation and solution of problems submitted by industrial and governmental clients.

2. HISTORY

The Clinic concept was initiated in Claremont in the undergraduate engineering curriculum at Harvey Mudd College (HMC). Later (1973) the HMC Mathematics Department conducted its first clinic project. Soon the Claremont Graduate School (CGS) saw the clinic as benefiting students in its applied mathematics master's program. Claremont McKenna College and Pomona College joined, and now clinic participation is available to all Claremont students. Further details on the evolution of the clinic is provided in the References.

To date there have been 48 clinic projects sponsored by about 30 different clients. Sponsors include companies from the oil, aerospace and electronic industries, but also diverse sources as the Pomona Water District, the Los Angeles Superior Court, the State Architect and the U.S. Forestry Service have submitted projects.

Two major grants from the National Science Foundation assisted the development of the Mathematics Clinic after its initial growth. One grant provided funds for (a) administrative support (b) visiting appointments for faculty members interested in the program, (c) publicity and, (d) support for graduate student assistance in developing notes and modules for a new course on modelling, with the clinic experience providing a source of recent applications. The other grant was mainly used to appoint a series of post-doctoral fellows who had received their training in pure mathematics and wished to broaden their experience. Their attachment as consultants to clinic projects gave them a unique insight into areas of mathematics previously believed remote, and substantially enhanced their job prospects in the universities and industry.

3. CLINIC ORGANIZATION

A Clinic Advisory Committee composed of senior-level industrialists and college administrators and faculty assists in long-range planning. The technical management of clinic operation is handled by the Director. The Director is a faculty member on partial release time whose duties are to

(a) contact potential clients so as to identify and develop clinic projects,
(b) administer the financial arrangements (the client is charged $21,500 for a year's project),
(c) set up the project teams at the beginning of each semester,

(d) monitor the progress of the projects throughout the year and see that report deadlines are met,
(e) report grades to the students' colleges.

4. CLINIC TEAM

The student team for each project consists of 2–5 undergraduates and 1–3 graduate students, each of whom receive course credit. A team leader, usually a graduate student chosen by the Director, is assigned to each project and receives a stipend equivalent to a T.A.'s remuneration. The faculty supervisor engages in clinic duties as with a course, including the assignment of grades. The team leader directs the day-to-day activities of the team, sets specific student tasks and oversees the operation as a project manager would. Such managerial experience, including the responsibility for team reporting both in oral and written modes, is of tremendous value in maturing students. We have many success stories of shy and introverted students gaining confidence in group participation and public speaking via their clinic experience. The faculty supervisor and consultants, if any, provide the general direction of the investigation and act as resource persons for the work carried out by the students.

5. CLINIC OPERATION

Even after the work done by the Director over the summer in capturing a project, the captive is often ill-defined and elusive. The team meets with the client's liaison(s) fairly often at the beginning, and tries for problem definition. Soon the supervisor is suggesting areas of mathematics (and physics, O.R., economics, etc.) that seem relevant and team members are given assignments to prepare and report on. Problem formulation may take a little time, and then there is analysis and computation. Monthly letter reports are sent to clients, and complete written reports (including computer programs) are prepared for the semester end, and year end. Oral presentations are given twice a semester to the whole clinic group, and the end-of-year presentation ('Projects' Day') is a grand affair to which all company liaisons and many other guests are invited.

6. PROJECTS

Since its inception in 1973, the Math Clinic has attracted projects from about 30 different clients. Nearly all the clients have been located in the greater Los Angeles area, the only exceptions being the California State Architect's Office in Sacramento, the Department of Energy's headquarters in Oakland, and Texas Instruments and ARCO in Dallas. These clients have sponsored a total of approximately fifty-two project-years of work. About half of the clients have sponsored repeat projects over the years. Most often these repeat projects are follow-on's from the previous year's work, but a significant number of new projects come from old clients.

Five project categories that have emerged so far are listed below:

MM + A(H): Math Modelling and Analysis (Hard)
MM + A(S) : Math Modelling and Analysis (Soft)
Anal. : Analysis
Econ. : Economics
Comp. : Computer-related Project

The modelling categories are used only if modelling is a significant component of the project, and the qualifiers 'hard' and 'soft' are used to indicate (respectively) that the modelling process uses continuous techniques (e.g. differential equations) or discrete techniques (e.g. discrete probabilities, game theory, programming, etc.). The 'analysis' category is used when the client more-or-less supplies the model and the goal of the project is to analyse the behavior of the model. The remaining two categories are self-explanatory. Like all classification schemes, however, it is not always an easy matter to assign to each project a unique label from the list—but since the purpose of the scheme is to aid planning and staffing, this ambiguity is not important.

Some typical examples:

MH + A(H) *Mathematical Modelling of Air Pollution Transport*
Rockwell International Science Center
A point source model based on the concept of a 'Gaussian Puff' was developed to describe pollution transport in the atmosphere. A novel mathematical model, intended for use in urban environments and based on the fractional calculus, was also developed.

MM + A(S) *Automatic Word Recognition*
Interstate Electronics
Based on a detailed investigation of the structure of an existing Interstate Electronics word recognition system, three methodologies were developed to improve the accuracy of the system: graphical methods, transition analysis, and distance tables. Suggested improvements ranged from changing single bits in a coding scheme to implementing a learning algorithm for improvement of recognition. Additional work involved many of the same considerations in moving from a speaker-dependent to a speaker-independent system.

ANAL *A Kinematic Handbook for Missile Design*
Northrop Corporation
The laser beam rider guidance system has the advantage

that the launcher rather than the missile tracks the target. While tracking the target, the launcher projects a coded laser beam of controlled width along the launcher/target line of sight. The missile follows this beam as it moves through space. The project team analysed alternative schemes for minimizing shortcomings of the laser beam rider system and assembled graphic data to facilitate the designing of systems to meet criteria.

ECON *Cash Flow Management in the Small, Private College Environment*
Harvey Mudd College, Office of Business Affairs

The College business office is required to manage operations that have continuing and regular expense profiles using cash generated on a variable and irregular basis. This project team developed an algorithm to optimize the management of the cash accounts of the college.

COMP *Search for Extraterrestrial Intelligence: Data Processing Techniques*
JPL-Cal Tech President's Fund, Pasadena

To discover evidence of intelligent extraterrestrial life, JPL performs broad frequency sky-surveys, using sophisticated receiving and data processing systems. The team worked with scientists from JPL to analyse the methodology associated with the design of very large fast Fourier transform data processing devices and conducted a mathematical study of sources and effects of errors in the data processing.

7. SPONSOR RELATIONS

The client's designated liaison(s) is the main line of communication between the team and the sponsor. The information received through this channel is very important, especially at the beginning of the project, since it determines the initial focus of the investigation. Group meetings with company representatives are sometimes necessary, but the project goes better if a good single channel of communication is developed.

There are some problems regarding confidentiality of data and patent rights. The sponsor's rights to protection of various kinds of data are recognized, and written down in the formal contract. Widespread use of clinic results is obtained a year after the final report is given to the company.

The major goal of the clinic is educational, and this is emphasized in its promotion in front of prospective clients. Nonetheless the client more often than not gets very good value for money in terms of results obtained. A secondary, important feature is that the company becomes known to a select group of students and its chances of hiring some of these are enhanced by this exposure.

8. EDUCATIONAL BENEFITS

Although students are bemused by the technical jargon at the beginning of a project, and sometimes show a prejudice against an applied area, they quickly become fascinated by the problem and their capacity for work and thought rises dramatically. Their appreciation of the complexities of other projects gains accordingly, and the art of the practical is viewed anew. This process of maturing spreads across many areas. Students learn to use the group, and being part of the group involves individual responsibility (no weak parts of the chain). Group participation encourages openness of views and enhances verbal skills. Speaking in front of the team eventually leads to speaking in front of large audiences. Writing skills are practised and the final arbiter is not an academic, but a user. Not only must the computer program run without bugs, but its answers must also be relevant and it must be coded in easily read language. These requirements place pressures on students different from the usual academic ones and they respond reassuringly well. We are rarely disappointed and most often we are thrilled by their successes.

Mathematically, the gains are not immediately obvious. Students learn how hard it is to translate an engineer's problem into something that can be approached by analysis. They learn the arts of approximation, computation and finally communication of their results. The section devoted to analysis is certainly always the easiest to write up. The usefulness of these skills is seen later when they take a second year of clinic, or when they leave to take a job. Perhaps the best test of the product of this training is the marketplace: our students are highly sought both by our sponsors and by employers of previous graduates.

REFERENCES

Borrelli, R., Busenberg, S., 1980, *The UMAP J.*, **1**, 3.
Busenberg, S., Tam, W. 1977, *SIGCSE Bul.*, **9**, 1.
Busenberg, S., Tam, W., 1979, Proc. of the ACM.
Henriksen M., 1977, *Am. Math. Mon.*
Henriksen, M., Orland, G., 1976, *Jurimetrics J.*
Spanier, J., 1976, *SIAM News*, **9**, 3.
Spanier, J., 1977, *SIAM Rev.*, **19**, 3.
Spanier, J., 1976, *Am. Math. Monthly*, **83**, 10.
Spanier, J., 1979, Proc. of the Second Int. Conf. on Math. Mod., St. Louis, July.
Spanier, J., 1980, *Int. J. of Math. Mod.*, **1**.
Spanier, J., 1981, in (Steen, L. (ed.), *Mathematics Tomorrow*, Springer-Verlag.
Woodson, T., 1973, *Engineering Education.*

Control Theory Ideas as a Component of a Modelling Course

D. J. G. James and N. C. Steele
Coventry (Lanchester) Polytechnic

1. INTRODUCTION

For a number of years the authors have been involved with teaching a course in mathematical modelling. The course is part of a second-year undergraduate mathematics programme so that students are already familiar with techniques for solving linear differential equations.

From the outset it is stressed that in modelling a real situation then there is *no unique* solution. Rather, any real system may be considered from several different points of view, all of which are equally acceptable. Any particular approach will give *some* information about the system although no model, however sophisticated, is an exact replica of the real situation. Consequently, students should become aware that all mathematical models are approximations with the degree of approximation permissible being a valid judgement made in the context of the system under consideration and the information required, striking a balance between sophistication and simplicity being the key to good modelling.

To reinforce these ideas the course is structured with various components designed to expose students to the various aspects of, and approaches to, a modelling exercise. Details of the course itself has been described elsewhere (James and Wilson, 1983) and this chapter considers only one aspect, the objective of which is to demonstrate some of the modelling techniques associated with classical control theory. These same techniques are now being applied in an increasing variety of areas including, for example, biological (Grodins, 1963) ecological (Patten, 1971) and socio-economics (Coyle, 1977) problems. Sometimes referred to as a systems approach to modelling, the technique involves formulation of a model via block diagrams leading to a visual model in which the various components of the system, and the interaction between them, are clearly identifiable. The approach provides a degree of insight and clarification not obtainable by other formulations.

The chapter gives a brief outline of how this particular aspect of the course is developed in the classroom.

2. OUTLINE OF THE PROGRAMME

This 'systems' approach to modelling is introduced midway through the course and consists of six three-hour sessions at weekly intervals. It is important that a lecture type approach be avoided and that the material is *developed interactively* with the class. At appropriate stages students must be given an opportunity to consider matters in small groups and then encouraged to suggest possible developments.

As a preliminary systems already familiar to the students are considered and open and closed loop control concepts are introduced and their advantages/disadvantages discussed.

Students are then introduced to system modelling using block diagrams, with illustrative examples. Although many of the students have little, if any, background knowledge in either physics or dynamics it is the authors' experience that physical systems provide the best vehicle for initial consideration. For such systems the individual components are readily identifiable and the variables involved, e.g. displacement, readily understood and visualized.

Most systems have a purpose and in order to achieve its objectives there is an element of control involved which, in man-made systems, is a matter for the designer to take into account. Consequently, the next stage of the course is devoted to considering ways of modelling the necessary control elements. This is best done by considering a case study interactively with the class. Again a case study involving a physical system appears most appropriate and it is particularly important for students to be given an opportunity to discuss ideas in small groups.

Having exposed the students to the general approach they are then encouraged to consider systems other than physical ones. This is achieved by concluding the programme with the students investigating further case studies in groups of two or three. During this final stage the lecturer should simply act as a consultant and give limited guidance, answering only specific questions when asked. It should be stressed on the students that in their report they should emphasize formulation, identification of the roles of various components and their interaction and any conclusions they wish to draw on validity, etc. Details of the mathematical manipulation required to solve any particular equation obtained is of secondary importance and, only if essential, included as appendices.

3. SYSTEM CONTROL STRATEGIES

The concept of a system is a general one. Here a system is taken to mean a collection of components arranged and interconnected in a definite way to achieve some objective. The components themselves may take various form, e.g. physical, economic, biological, sociological, or possibly, combinations of various forms. A distinguishing feature of such a system is that there is an input and an output and the input–output relation represents the cause–effect relationship of the system. The input represents the reference data or command signal and conveys instruction of what we are trying to achieve. The output represents the controlled variable and is a measure of the success of the system in achieving its goal. A basic control system may thus be represented pictorially by the simple block diagram of Fig. 1.

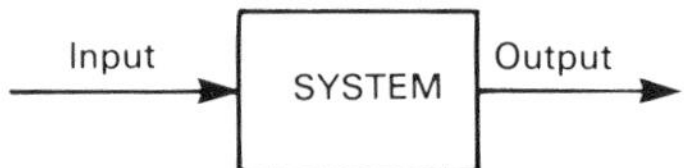

Fig. 1. System block diagram

The dependence of the output on the input is defined by the laws of the system which, ideally, may be modelled in terms of mathematical equations. The main objective of the system is to control the output variable in some prescribed manner by using the available input signal. In determining the method of control two basic control strategies are available. To illustrate consider a gas centrally heated room and possible alternatives for controlling room temperature.

3.1 Open-loop control

A simple control strategy, used in older systems, is to install a thermostat in the boiler which controls the temperature of the water in the boiler. The thermostat is set at a particular value and when the water temperature in the boiler rises above this value the gas supply is turned off and when it falls below this set value the gas supply is turned on; the turning on and off is done by a control unit, being a simple on/off switch in this case. Based on experimental work, under average conditions, the thermostat is calibrated to indicate room temperature corresponding to various boiler/radiator water temperature values. The configuration may be represented pictorially by the flow diagram of Fig. 2 where, in this case, the system consists of boiler, pipes, radiators, rooms, etc., whilst the goal is to achieve the desired room temperature.

The strategy adopted here is referred to as *open-loop control* and essentially it requires inclusion in the controller a plan which predicts the consequences of various actions; that is, the controller must include some form of calibration so

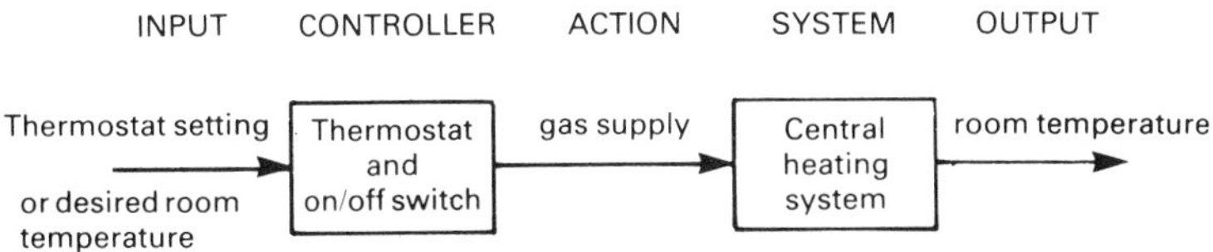

Fig. 2. Open-loop temperature control

that the value of the output produced by any particular value of the input would be known. A general open-loop control system has the configuration of Fig. 3.

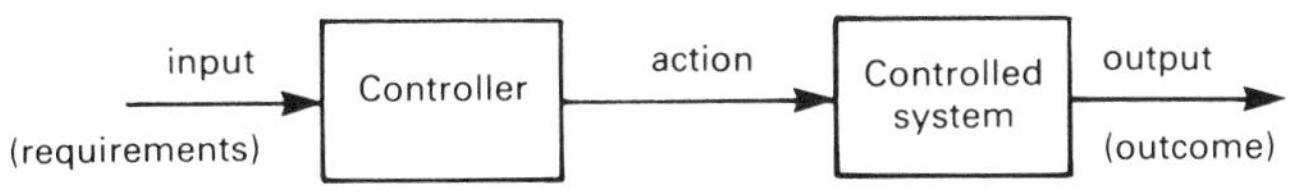

Fig. 3. Open-loop control system

At this stage the students should be given the opportunity to consider both
(i) other examples of open loop control systems, and
(ii) advantages and disadvantages of such systems.
Clearly such a control strategy represents complete faith in the accuracy of design of the controller as the actual outcome depends on the validity of the calibration. A distinctive feature is that no checks are made on the actual outcome and it is assumed that the effects of outside influences are negligible. In particular, for the central heating problem no account has been taken of factors such as, possible variations in the outside temperature, variations in the performance of the central heating system, or of possible disturbances such as intermittent opening of windows or doors.

If changes in outside circumstances are known to influence the system that is to be controlled then clearly open-loop control is not a satisfactory control strategy and alternatives need to be considered. Students should be given an opportunity to propose such alternative strategies.

3.2 Closed-loop control

To correct the above deficiency in the system, the outcome (output) should be continuously monitored and compared with the desired value (input) and then action taken based on the deficiencies. To achieve this we need a link or *feedback path* between the output and input of the system as shown in Fig. 4.

Such a control strategy is referred to as *closed-loop control* or feedback control. Here the controller is no longer activated by the input but by the 'error', where the error is a measure of the deviation of the output (actual outcome) from the input (desired outcome). In this way disturbances within the

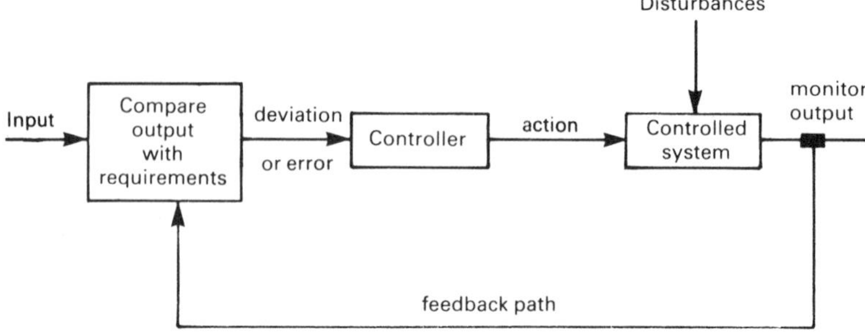

Fig. 4. Feedback control system

system will be automatically accounted for; a disturbance will effect the output causing a deviation from the desired value, the feedback path will initiate the necessary corrective action in order to attempt to restore the system output to the desired value.

Mathematically the comparator may be regarded as a subtractor so that symbolically it may be represented as in Fig. 5, giving

error = input − output

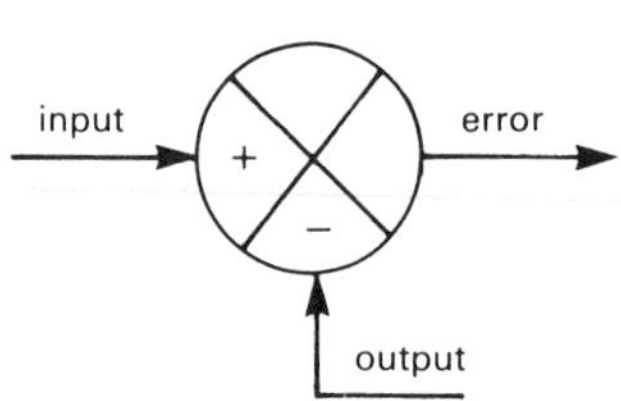

Fig. 5. Comparator as a subtractor

Since the output is subtracted from the input to create the error signal the systems are frequently referred to as negative feedback systems.

In the case of our central heating system a closed-loop control strategy may be implemented by installing the thermostat, which controls the on/off switch in the boiler, in the room to be heated.

At this stage time should be spent discussing other examples of closed-loop control systems with students being encouraged to consider possible applications in various fields including biological, sociological, economic systems. In each case they should be encouraged to identify the comparator and controller components.

It now appears that feedback control solves the design problem. Discussion should take place as to whether this is actually the case with students being encouraged to consider potential problems with perhaps a hint being given to consider possible effect of overcorrection.

3.3 Oscillating response

In order to investigate what could go wrong let us consider another example of everyday use of the feedback control strategy. Imagine a person taking a shower; a situation depicted in Fig. 6. The various elements, such as input, controller, etc. are readily identifiable.

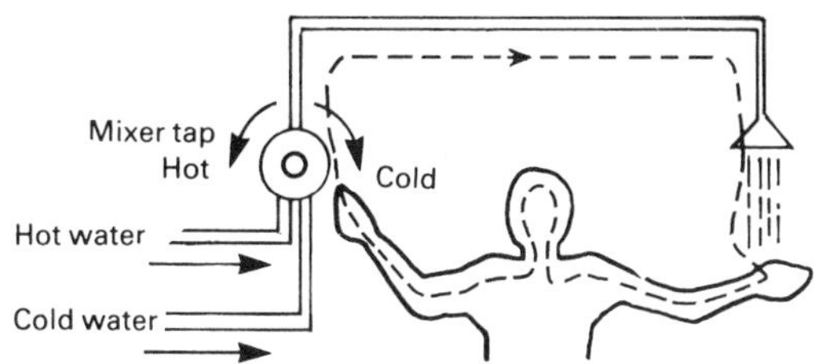

Fig. 6. Shower as feedback control system

Students should be given an opportunity to consider the questions:

(i) Do they always find that the system works satisfactory or do they have trouble getting the correct water temperature?
(ii) What can go wrong and what are the causes?

After some discussion it should become apparent that a possible cause of problems is the delay due to the water travelling along the pipes from the mixer tap to the shower head. Thus, if the water at the shower head is too cold the mixer tap is turned to allow more hot water to enter; this will not immediately produce warmer water at the shower head as it takes time to travel along the pipes. Impatience coupled with an attempt to 'hurry' the process up causes the mixer tap to be turned too far, first in one direction and then the other. As a result the water temperature at the shower head will tend to fluctuate about the desired value as shown in Fig. 7. If the mixture tap were to be controlled automatically, and not by hand, using sensing elements at the shower head then the effect would be more drastic.

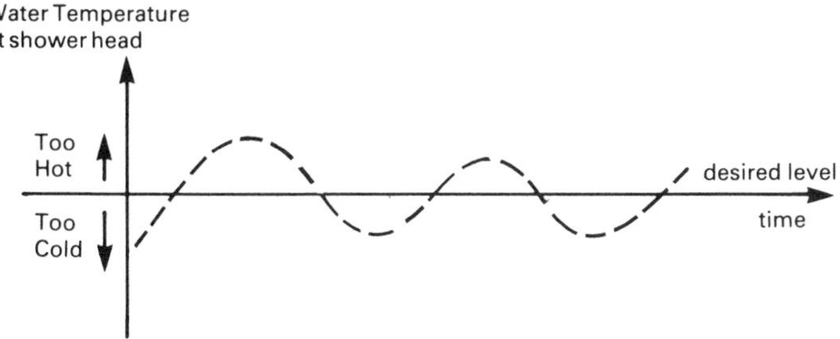

Fig. 7. Oscillations in temperature due to delay

Such an oscillatory response, sometimes referred to as hunting, is a common feature of feedback control systems. Clearly the controller needs to be designed so that such oscillations die out fairly quickly; in such cases the overall system is said to be *stable*. Unless great care is taken in designing the controller it is quite possible for the oscillations to grow; in such cases the situation is continually getting worse and the overall system is said to be *unstable*, the consequence of which can be serious—for example, bankruptcy in an economic system. Designing a stable system is obviously a major role for a system designer.

In the case of the shower problem the delay was due to the time taken for the water to travel along the pipes. Students should be asked to consider other sources of delay in the context of the system being discussed; for example, in a manpower planning system the delay could be due to the time difference from taking on new recruits to when they were trained.

4. MATHEMATICAL MODELLING OF A SYSTEM

4.1 Block diagrams and transfer functions

A first step in designing an overall system is to formulate a mathematical model of the system, which could be a large and complex one, to be controlled. The approach adopted is to isolate the individual components making up the system and represent the functional operation of each by a block such as that shown in Fig. 8. The symbol G_i appearing in the block represents the mathematical operation carried out on the input in order to produce the output, with the arrows indicating the direction of information, or signal, flow.

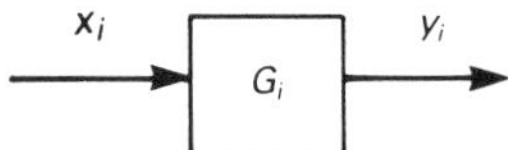

Fig. 8. Block diagram of an individual component

In general, the individual components are likely to have dynamics associated with them so that the mathematical relation between y_i and x_i involves a differential equation. As a consequence, it is convention to take Laplace transforms in order that the relation becomes algebraic, namely

$$y_i(s) = G_i(s)x_i(s)$$

where $x_i(s)$ and $y_i(s)$ denote the Laplace transforms of the input and output terms respectively and $G_i(s)$ is the *transfer function* representing the role of the component.

Background knowledge in Laplace transforms, although desirable, is not essential for adopting the approach to a modelling exercise. The problem can be overcome by using the symbol s to denote differentiation and $1/s$ to denote integration. Having carried out algebraic manipulations using the s variable the model can readily be reformulated as a differential equation.

Having determined the functional block for the various components making up a system these are then connected by lines, with appropriate arrows, indicating the flow of information. This leads to an overall block diagram representation of the system; this is a pictorial model which clearly illustrates the role played by each component and the interrelationship that exists between the various components.

4.2 Block diagram algebra

A block diagram representation of a system consisting of a number of components may be reduced to a single block, representing the functional relationship between the input and output of the overall system by making use of block diagram algebra rules. For an introductory course in modelling the following basic rules will be sufficient.

Rule 1: blocks in cascade

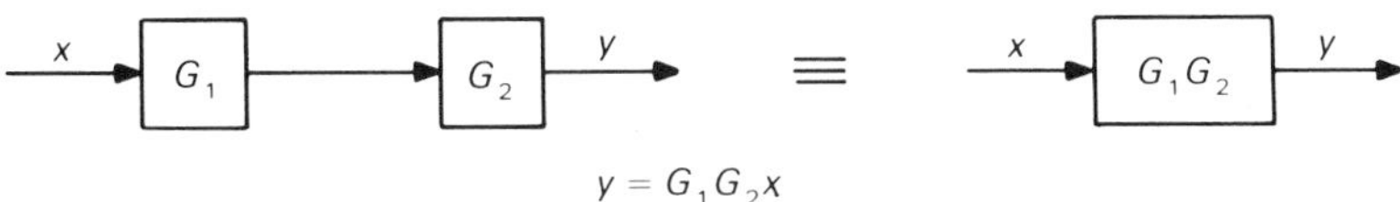

Rule 2: Blocks in parallel

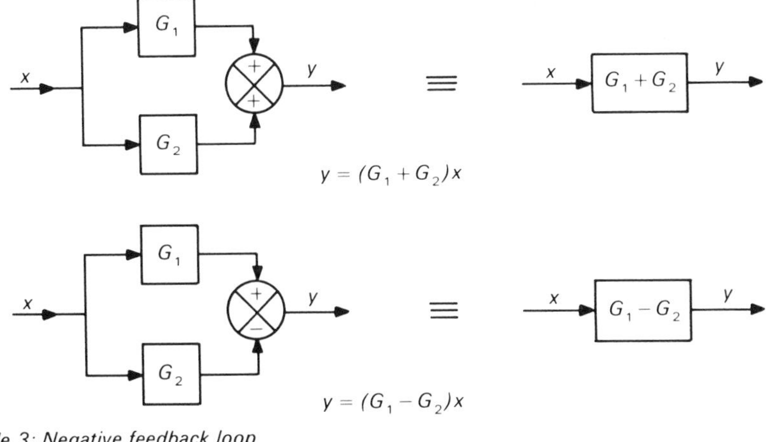

Rule 3: Negative feedback loop

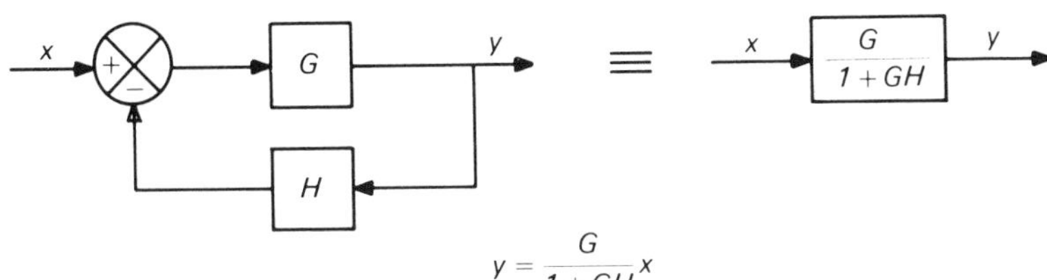

By the use of these three rules, complicated block diagrams may be readily simplified. It should be noted in discussion with students that in fact intermediate variables are being eliminated graphically rather than algebraically. An example is given here for illustrative purposes; examples for use with students may be found in any textbook on feedback Control Theory (Ogata, 1970; Dorf, 1976; Di Stefano *et al.*, 1967).

Example

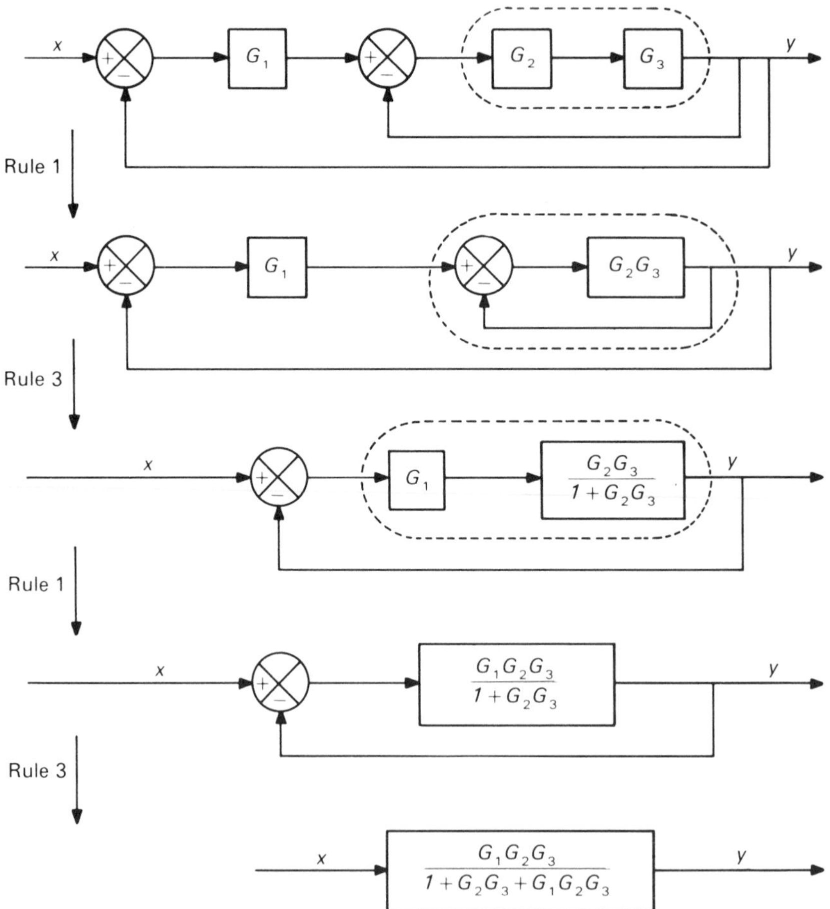

4.3 Some transfer functions

As mentioned in section 2, it is the authors' experience that physical systems provide the best source for illustrative purposes. In particular, mechanical translational systems have proven particularly useful. Such systems involve three basic components, namely masses, springs and dampers and at this stage the role of each of these should be discussed as many of the students are likely to have little, if any, relevant background knowledge. A brief discussion is included here for continuity.

(i) *Mass.* When a force is applied to a mass M it produces an acceleration of the mass which, by Newton's law is given by Force = (Mass) × (Acceleration)

The displacement of the mass is therefore represented by the flow diagram of Fig. 9.

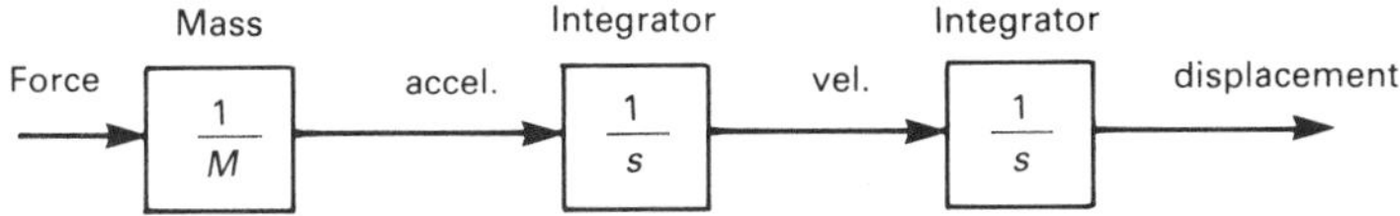

Fig. 9. Force–displacement relationship for a mass

Thus, by block diagram algebra, rule 1, the functional block representation of the mass is shown in Fig. 10.

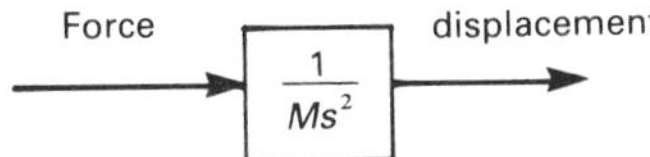

Fig. 10. Functional block for mass M

(ii) *Spring.* If a force is applied to a spring, Fig. 11, and stretches it then the spring tries to contract, and under compression it tries to expand. This reaction is due to the elastic property of stiffness which attempts to act as a restoring force. According to Hooke's law this restoring force, for an ideal spring, is proportional to the extension; the constant of proportionality being called the spring stiffness. Thus, a spring may be represented by either of the functional blocks shown in Fig. 12.

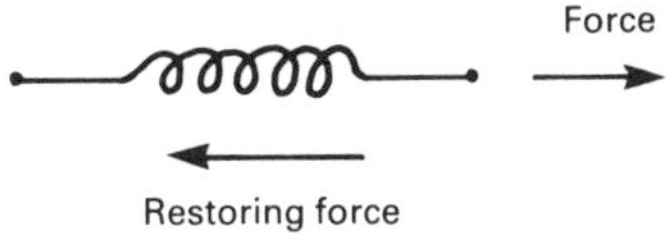

Fig. 11. Force applied to a spring

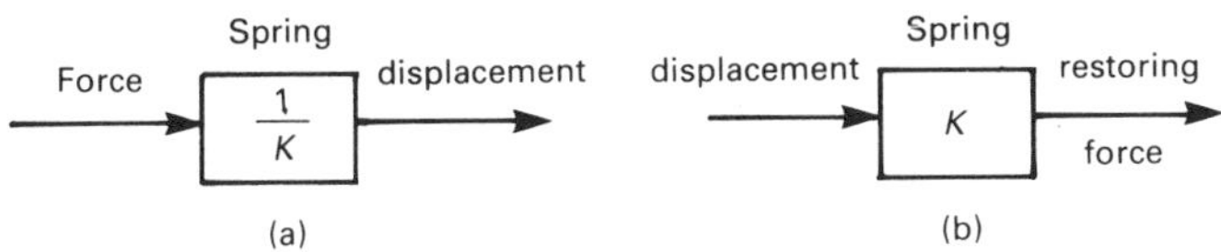

Fig. 12. Functional boxes for spring, stiffness K

(iii) *Damper*. A damper, Fig. 13, is a device that dissipates energy in the form of heat rather than store it as in the case of a mass or spring. Basically it can be thought of as a piston moving inside a housing containing a fluid, a typical example being the shock absorber of a car. When a force is applied a reaction force is developed which, in the case of an ideal damper, is proportional to the relative velocity of the two components comprising the damping device; the proportionality constant representing the damping coefficient (If the housing is fixed then the reaction force will be

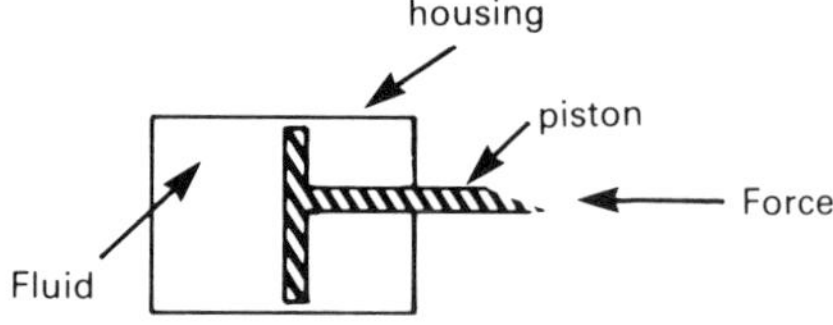

Fig. 13. A damper

proportional to the velocity of the piston.) Thus, the action of a damper may be represented by the flow diagram of Fig. 14, leading to the function block representation shown in Fig. 15.

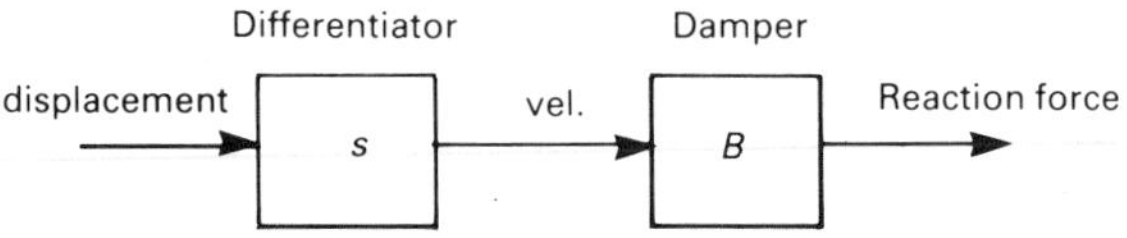

Fig. 14. Action of a damper, damping coefficient B

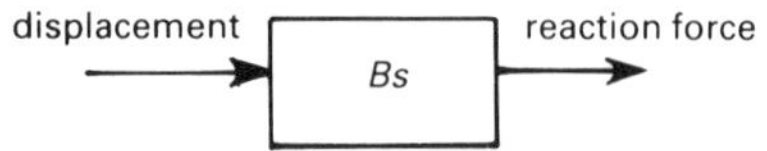

Fig. 15. Function block a damper, coefficient B

4.4 Modelling of mechanical systems

Having established the functional blocks for the individual components we are now in a position to develop block diagram models for various combinations. In the first instance a simple system such as the mass-spring-damper system of Fig. 16 is considered. This could be regarded as a simplified model of a car suspension system with the wheels being considered solid (i.e. they have no springiness). As the car moves along the road, the profile of the load will cause a displacement of the spring and damper which will result in a force being transmitted through to the car body. The input, in this case, is the vertical displacement x of the wheel and the output of interest is the vertical displacement y of the car body. The functional boxes for the individual components and the interaction between them are shown in Fig. 17.

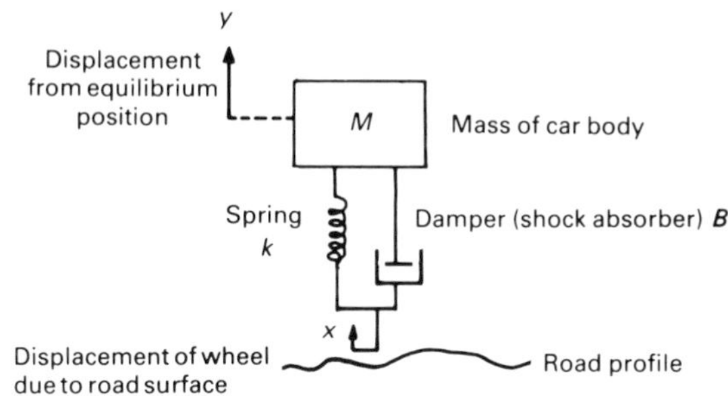

Fig. 16. Simple model of a car suspension system

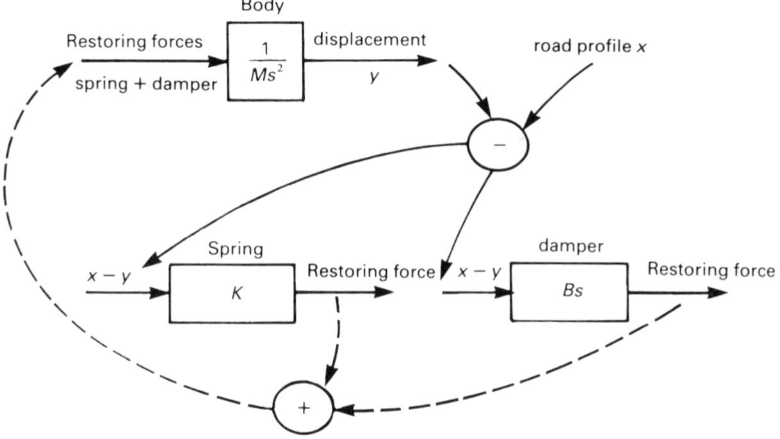

Fig. 17. Flow diagram for simple car suspension system

Using the functional blocks developed above for the individual components leads to the system block diagram of Fig. 18. It should be pointed out to students that although the system has an internal feedback loop it is not a feedback control system as the output is not used to monitor the input.

Applying the block diagram rules leads to the single functional block representation of Fig. 19 indicating that the mathematical model representing the system is

$$(Ms^2 + Bs + k)y = (Bs + k)x$$

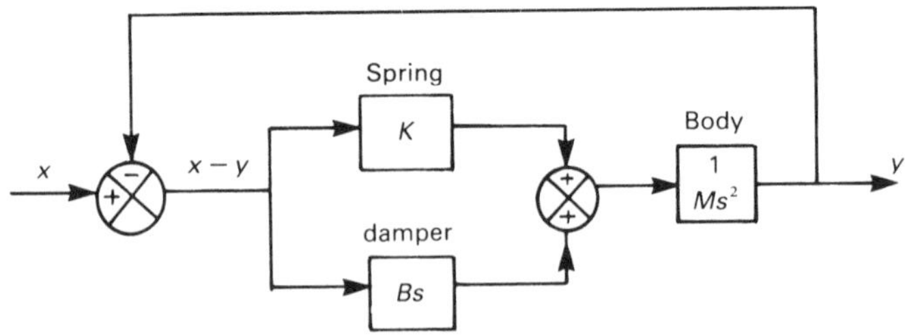

Fig. 18. Block diagram for simple car suspension system

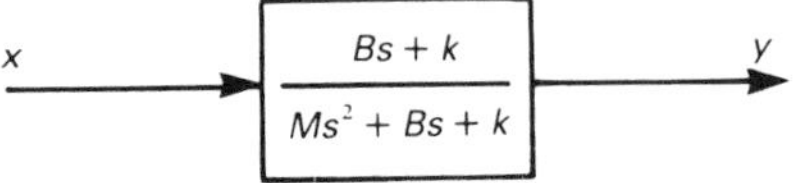

Fig. 19. Transfer function for simple car suspension system

or, in differential equation form.

$$M\frac{\mathrm{d}^2y}{\mathrm{d}t^2}+B\frac{\mathrm{d}y}{\mathrm{d}t}+ky=kx+B\frac{\mathrm{d}x}{\mathrm{d}t}$$

This model can now be used to consider the effects on the car body of various road profiles described by $x(t)$.

In the case of this simple system the differential model could be easily written down by direct application of Newton's law to the motion of the mass. However, experience with students, particularly those with little relevant background knowledge, has shown that they readily accept the block diagram approach and are able to go on to model more complex systems. At this stage a further example should be considered interactively with the class before asking them to investigate some systems for themselves.

A system found suitable for class discussion is shown in Fig. 20. This may be regarded as an enhanced model of a car suspension system with M_1 denoting the mass of the axle and k_1 the stiffness of the tyres. Again all displacements are measured from equilibrium positions.

The functional boxes for the individual components and the interaction between them are shown in Fig. 21, leading to the block diagram model of Fig. 22. Using block diagram algebra this may be reduced to the single block of Fig. 23.

Expanding the denominator and reverting to the time domain leads to the differential equation model

$$M_1\frac{M\mathrm{d}^4y}{\mathrm{d}t^4}+B(M_1+M)\frac{\mathrm{d}^3y}{\mathrm{d}t^3}+(K_1M+KM_1+KM)\frac{\mathrm{d}^2y}{\mathrm{d}t^2}+K_1\frac{B\mathrm{d}y}{\mathrm{d}t}$$

$$+K_1Ky=K_1K_2+K_1B\frac{\mathrm{d}x}{\mathrm{d}t}.$$

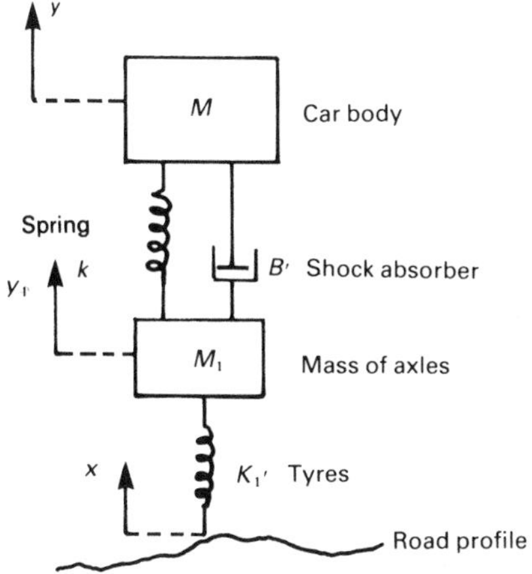

Fig. 20. Enhanced model of car-suspension system

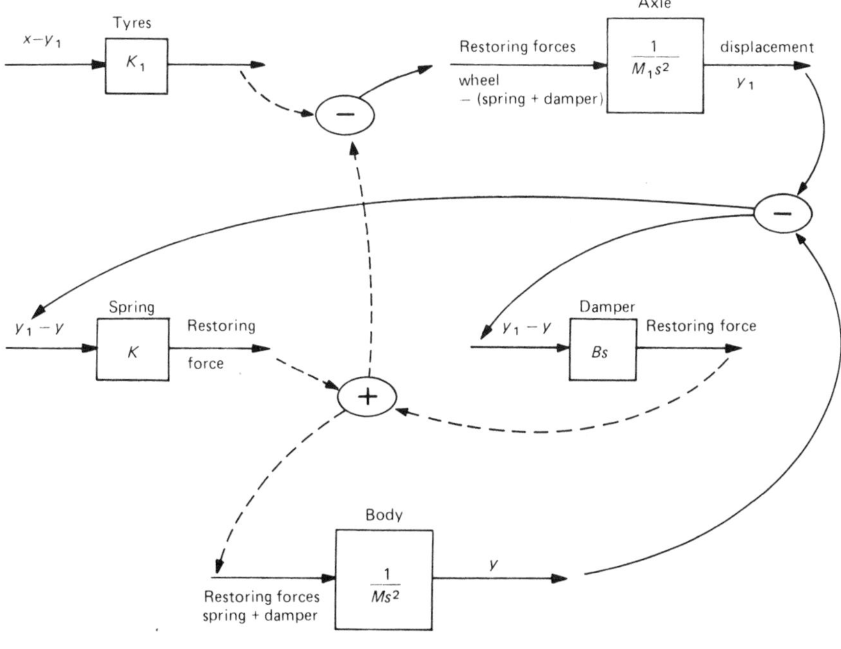

Fig. 21.

Attention should be given to the fact that the order corresponds to the number of energy stores in the system which are the two masses and two springs in this case.

A distinct advantage of the pictorial model of Fig. 22 over the differential

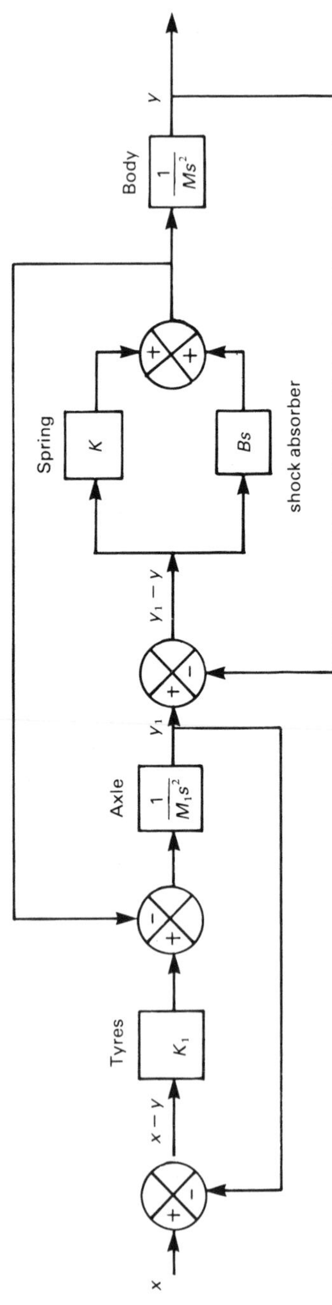

Fig. 22. Block diagram for enhanced car suspension system

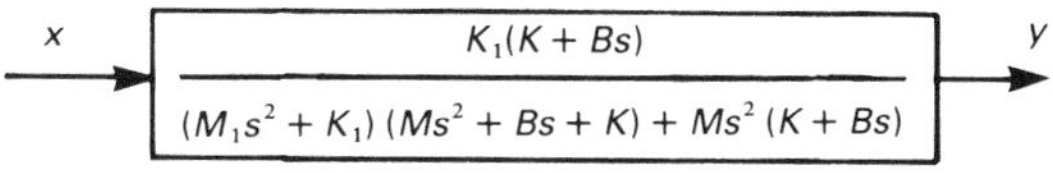

$$x \longrightarrow \boxed{\dfrac{K_1(K+Bs)}{(M_1s^2+K_1)(Ms^2+Bs+K)+Ms^2(K+Bs)}} \longrightarrow y$$

Fig. 23. Transfer function for enhanced car suspension system

equation model is that it clearly indicates the role played by each component and the interaction between them.

4.5 Modelling of non-mechanical systems

At this stage in the progression there are definite advantages in discussing at least one system of a non-mechanical nature. This should lead to an appreciation that different physical systems can be represented by the same mathematical model. In this context liquid level systems, involving flow through connecting pipes and tanks, form a fruitful area for investigation as they also have the advantage in that they can be used to model some real physical systems.

Initially some time should be spent developing a suitable mathematical model for the arrangment of Fig. 24. Ideally students should be given time to consider this for themselves with possibly some experimental work being suggested (James, 1981). Here the vessel has uniform cross-sectional area Am^2, the input and output flows are $q_i(t)$, $q_0(t)$ $\mathrm{m}^3\,\mathrm{s}^{-1}$ respectively, and the liquid level being $h(t)m$. Since

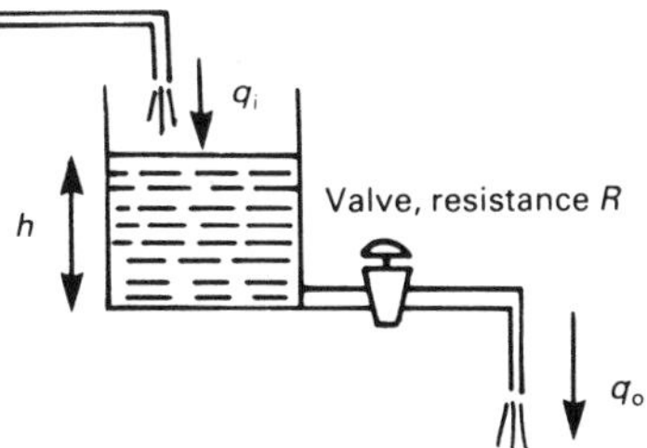

Fig. 24. Single water tank

Rate of increase of volume of liquid = Rate of inflow − Rate of outflow

we have that

$$\frac{\mathrm{d}}{\mathrm{d}t}(Ah) = q_i - q_0$$

or $$A\frac{\mathrm{d}h}{\mathrm{d}t} = q_i - q_0 \qquad \text{(i)}$$

when the cross-sectional area is constant.

Clearly the rate of outflow will increase as the liquid level in the tank increases so that $q_0(t)$ is an increasing function of water level $h(t)$. It is also reasonable to assume that the outflow is inversely proportional to the resistance R of the valve. The relationship between $q_0(t)$ and $h(t)$ should be fully discussed, leading eventually to the conclusion that provided variations in the variables are small in comparison to steady state values then a realistic model for the relationship is

$$q_0 = \frac{1}{R}h \qquad \text{(ii)}$$

Using (i) and (ii) the single tank arrangement of Fig. 24 may be represented by the block diagram of Fig. 25.

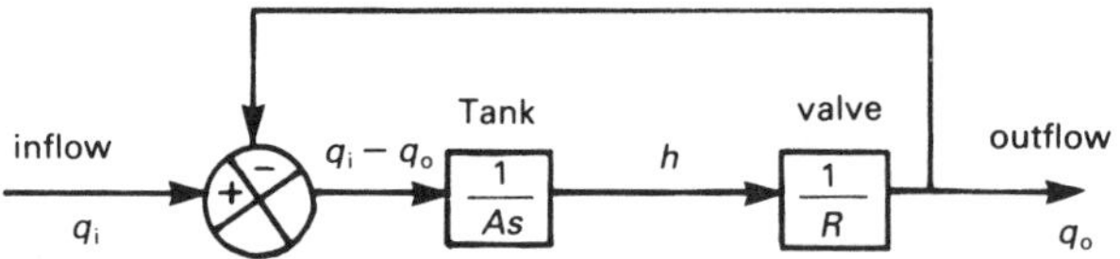

Fig. 25. Block diagram for single water tank

As in the case of the mechanical system this may now be readily extended to include a combination of tanks. As an illustrative example the system of Fig. 26 could be considered which leads to the block diagram model of Fig. 27. Block diagram algebra reduces this to the single block of Fig. 28 equivalent to the differential equation model

$$R_1A_1R_2A_2\frac{d^2q_0}{dt^2} + (R_1A_1 + R_2A_2 + R_2A_1)\frac{dq_0}{dt} + q_0 = q_i$$

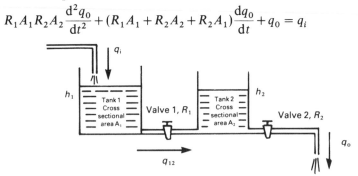

Fig. 26. Two tanks with interaction

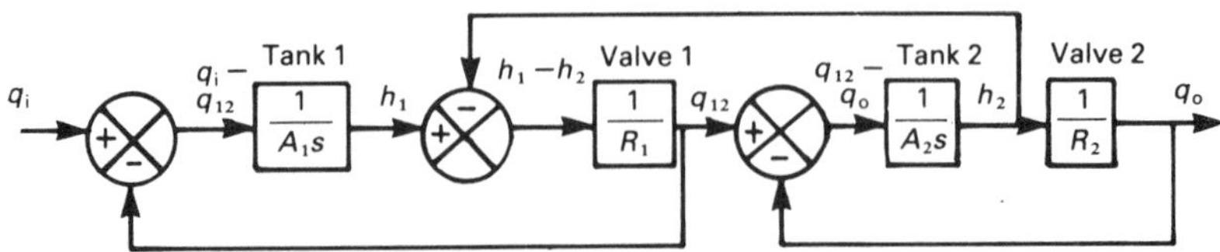

Fig. 27. Block diagram for tanks with interaction

5. CASE STUDY TO ILLUSTRATE CONTROL STRATEGIES

Having exposed students to the block diagram approach to modelling a basic system the final part of the 'classroom' component of this section of the course is devoted to considering ways of incorporating the control strategies, outlined in section 3, within the system models. This is achieved by developing a further case study interactively with the class. Again physical systems have proved to be excellent in this context and an outline of how such a system may be considered is given here. At each stage students should be encouraged to discuss possible developments or strategies. An alternative example, which has also been found effective at this stage, is the flow control problem discussed in James (1981).

Consider the position control system illustrated in Fig. 29, which represents a simple form of quay-side loading platform. The objective here is to ensure that the applied force F, which may be applied hydraulically, is such that the platform position is continuously in line with the level of the ship's deck.

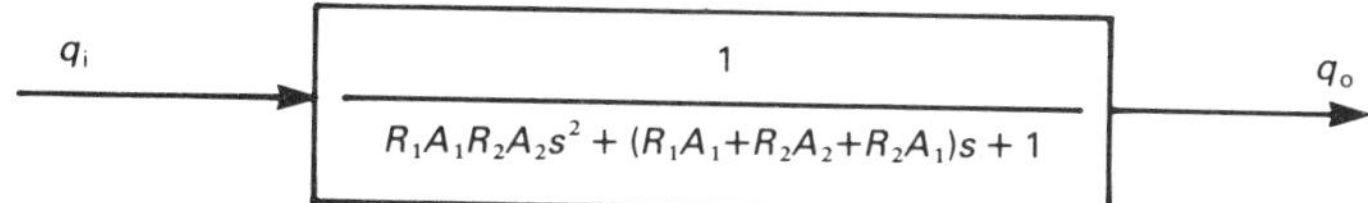

Fig. 28. Transfer function for tanks with interaction

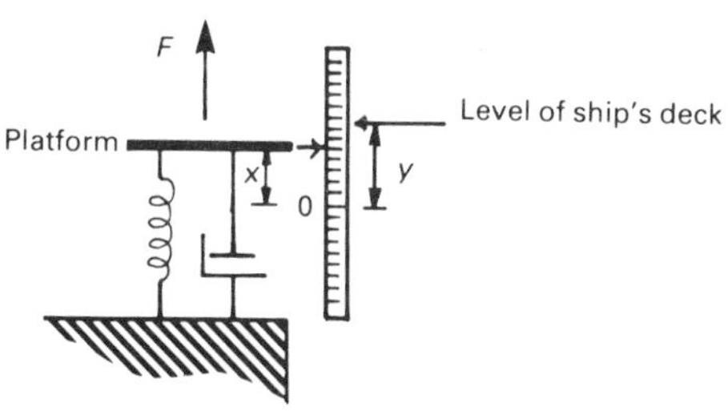

Fig. 29. Loading platform

If the mass of the loading platform under normal loading is M, the spring stiffness is K and the damping coefficient is B then the loading platform may be modelled by the block diagram of Fig. 30. Thus, the position $x(t)$, measured

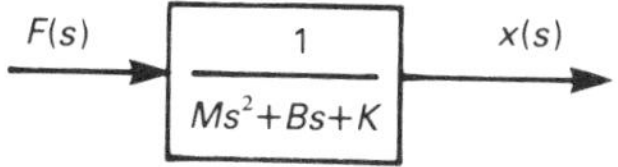

Fig. 30. Transfer function for loading platform

from an equilibrium position, of the platform is related to the applied force F by the second order differential equation

$$M\frac{d^2x}{dt^2}+B\frac{dx}{dt}+Kx=F(t).$$

At this stage comparison should be made with the 'standard' second order differential equation

$$\frac{d^2x}{dt^2}+2\xi\omega_n\frac{dx}{dt}+\omega_n^2x=r(t)$$

and responses for various values of the damping ratio ξ considered (James, 1981), leading to a discussion on suitable values for B and K to assist the designer in his choice of spring and damper.

Having chosen suitable values for B and K it could be argued that all that is necessary is the solution of the differential equation model for various constant values of $F(t)$. In this way a table of values of $x(t)$ versus $F(t)$ could be drawn up and then used to calibrate the force supply mechanism.

It appears that a suitable system may have been designed by this approach and a discussion as to whether this is the case should be initiated. Clearly, what is proposed is an *open-loop control system* which suffers from the shortcomings discussed earlier in the course. Briefly, such a system has no way of monitoring its own behaviour and is, therefore, not capable of reacting to shortcomings such as, for example, efficiency of the springs and dampers, or sudden changes in platform loads.

These shortcomings lead to the question of how the system can be improved and the ideas of feedback control considered earlier should readily come to mind generating the enhanced system of Fig. 31. Here the actual position $x(t)$ of the platform is continuously compared with the desired position, namely the position of the ship's deck. If these are different then a change must be made in the applied force F in order to move the error, or deviation, to zero.

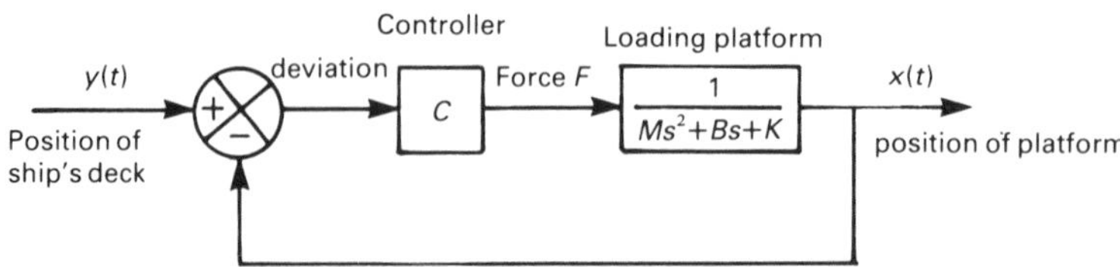

Fig. 31. Feedback control of platform

The obvious question at this stage is, how should the applied force be related to the error; that is, what is the nature of the controller C in Fig. 31? Initially students should be asked to consider this for themselves and generally they suggest that the force should be proportional to the error so that the controller is simply a gain; that is,

$$C = K_1, \text{ a constant}$$

This type of control is normally referred to as *proportional control* and use of block diagram algebra leads to the differential equation model

$$M\frac{d^2x}{dt^2} + B\frac{dx}{dt} + (K + K_1)x = K_1 y$$

the behaviour of which should be discussed. Discussion should highlight the following matters

(i) introducing proportional feedback control does not increase the order of the differential equation model;
(ii) proportional control is characterised by a steady state error;
(iii) there are two conflicting requirements in choosing the control gain K_1; on the one hand it is desirable to choose K_1 as large as possible in order to reduce the steady state error (i.e. the shortfall) whilst on the other hand large values of K_1 implies a smaller damping ratio thus showing an oscillatory response, leading to a long settling down time. Thus, when using proportional control there must be some compromise to ensure that the steady state error is as small as possible without undue oscillations.

Consideration is then directed towards the question of whether or not this compromise can be improved upon. Discussion with the students should lead to the conclusion that the rate at which the platform is approaching its desired level should be taken into account. This leads to the idea that the applied force be proportional to both the error and the rate of change of error, giving the controller model as

$$C = K_1 + K_2 s,$$

which is termed proportional plus derivative (PD) control. Block diagram algebra then leads to the differential equation model

$$M\frac{d^2x}{dt^2} + (B + K_2)\frac{dx}{dt} + (K + K_1)K = K_1 y + K_2\frac{dy}{dt}$$

The advantages of introducing the derivative element should be investigated. A distinct advantage to emerge is that the effective damping has been increased so that the proportional control gain K_1 may be chosen large enough to reduce the steady state error whilst simultaneously maintaining adequate damping by increasing the derivative control gain K_2. It should be pointed out that the derivative control cannot be used in isolation as it is a control action which is only effective during transient periods.

In conclusion it should be noticed that a PD controller still leads to a steady state error and the students should be asked to consider how to overcome this deficiency. Having seen that increasing the value of K_1 results in a decrease in the steady state error with the error, in theory, eliminated by an infinite value, the students should become aware that the cause of the problem is that realistic values of K_1 and K_2 do not produce a sufficiently large output, to enable correction to take place, when the error is small. A possible solution is to build up these small errors in some way in order that an effective signal can be produced. This leads to the idea of summation or integration and the introduction of an integrator into the controller, so that

$$C = K_1 + K_2 s + \frac{K_3}{s},$$

termed a proportional plus integral plus derivative (PID) controller. Block diagram algebra then leads to the differential equation model

$$M\frac{\mathrm{d}^3x}{\mathrm{d}t^3} + (B+K_2)\frac{\mathrm{d}^2x}{\mathrm{d}t^2} + (K+K_1)\frac{\mathrm{d}x}{\mathrm{d}t} + K_3x = K_3y + K_1\frac{\mathrm{d}y}{\mathrm{d}t} + K_2\frac{\mathrm{d}^2y}{\mathrm{d}t^2}$$

A clear consequence of introducing the integrator is to increase the order of the differential equation model. Discussion should take place as to the effect of this on both the stability and performance of the system.

6. CONCLUDING REMARKS

The main criticism of a case study approach to teaching modelling is that it concentrates on the particular and ignores the possible generalization. This situation should, and can, be avoided by the consideration at the later stages of such a study of possible wider applications of the ideas developed. In this way students may become aware of feed-back control ideas in non-physical systems and the course should encourage this development.

Perhaps for most students the new skill acquired, by pursuing the programme outlined here, is the representation of systems by block diagram models. This, in itself, should be highlighted as a powerful approach to modelling since in developing such a model the interactions between component parts of the system are clearly exhibited and thus a major step in a model formulation exercise is taken account of.

The systems examined here are all of the physical type. However, great care must be taken to ensure that students do not see the approach as applicable only to such systems and the exploration of non-physical system must be encouraged. This is achieved by presenting students with short exercises throughout the development stages of the course and by the case studies being investigated during the concluding stage of the course. The framework for a number of suitable case studies may be found in Dorf (1976) and in the other references cited.

REFERENCES

Coyle R. G., 1977, *Management System Dynamics*, John Wiley.

Di Stefano J. J., Stubberud A. R., Williams I. J., 1967, *Feedback and Control Systems*, Schaum Outline Series, McGraw-Hill.

Dorf R. C., 1976, *Modern Control Systems*, Addison Wesley.

Grodins F. S., 1963, *Control Theory and Biological Systems*, Columbia University Press.

James D. J. G., 1981, Hydroelectric power generation systems, in *Case Studies in Mathematical Modelling*, Ed. D. J. G. James and J. J. McDonald, Stanley Thornes, p. 137.

James D. J. G., Wilson M. A., 1983, Vol. 19, 9/10, pp. 180–182 *Bulletin Inst. Maths. Applics.*

Ogata K., 1970, *Modern Control Engineering*, Prentice-Hall.

Patten B. C. (Editor), 1971, *Systems Analysis and Simulation in Ecology*, Vols. I, II, Academic Press.

Magic Squares Leading to Vector Spaces

Paul Bungartz
Universität Bonn

1. INTRODUCTION

Obviously we can't teach 'linear algebra' beginning with the axioms of a vector space, deducing theorems and constructing proofs. On the other hand we should avoid arrows as the only model for vectors. In middle-band classes we often introduce arrows to symbolize a translation in geometry or physical forces and call them 'vectors'. But one arrow, that means a 'directed straight line of finite length', is no vector, so we try to explain it by introducing 'classes of arrows'. But nobody can imagine what a class is!

In my country, at least in the upper classes of the German Gymnasium, we are obliged to introduce vector spaces. So I developed in cooperation with some teachers a sequence of lessons with the learning goal: pupils, young students should discover, should learn by doing, that a vector is only an element of a structured set, called a vector space. We only can tell what a vector is if we know which model we mean. So we have to teach a lot of models of vector spaces and at the end of our lessons we should discover that finally all models of finite dimension n are in some sense 'essentially the same', and are 'essentially equal to' the arithmetic vector space $\mathbb{R}^n$.

But when introducing vector spaces at school-level I think we should not treat a lot of models abstracting the common structure. On the contrary, we should study intensively only one model, a suitable one. Of course if we have a very good model, attractive for pupils, they can discover all properties of a vector space by handling this one model. I think a very good model is 'magic squares'. We found that our young students of age 16 or 17 are very interested in searching for new squares. There is a great motivation in dealing with magic squares because it seems to be something quite different from usual mathematical teaching. Even teacher students at the university are very interested in searching for magic squares and I have used magic squares in teacher training courses.

2. MAGIC-SQUARES—DÜRER-SQUARES

We began our lesson looking at the copperprint (Fig. 1) of the famous German artist Albrecht Dürer (1471–1521), it is titled 'Melencolia I' and was constructed in 1514. The picture is full of mathematical symbols, signs and geometric figures. We are not interested in geometric figurations (Schaal) but in the numbers in the upper right corner beyond the bell. We see that there is a square of natural numbers, which we call a *Dürer-square*:

16	3	2	13
5	10	11	8
9	6	7	12
4	15	14	1

We try to find out why that is called a magic square, what are its properties? Students will find:

The sum of all numbers in each row equals 34.
The sum of all numbers in each column is 34.
Even if we add the numbers in both diagonals, we get 34.

If we divide the square into 4 subsquares by vertical and horizontal lines and add up the numbers of each subsquare we get 34.

Finally we find only the natural numbers 1, . . . , 16; and the date of origination, 1514, is written down in the last row; and add the four numbers at the corners—you again get 34.

That's fantastic, very fascinating, we are right to call it a 'magic square'.

Remark: Probably Albrecht Dürer didn't find that square himself, he constructed it by transformation of the 'tabula iovis':

4	14	15	1
9	7	6	12
5	11	10	8
16	2	3	13

tabula iovis

He got his own square by axial reflexion at the middle line and permutation of the two middle columns (Kracke, 1970).

Students now should become magicians, witches, they should try to create comparable 'Magic Squares'. There are some of course, that the teacher will write on the blackboard:

Fig. 1.

e.g.

13	3	2	16
11	8	5	10
6	9	12	7
4	15	14	1

13	4	1	16
11	6	7	10
8	9	12	5
2	15	14	3

Perhaps they discover that they get new squares by skilful transformations of a known one. By rotation (90°, 180°, 270°) they get 3 new squares and by reflexions at the middle lines and diagonals they get 4 new squares. (The Dieder-group of a square is of order 8.)

If we search for new squares, not all properties of the original magic square of A. Dürer are reproduced: the number 1514 is not in the last row, the sum of the numbers is not exactly 34, we use not only natural numbers between 1 and 16 but also real numbers. There is no restriction, we use all numbers we know, that are the real ones. So we call a 4×4 number-square '*Dürer-square*' if it has the following properties: we get the same sum if we add up the numbers in each row, each column, each diagonal and each subsquare ($R = C = D = S$).

$\mathscr{D}$ is the set of all Dürer-squares. The homework for students should be the construction of new magic squares. Each student should have his own magic square. e.g.

10	3	11	4
9	6	8	5
1	8	6	13
8	11	3	6

($R = C = D = S = 28$)

1	14	4	15
12	7	9	6
13	2	16	3
8	11	5	10

($R = C = D = S = 34$)

Exercise. Complete the following square to a Dürer-square: (You can easily find 'whole-number-squares' but only one Dürer-square with natural numbers):

6	7	9	8
	6		
		9	
		7	

Complete a Dürer-square with $R = C = D = S = 100$:

6			
		14	
9		48	
81	7		11

It is fascinating to find so many Dürer-squares. But do we get all D-squares?

3. A GENERATING SET FOR DÜRER-SQUARES

If we have a lot of squares, there are some questions: How many squares do exist? Is there a way to construct all squares? Perhaps we should try to find the most simple square. What is a simple square?
We find by discussion, the zero-square 0 and the 1-square E:

$$0 = \begin{array}{|cccc|} \hline 0 & 0 & 0 & 0 \\ 0 & 0 & 0 & 0 \\ 0 & 0 & 0 & 0 \\ 0 & 0 & 0 & 0 \\ \hline \end{array} \qquad E = \begin{array}{|cccc|} \hline 1 & 1 & 1 & 1 \\ 1 & 1 & 1 & 1 \\ 1 & 1 & 1 & 1 \\ 1 & 1 & 1 & 1 \\ \hline \end{array}$$

$(R = C = D = S = 0)$ $(R = C = D = S = 4)$

The next exercise should be: find all D-squares with $R = C = S = D = 1$, in other words all D-squares using only the two numbers 0 and 1.

Hint: You can find them combinatorically. Put 1 into a corner and regard the condition $R = C = D = S = 1$, two possibilities remain to put 1 into other rows, so you get $4 \times 2 = 8$. Basic-squares: $Q_1, \ldots, Q_8$.
Or you will apply the Dieder-group of a square: If you find one D-square with $R = C = D = S = 1$, the Dieder-group gives you the other squares. Again you get 8 Basic-squares:

$$Q_1 = \begin{array}{|cccc|} \hline 1 & 0 & 0 & 0 \\ 0 & 0 & 1 & 0 \\ 0 & 0 & 0 & 1 \\ 0 & 1 & 0 & 0 \\ \hline \end{array} \qquad Q_2 = \begin{array}{|cccc|} \hline 1 & 0 & 0 & 0 \\ 0 & 0 & 0 & 1 \\ 0 & 1 & 0 & 0 \\ 0 & 0 & 1 & 0 \\ \hline \end{array}$$

$$Q_3 = \begin{array}{|cccc|} \hline 0 & 0 & 0 & 1 \\ 1 & 0 & 0 & 0 \\ 0 & 0 & 1 & 0 \\ 0 & 1 & 0 & 0 \\ \hline \end{array} \qquad Q_4 = \begin{array}{|cccc|} \hline 0 & 0 & 0 & 1 \\ 0 & 1 & 0 & 0 \\ 1 & 0 & 0 & 0 \\ 0 & 0 & 1 & 0 \\ \hline \end{array}$$

$$Q_5 = \begin{array}{|cccc|} \hline 0 & 0 & 1 & 0 \\ 1 & 0 & 0 & 0 \\ 0 & 1 & 0 & 0 \\ 0 & 0 & 0 & 1 \\ \hline \end{array} \qquad Q_6 = \begin{array}{|cccc|} \hline 0 & 1 & 0 & 0 \\ 0 & 0 & 1 & 0 \\ 1 & 0 & 0 & 0 \\ 0 & 0 & 0 & 1 \\ \hline \end{array}$$

$$Q_7 = \begin{array}{|cccc|} \hline 0 & 0 & 1 & 0 \\ 0 & 1 & 0 & 0 \\ 0 & 0 & 0 & 1 \\ 1 & 0 & 0 & 0 \\ \hline \end{array} \qquad Q_8 = \begin{array}{|cccc|} \hline 0 & 1 & 0 & 0 \\ 0 & 0 & 0 & 1 \\ 0 & 0 & 1 & 0 \\ 1 & 0 & 0 & 0 \\ \hline \end{array}$$

Now we try to construct systematically Dürer-squares:

1. If we put instead of 1 an other real number r into the 'Basic-squares' we get Dürer-squares with $R = C = D = S = r$. We call this operation the '*multiplication of a D-square with a real number r.*

   So each D-square can be multiplied by a real number r by multiplication of each number.
2. Students find the *addition of D-squares*: cut out two squares, superpose them and add the superposing numbers. You get a new D-square and the new sum is the added sum of the old squares:

$$\begin{array}{|cccc|}\hline 4 & 0 & 7 & 1\\ 3 & 5 & 4 & 0\\ 0 & 2 & 1 & 9\\ 5 & 5 & 0 & 2\\ \hline\end{array} + \begin{array}{|cccc|}\hline 1 & 14 & 4 & 15\\ 12 & 7 & 9 & 6\\ 13 & 2 & 16 & 3\\ 8 & 11 & 5 & 10\\ \hline\end{array} = \begin{array}{|cccc|}\hline 5 & 14 & 11 & 16\\ 15 & 12 & 13 & 6\\ 13 & 4 & 17 & 12\\ 13 & 16 & 5 & 12\\ \hline\end{array}$$

$Z = S = D = T = 12$ $Z = S = D = T = 34$ $Z = S = D = T = 12 + 34 = 46$

So we are able to construct new D-squares by '*linear combinations*' of just known squares, perhaps the 8 Basic-squares.

Again there are two questions:

1. Do we get all D-squares by linear combinations only of the 8 Basic-squares? Then we call the set $\{Q_1, \ldots, Q_8\}$ a *generating set.*
2. Do we need all Basic-squares to generate all other D-squares? Perhaps we only need 6 or 7 Basic-squares? We search for *a minimal generating set*!

Remark: You see the benefit of treating 'Magic squares': Contrary to the arithmetic vector spaces $\mathbb{R}^2$, $\mathbb{R}^3, \ldots$ you get no canonical basis. The above-mentioned 'Basic-squares' $Q_1, \ldots, Q_8$ generate all D-squares but they are not linear independent. The dimension of the vector space $\mathscr{D}$ is not 8 but 7! So students discover that we need two properties for a basis: It has to be a generating set and a set of linearly independent vectors!

We answer question 2:
We have a lot of linear combinations of D-squares; in particular we look at the following:

$$\begin{aligned} D_1 &= 10Q_1 + 8Q_4 + 6Q_2 + 5Q_3 + 2Q_8 + 2Q_7 + Q_6 \\ D_2 &= 11Q_1 + 5Q_2 + 4Q_3 + 9Q_4 + Q_5 + Q_7 + 3Q_8 \\ E_1 &= Q_1 + Q_4 + Q_5 + Q_8 \\ E_2 &= Q_2 + Q_3 + Q_6 + Q_7 \\ N &= Q_1 - Q_2 + Q_4 - Q_3 + Q_5 - Q_6 + Q_8 - Q_7 \end{aligned}$$

We get the relations: $D_1 = D_2$, that's the D-square on the copperprint of A. Dürer, $E_1 = E_2$, that's the 1-square, and N is the 0-square.

$E_1 = E_2$ gives us

$$Q_1+Q_4+Q_5+Q_8 = Q_2+Q_3+Q_6+Q_7,$$

that means

$$0 = Q_1+Q_4+Q_5+Q_8-Q_2-Q_3-Q_6-Q_7,$$

or we compute

$$\begin{aligned} Q_1 &= Q_2+Q_3+Q_6+Q_7-Q_4-Q_5-Q_8, \\ Q_2 &= Q_3+Q_6+Q_7-Q_1-Q_4-Q_5-Q_8, \\ Q_3 &= \ldots \\ Q_4 &= \ldots \end{aligned}$$

Each Basic-square Q_j is a linear combination of the other. We only need 7 Basic-squares to generate the 8th. Perhaps we only need 6? Look at the set $(Q_1, \ldots, Q_7)$ and try to generate Q_7 by linear combination of the rest. You must find real numbers $r_1, \ldots, r_6$ so that

$$Q_7 = r_1Q_1+r_2Q_2+r_3Q_3+r_4Q_4+r_5Q_5+r_6Q_6.$$

Explicitly you have to solve the equation:

$$\begin{array}{|cccc|}\hline 0 & 0 & 1 & 0 \\ 0 & 1 & 0 & 0 \\ 0 & 0 & 0 & 1 \\ 1 & 0 & 0 & 0 \\ \hline \end{array} = \begin{array}{|llll|}\hline r_1+r_2 & r_6 & r_5 & r_3+r_4 \\ r_3+r_5 & r_4 & r_1+r_6 & r_2 \\ r_4+r_6 & r_2+r_5 & r_3 & r_1 \\ 0 & r_1+r_3 & r_2+r_4 & r_5+r_6 \\ \hline \end{array}$$

Compare the elements of both squares and you see that there is no solution because $0 \neq 1$! We find no real numbers so that Q_7 is a linear combination of $Q_1, \ldots, Q_6$. Therefore we need at least 7 Basic-squares to generate all other D-squares. These considerations are the same for each subset of 7 Basic-squares.

We see there are no real numbers except $r_1 = \ldots r_7 = 0$ so that $r_1Q_1+\ldots+r_7Q_7 = 0$ is correct. Explicitly there is one and only one solution of the equation:

$$\begin{array}{|llll|}\hline r_1+r_2 & r_6 & r_5+r_7 & r_3+r_4 \\ r_3+r_5 & r_4+r_7 & r_1+r_6 & r_2 \\ r_4+r_6 & r_2+r_5 & r_3 & r_1+r_7 \\ r_7 & r_1+r_3 & r_2+r_4 & r_5+r_6 \\ \hline \end{array} = \begin{array}{|cccc|}\hline 0 & 0 & 0 & 0 \\ 0 & 0 & 0 & 0 \\ 0 & 0 & 0 & 0 \\ 0 & 0 & 0 & 0 \\ \hline \end{array}$$

There is no linear combination except the trivial one. We get the definition of linear independence:

Definition: D-squares $D_1, \ldots, D_n$ are linear independent if and only if

$$r_1D_1 + \ldots + r_nD_n = 0 \Leftrightarrow r_1 = \ldots = r_n = 0$$

(for real numbers r_j).

Our so-called 'Basic-squares' are not linearly independent, they are linearly dependent. Using the definition, we find that the 0-square is linearly dependent and each D-square $D \neq 0$ is linearly independent itself. Each subset of 7 squares of $\{Q_1, \ldots, Q_8\}$ is linearly independent.

Finally we answer the question 1. Take 7 Basic-squares, i.e. $\{Q_1, \ldots, Q_7\}$ and show that this is a generating set for all D-squares!

If we have 7 real numbers $r_1, \ldots, r_7$ then the linear combination $r_1Q_1 + \ldots + r_7Q_7 = L$ is a D-square, of course, all the Basic-squares are D-squares. The R-, C-, D-, S-sum of the D-squares L is $R = C = D = S = r_1 + \ldots + r_7$, i.e. the sum of the coefficients of that linear combination.

On the other hand if we take some special D-square, D say, we must search for real numbers $d_1, \ldots, d_7$ so that

$$D = d_1Q_1 + \ldots + d_7Q_7.$$

i.e. if we take the D-square of Albrecht Dürer we must solve the following equation:

$$\begin{array}{|cccc|}\hline 16 & 3 & 2 & 13 \\ 5 & 10 & 11 & 8 \\ 9 & 6 & 7 & 12 \\ 5 & 15 & 14 & 1 \\ \hline \end{array} = \begin{array}{|llll|}\hline d_1+d_2 & d_6 & d_5+d_7 & d_3+d_4 \\ d_3+d_5 & d_4+d_7 & d_1+d_6 & d_2 \\ d_4+d_6 & d_2+d_5 & d_3 & d_1+d_7 \\ d_7 & d_1+d_3 & d_2+d_4 & d_5+d_6 \\ \hline \end{array}.$$

Comparing the elements of both squares we get

$$D = 8Q_1 + 8Q_2 + 7Q_3 + 6Q_4 - 3Q_5 + 3Q_6 + 5Q_7$$

$(8 + 8 + 7 + 6 - 3 + 3 + 5 = 34 = R = C = S = D)$

That's the way to prove that each D-square has an unique decomposition into a linear combination of the seven linearly independent Basic-squares $Q_1, \ldots,$ Q_7. We only write down the D-square $Q = (q_{ij})$ and the relation $Q = q_1Q_1 + \ldots q_7Q_7$, so the real numbers $q_1, \ldots, q_7$ are uniquely determined and the sum of all these q_j is the R-, C-, D-, S-sum.

Summary. To get all D-squares we need a minimal generating set of D-squares: i.e. each subset of 7 Basic-squares of $\{Q_1, \ldots, Q_8\}$ is a generating set for D-squares.

Remark: We have reached our first goal: we know how to construct D-squares by linear combinations of some basic squares and we know what 'linear independence' means.

Now we look at D-squares mathematically to discover arithmetic rules and the structure of all D-squares.

4. VECTOR SPACE $\mathscr{D}$

Looking back at the calculations, we see that D-squares can be added like numbers, we find the same rules, D-squares are an abelian group: ($\mathscr{D}$, +). But the multiplication of a real number and a D-square is not the same as the multiplication of two numbers.

We list the rules:

S_1: $r.(s.Q) = (r.s).Q$ $r, s, \ldots$ real numbers,
S_2: $(r+s).Q = r.Q + s.Q$ $Q, R, \ldots$ D-squares.
S_3: $r.(Q+R) = r.Q + r.R$
S_4: $1.Q = Q$

Summing up we get the usual 'axioms' of a vectorspace (Fischer, 1978; Toussaint): we have a set $\mathscr{D}$ and have defined addition and multiplication on $\mathscr{D}$ but it's different from real numbers, so we create a special name: *a vector space*. The elements of a vector space are called *vectors*. That's the denomination of mathematicians, so we will use it.

Each D-square Q is a vector!

Students may be surprised. They know perhaps that a vector is a class of arrows or is an arrow or denotes directed forces in physical teaching. But 'Magic-squares', Dürer-squares are vectors?

Now we discuss some other models of vector spaces we already know:

—In the Euclidean plane points are given by two real numbers. All these couples of real numbers are the vector space $\mathbb{R}^2$, each couple is a vector!

—We know the model of the surrounding space: $\mathbb{R}^3$, each triplet of three real numbers is a point in $\mathbb{R}^3$ if we fix coordinates. We can add these triplets, these points and get new ones. All axioms of a vector space are fulfilled. So each triplet of three real numbers is a vector!

—Sequences of real numbers can be added and multiplied by real numbers; sequences might be vectors!

Arithmetic sequences: $AF = \{(a_n) | a_n = a_n + d,\ a_j,\ d \in \mathbb{R}\}$
AF is a vector space, each sequence is a vector!
Geometric sequences: $GF = \{(b_n) | b_n = b_0 . q^n,\ b_0,\ q \in \mathbb{R}\}$
GF is a vector space, each geometric sequence is a vector!
(Artmann, 1978)

The famous Fibonacci-sequences are a structured set:

$$FF = \{(f_n) | f_n = f_{n-1} + f_{n-2}, f_1 = a, f_2 = b,\ a,\ b \in \mathbb{R}\}$$

is a vector space, each Fibonacci-sequence is a vector! (Grassle, Stowasser, 1978).

—Consider polynomials:

$$P(x) = \{a_n x^n + a_{n-1} x^{n-1} + \ldots + a_1 x + a_0 \quad | a_j \in \mathbb{R},\ n \in \mathbb{N}\}.$$

The set of all polynomials of some degree n is a vector space. Each polynomial is a vector!

—Other functions could be vectors. Look at geometric movements in the Euclidean plane, i.e. translations; they are often symbolized by arrows. But not the arrows of vectors but the functions!

—Look at the solution set of linear equations, rows or columns of real numbers. All these rows or columns of the same length n build a vector space, each row or column is a vector!

—Look at the real numbers itself. All rules for vectors are good for numbers, so $\mathbb{R}$, the real numbers, is a vector space and each real number is a vector!

There are many other models for vector spaces and vectors. All these sets are structured in the same way as Dürer-squares! If you would like to answer the question: What is a vector? You first have to say which model you mean! I think we have reached a main goal of our course: we now know what a vector is !

Further on we search for communities of vector spaces:

5. DIMENSION, BASIS, SUBSPACES

Looking at the Dürer-squares we recognize the importance of the generating set, i.e. $\{Q_1, \ldots, Q_7\}$. We define:

Definition: Generally we call a *generating set of linearly independent vectors* a *basis* of the vector space. The number of elements in a basis is called the *dimension* of that vector space.

If we have any subset of a vector space and want to know whether that is a basis or not we must test two conditions: (1) Are these vectors linearly independent? (2) Do they generate all other vectors? Perhaps we now will explain or in some cases we will prove theorems like:

—Every vector space has a basis.

—All bases of a vector space have the same length (this number is called the dimension of the vector space).

—We get one and only one representation of a vector according to a fixed basis.

—. . . .

Remark. In our course we taught on two levels: on one level are the models; we compute something; we can imagine our vectors and the connections between them; we interpret the theorems in a geometric-visual way.

On the other level we see the axioms of a vector space, definitions, theorems and proofs without any connection to some model. Abstractly we formulate and prove our theorems. I think we should not teach linear algebra in schools in a universal way, neither should we do it by only treating some models. Students should discover the main facts by studying models; on the other hand we have to show them abstract mathematical theory. Perhaps we only refer to some book, perhaps we write down all the axioms including exactly formulated proofs and let them have a look at that paper. So each student should get an idea, a first impression of what a mathematical theory is.

In exercises we now try to find a basis for the above-mentioned vector spaces. That's a little difficult regarding the vector space of all polynomials and perhaps the vector space of Fibonacci sequences.

If we look at a subset of a vector space we get the definition of a subspace:

Definition: Some subset U of a vector space V is a subspace if and only if U is itself a vector space.

By treating some subspaces, e.g. look at the polynomials of degree less than or equal to n, and other vector spaces, we find the well-known criterion for subspaces: The subset U of the vector space V is a subspace if and only if

$$u, v \in U \Rightarrow u+v \in U,$$

$$r \in \mathbb{R},\ u \in U \Rightarrow r.u \in U.$$

i.e. the set which contains only one vector v: $\{v\}$, $(v \neq 0)$, is no subspace but all linear combinations of one vector is a subspace. We have a lot of subspaces in the above-mentioned models.

Now we return to our 'Magic squares'. By specialization of the properties we get subspaces and larger spaces of magic squares:

1. We only demand the sum of the numbers in all rows, columns and diagonals should be the same, not the sum in the subsquares:

i.e.

6	7	9	8
12	6	5	7
5	10	9	6
7	7	7	9

$R = C = D = 30,\ S \neq 30$

If we try to find a basis for those squares, i.e. if we try to decompose the squares relative to the Basic-squares $\{Q_1, \ldots, Q_7\}$ of Dürer-squares we always get some multiplicity of N_0:

$$N_0 = \begin{array}{|cccc|} \hline 0 & 1 & -1 & 0 \\ 0 & 0 & 0 & 0 \\ 0 & 0 & 0 & 0 \\ 0 & -1 & 1 & 0 \\ \hline \end{array} \qquad R = C = D = 0,\ S \neq 0$$

If we call the set of such squares $\mathfrak{Q}$, it's a vector space of dimension 8 with basis $B_{\mathfrak{Q}} = \{Q_1, \ldots, Q_7, N_0\}$. $\mathscr{D}$ is a 7-dimensional subspace of $\mathfrak{Q}$.

2. If we demand no properties, i.e. if we are looking at all 4×4 number-squares we get the vector space of 4×4 matrices called $\mathfrak{M}$. We find the canonical basis and dimension 16.
3. The strongest demand is: all elements of a square are equal. That is the set $\mathfrak{E} = \{r.E | r \in \mathbb{R}\}$, a 1-dimensional vector space with basis $B_{\mathfrak{E}} = \{E\}$, a 1-dimensional subspace of $\mathscr{D}$.
4. The vector space containing only the 0-square is a vector space of dimension 0 by definition.
5. We are able to construct a lot of subspaces or 'overspaces' of our vector space $\mathscr{D}$. For each number k between 1 and 16 there exists a vector space of 4×4 squares of dimension k. (Botsch, 1967).

   We only look at two other vector spaces: We demand the properties row-sum = column-sum ($R = C$) and get a 10-dimensional vector space of squares: $\mathfrak{Z}$, with basis $B_{\mathfrak{Z}} = \{Q_1, \ldots, Q_7, N_1, N_2, N_3\}$ ($Q_1, \ldots, Q_7$ are Basic-squares of $\mathscr{D}$)

$$N_1 = \begin{array}{|rrrr|} 0 & 0 & 0 & 0 \\ 1 & 0 & 0 & -1 \\ -1 & 0 & 0 & 1 \\ 0 & 0 & 0 & 0 \end{array} \qquad N_2 = \begin{array}{|rrrr|} 0 & 1 & 0 & -1 \\ 1 & 0 & -1 & 0 \\ -1 & 0 & 0 & 1 \\ 0 & -1 & 1 & 0 \end{array}$$

$$N_3 = \begin{array}{|rrrr|} 0 & 1 & 0 & 0 \\ 1 & 0 & 0 & 0 \\ 0 & 0 & 0 & 1 \\ 0 & 0 & 1 & 0 \end{array}$$

If we demand the row-sum, column-sum and the sum over all possible diagonals should be the same, we get the 5-dimensional vector space of 'pandiagonal' squares $\mathfrak{P}$:

i.e.:

$$P = \begin{array}{|rrrr|} 17 & 2 & 11 & 16 \\ 16 & 11 & 22 & -3 \\ 12 & 7 & 6 & 21 \\ 1 & 26 & 7 & 12 \end{array} = \begin{array}{|rrrr|} 17 & 2 & 11 & 16 \\ 16 & 11 & 22 & -3 \\ 12 & 7 & 6 & 21 \\ 1 & 26 & 7 & 12 \end{array}$$

$H = 46 = R = C$ $\qquad$ $N = 46 = R = C$

Basis $B_{\mathfrak{P}}$:

$$P_1 = \begin{array}{|rrrr|} 1 & 0 & 1 & 0 \\ 1 & 0 & 1 & 0 \\ 0 & 1 & 0 & 1 \\ 0 & 1 & 0 & 1 \end{array} \quad P_2 = \begin{array}{|rrrr|} 1 & 0 & 0 & 1 \\ 0 & 1 & 1 & 0 \\ 1 & 0 & 0 & 1 \\ 0 & 1 & 1 & 0 \end{array} \quad P_3 = \begin{array}{|rrrr|} 0 & 1 & 1 & 0 \\ 1 & 0 & 0 & 1 \\ 0 & 1 & 1 & 0 \\ 1 & 0 & 0 & 1 \end{array}$$

$$P_4 = \begin{array}{|cccc|} \hline 0 & 1 & 0 & 1 \\ 1 & 0 & 1 & 0 \\ 1 & 0 & 1 & 0 \\ 0 & 1 & 0 & 1 \\ \hline \end{array} \qquad P_5 = \begin{array}{|cccc|} \hline 1 & 1 & 0 & 0 \\ 0 & 0 & 1 & 1 \\ 1 & 1 & 0 & 0 \\ 0 & 0 & 1 & 1 \\ \hline \end{array}$$

Summary. We have the following implications of vector spaces:

VS: $\{0\} \subseteq \mathfrak{E} \subseteq \mathfrak{P} \subseteq \mathscr{D} \subseteq \mathfrak{P} \subseteq \mathfrak{Z} \subseteq \mathfrak{M}$
Dim: 0 1 5 7 8 10 16

6. ISOMORPHIC VECTOR SPACES

The learning goal is that students should discover by doing that finite dimensional vector spaces of dimension n all are 'essentially the same', that means in some sense they are all the same as the well-known vector space $\mathbb{R}^n$. In exercises during our lessons we shortened the representation of D-squares by finite number-sequences: i.e. if we take the basis $(Q_1, \ldots, Q_7)$ and the D-square $D = 8Q_1 + 8Q_2 + 7Q_3 + 6Q_4 - 3Q_5 + 3Q_6 + 5Q_7$ we only write down $D = (8, 8, 7, 6, -3, 3, 7)$.

So respectively to the ordered basis $B_{\mathscr{D}} = (Q_1, \ldots, Q_7)$ we get a uniquely determined square if we have a sequence of 7 real numbers.

Accordingly we get a fixed pandiagonal square with respect to the basis $B_{\mathfrak{P}} = (P_1, \ldots, P_5)$ if we write a sequence of 5 numbers $(b_1, \ldots, b_5)$.

On the other hand we may interpret a sequence of 7 numbers as a point in $\mathbb{R}^7$ or 5 numbers as a point in $\mathbb{R}^5$, 7 numbers perhaps as the coefficients of a polynomial of degree 6 or less? If I write a couple of two real numbers it may stand for a point in the Euclidean plane or a sequence in the vector space AF (arithmetic s.) or a sequence in the vector space FF (Fibonacci-s.) or a polynomial of degree less or equal 1.

So if we write down on the blackboard a finite sequence of real numbers of the length $n = 1, 5, 7, 8, 10, 16$ it has the interpretation as a magic-square of $\mathfrak{E}$, $\mathfrak{Q}$, $\mathscr{D}$, $\mathfrak{P}$, $\mathfrak{Z}$, $\mathfrak{M}$ or a point in the arithmetic vector spaces $\mathbb{R}^1$, $\mathbb{R}^5$, $\mathbb{R}^7$, $\mathbb{R}^8$, $\mathbb{R}^{10}$, $\mathbb{R}^{16}$, or as polynomials of degree less or equal to 0, 4, 6, 7, 9, 15, always regarded to a fixed basis of these vector spaces.

We compare the models of vector spaces of the same finite dimension n and discover that each vector is fixed by a sequence of n real numbers with regard to a fixed basis. So all these models are 'essentially the same', we only take another basis and get another model but essentially we must only regard the n real numbers $(a_1, \ldots, a_n)$. We understand the fundamental idea of isomorphism, we recognize that all these models are represented in the arithmetic vector space $\mathbb{R}^n$.

Again we are teaching on two levels: models on the one hand—mathematical theory on the other. Mathematically, we can define what that means: the vector spaces U and V are isomorphic and we may proof the well-known theorem. But in models we find the understanding and the idea for the proof, we find the correlations:

Magic squares:	$\{0\}$	$\mathfrak{E}$	$\mathfrak{Q}$	$\mathcal{D}$	$\mathfrak{P}$	$\mathfrak{Z}$	$\mathfrak{M}$
Arithmetic VS:	$\{0\}$	$\mathbb{R}$	$\mathbb{R}^5$	$\mathbb{R}^7$	$\mathbb{R}^8$	$\mathbb{R}^{10}$	$\mathbb{R}^{16}$
Polynomial VS:	$\{0\}$	P_0	P_4	P_6	P_7	P_9	P_{15}

Our young students should understand the meaning of: you only need the vector space $\mathbb{R}^n$ if you want to demonstrate some facts of vector space theory, because all other models of vector spaces are isomorphic to $\mathbb{R}^n$. Pupils do not have to know the word 'isomorphism' but they should understand the connections (Artmann, 1978; Stowasser, 1978).

7. FINAL REMARK

In this way we have taught Linear Algebra at high-school (Gymnasium). We found that students, teachers and teacher-students like magic squares and that this model is very suitable to introduce vector spaces. Students have been surprised to find a lot of interesting facts about squares and they got a first idea of a mathematical theory. We put the elementarization by studying concrete models versus the exactification in mathematical theory (Fischer, 1978). I think we have had very attractive mathematical lessons.

REFERENCES

Artmann, B., 1978, Arithmetische und geometrische Folgen als zweidimensionale Vektorräume, P.M. 20.

Botsch, O., 1967, *Spiel mit Zahlenquadraten*, Frankfurt

Fischer, R., 1976, Fundamentale Ideen bei den reellen Funktionen, *ZDM*, **4**.

Fischer, G., 1978, *Lineare Algebra*, Braunschweig

Grassle, W., Basiswahl in Vektorräumen—explizite Darstellung von Fibonacci-Folgen, *MNU*, **4**, 78.

Kracke, H., 1970, *Aus eins mach zehn und zehn ist keins*, Hamburg.

Schaal, H., Geometrische Studien zu Dürers Melencolia, *MU*, **2**, 82.

Stowasser, R., Mohry, B., 1978, *Rekursive Verfahren*, Hannover.

Toussaint, M., Lineare Algebra auf der Oberstufe, *MU*, **5**, 73.

Ideal Fluid Flow and Finite Elements

E.-M. Salonen,
Helsinki University of Technology
A. Pramila
The University of Oulu
and
P. Lehtonen
Finnish Broadcasting Company

SUMMARY

Two-dimensional ideal fluid flow is employed as an example problem to illuminate numerical modelling in the finite element method. The basic unknowns are the velocity components $u(x, y)$ and $v(x, y)$. Both the weighted residual and the variational formulation are employed.

Special topics emphasized are: (1) The difficulties in obtaining the discrete equations systematically in the weighted residual formulation; (2) the need for using transformations of nodal variables at the boundary nodes and the derivation of the transformation formulae; (3) simple explanation for the ill-conditioning in the penalty method when the value of the penalty number is high; (4) the possibility of combining the classical least squares method with the other versions of the method of weighted residuals.

Some numerical results obtained by the different formulations are given.

1. INTRODUCTION

We consider the two-dimensional ideal fluid flow as an example problem to illuminate several features of the finite element method in a simple way.

The governing equations are in the domain A

$$\frac{\partial u}{\partial x} + \frac{\partial v}{\partial y} = 0 \tag{1}$$

$$\frac{\partial v}{\partial x} - \frac{\partial u}{\partial y} = 0, \tag{2}$$

on the boundary s_1

$$u' \equiv n_x u + n_y v = \bar{u}' \tag{3}$$

and on the boundary s_2

$$v' \equiv -n_y u + n_x v = \bar{v}'. \tag{4}$$

Here (Fig. 1) u and v are the velocity components in the x and y directions, respectively, n_x and n_y are the components of the outward unit normal vector to the boundary and u' and v' are the normal and tangential velocity components at the boundary. Parts s_1 and s_2 form the whole boundary. A bar refers to a given quantity.

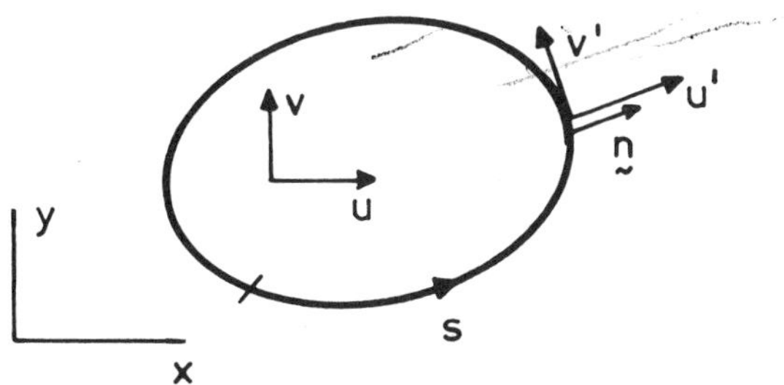

Fig. 1. Some notations

Equation (1) is the condition of incompressibility and equation (2) that of irrotationality. Equation (3) states that the volume flux intensity calculated from u and v must be consistent with the prescribed value $\bar{u}'$. Equation (4), usually, appears from symmetry reasons, when for instance only a half of a symmetric domain is analysed. The given tangential velocity component $\bar{v}'$ is then zero.

The normal way to proceed is to introduce either the velocity potential function $\phi(x, y)$ with $u = -\partial\phi/\partial x$, $v = -\partial\phi/\partial y$ or the stream function $\psi(x, y)$ with $u = \partial\psi/\partial y$, $v = -\partial\psi/\partial x$. In both cases we then end up with only one dependent function, ϕ or ψ, obeying the Laplace equation with some Dirichlet and Neumann boundary conditions. How this problem can be solved with finite elements is standard knowledge and is well documented in textbooks.

For instructive purposes, however, it is much more rewarding to apply the finite element method directly to the set of equations (1) to (4). The equations are simple in form (for instance no material parameters appear in them), but they still represent a boundary value problem with two independent variables x and y and two dependent variables $u(x, y)$ and $v(x, y)$. The problem statement is wide enough for the student to grasp most of the essential features of the finite element method from formulations built around it. First order derivatives only appear in the field equations. Thus simple C^0-continuous elements can be used and for instance no integrations by parts are necessary in the Galerkin method.

One can say that there are two levels in the assimilation of the finite element method. The first one consists of the general ideas and principles on which the discretizations are based. The second one deals with the detailed bookkeeping and programming effort which is necessary when the method is actually used in a computer.

In this paper we try to illuminate some aspects of the first level. There are four main themes. (1) The difficulties in obtaining the discrete equations systematically in the method of weighted residuals. These are not usually discussed in the literature as the model example problems normally deal with cases where the number of boundary conditions at a point on the boundary is equal to the number of unknown functions. Here this is not the case as these numbers are one and two, respectively. (2) The need for using transformation of nodal variables at the boundary nodes and the derivation of the transformation formulae. (3) Simple explanation for the ill-conditioning in the penalty method when the value of the penalty number is high. (4) The possibility of combining the classical least squares method with the other versions of the method of weighted residuals.

2. FINITE ELEMENT APPROXIMATION

We adhere closely to the notation of the finite element text by Zienkiewicz (1977) and define

$$\left.\begin{aligned} A_1(\mathbf{u}) &\equiv \frac{\partial u}{\partial x}+\frac{\partial v}{\partial y}, \\ A_2(\mathbf{u}) &\equiv \frac{\partial v}{\partial x}-\frac{\partial u}{\partial y}, \\ B_1(\mathbf{u}) &\equiv n_x u+n_y v-\bar{u}', \\ B_2(\mathbf{u}) &\equiv -n_y u+n_x v-\bar{v}', \end{aligned}\right\} \tag{5}$$

so that all equations (1) to (4) can be written in short as

$$\left.\begin{aligned} A_1(\mathbf{u}) &= 0 \quad \text{in } A, \\ A_2(\mathbf{u}) &= 0 \quad \text{in } A, \\ B_1(\mathbf{u}) &= 0 \quad \text{on } s_1, \\ B_2(\mathbf{u}) &= 0 \quad \text{on } s_2. \end{aligned}\right\} \tag{6}$$

These standard forms ‘something on the left is equal to zero on the right’ are most useful for performing the necessary derivations in a consistent way. In addition, by employing notations A_1, A_2, etc. instead of specific detailed expressions on the right-hand sides in (5), the student can probably more easily visualize the generalizations of the formulations to come for other sets of equations.

The finite element approximation for $\mathbf{u} = [u, v]^{\mathrm{T}}$ is

$$\hat{\mathbf{u}} = \mathbf{N}\mathbf{a} \tag{7}$$

where $\mathbf{N}(x, y)$ is the given shape function matrix and $\mathbf{a}$ is the list of unknown nodal variables.

Formally similar approximations abound in mathematics and physics; for instance in Fourier series, in power series in the Ritz method, etc. The only—and in practice extremely important—difference is that in the finite element method the functions $\mathbf{N}$ are defined in a piecewise manner and that the unknowns $\mathbf{a}$ are usually values of the unknown functions or of their derivatives at the nodes.

After we have decided on expression (7), the only thing left in the finite element method in principle is the determination of suitable values for $\mathbf{a}$. The criterion for this naturally comes from the fact that $\hat{\mathbf{u}}$ must satisfy equations (6) in some sense as well as possible.

As is well known, there are two main possibilities to determine $\mathbf{a}$: the weighted residual formulation and the variational formulation. The most usual versions of the former formulation are the Galerkin method, the subdomain collocation method, the point collocation method, and the least squares method. However, in this paper we look upon the least squares method as a variational formulation. If wanted, a variational formulation in the classical sense is easily obtained for the problem which follows from the set (1) to (4) when either the velocity potential or the stream function is employed.

3. WEIGHTED RESIDUAL FORMULATION

3.1 General

Substitution of approximation (7) into expressions (5) gives the so-called residuals

$$\left.\begin{aligned} \hat{A}_1(\mathbf{a}) &\equiv A_1(\hat{\mathbf{u}}), \\ \hat{A}_2(\mathbf{a}) &\equiv A_2(\hat{\mathbf{u}}), \\ \hat{B}_1(\mathbf{a}) &\equiv B_1(\hat{\mathbf{u}}), \\ \hat{B}_2(\mathbf{a}) &\equiv B_2(\hat{\mathbf{u}}). \end{aligned}\right\} \tag{8}$$

Actually the field residuals $\hat{A}_1$ and $\hat{A}_2$ depend in addition to $\mathbf{a}$ on position $\mathbf{x} = [x, y]^{\mathrm{T}}$ and similarly the boundary residuals $\hat{B}_1$ and $\hat{B}_2$ depend on s. These dependencies are usually, however, not shown in expressions like (8).

We write the equation

$$\int_A w_1 \hat{A}_1 \mathrm{d}A + \int_A w_2 \hat{A}_2 \mathrm{d}A + \int_{s_1} w_1^* \hat{B}_1 \, \mathrm{d}s + \int_{s_2} w_2^* \hat{B}_2 \, \mathrm{d}s = 0 \tag{9}$$

or equivalently the equations

$$\left.\begin{aligned}
&\int_A w_1 \hat{A}_1 \,\mathrm{d}A = 0,\\
&\int_A w_2 \hat{A}_2 \,\mathrm{d}A = 0,\\
&\int_{s_1} w_1^* \hat{B}_1 \,\mathrm{d}s = 0,\\
&\int_{s_2} w_2^* \hat{B}_2 \,\mathrm{d}s = 0.
\end{aligned}\right\} \tag{10}$$

Quantities $w_1(\mathbf{x})$, $w_2(\mathbf{x})$, $w_1^*(s)$, and $w_2^*(s)$ are called test functions or weight functions. We try to get the discrete system equations for **a** by selecting suitable combinations of weight functions n times (n is the number of nodal variables in **a**) and by demanding equation (9) to be valid for these functions. Alternatively, we select one weight function at a time and demand the corresponding equation in set (10) to be valid and continue in this fashion until we have got enough equations. The two forms (9) or (10) appear alternatively in the literature, but we can easily see that both of them can be used equivalently to generate the same discrete set. For instance, by taking the special case $w_1 \neq 0$, $w_2 = w_1^* = w_2^* = 0$, we obtain from equation (9) the first equation in set (10). On the other hand, by summing the both sides of the four equations (10), we arrive at equation (9). In what follows we shall employ set (10).

The idea behind the weighted residual formulation relies on the reasonable hope that if the weighted integrals of the residuals disappear at least with respect to some weight functions, the residuals themselves must be small. It is on the responsibility of the user to select the weight functions in a 'good' way. Fig. 2 describes schematically the most usual prescriptions given in the literature.

These prescriptions are regrettably not always quite adequate even for a reasonable expert, not to mention a student. Let us consider the simplest case, where three-noded linear triangular elements are used for the approximation (7). Each node contains two unknown nodal variables: the components u and v. The field residuals $\hat{A}_1^e$ and $\hat{A}_2^e$ are found to be constants in each element e. Let there be m elements. We realize that the satisfaction of the $2m$ equations

$$\left.\begin{aligned}
&\hat{A}_1^e(\mathbf{a}) = \text{constant} = 0,\\
&\hat{A}_2^e(\mathbf{a}) = \text{constant} = 0
\end{aligned}\right\} \quad e = 1, 2, \ldots, m \tag{11}$$

would be the best possible basis for the determination of **a** if we consider the field equations only. Equations (11) are obtained from the first two equations (10) both by the subdomain collocation method by taking each element as a

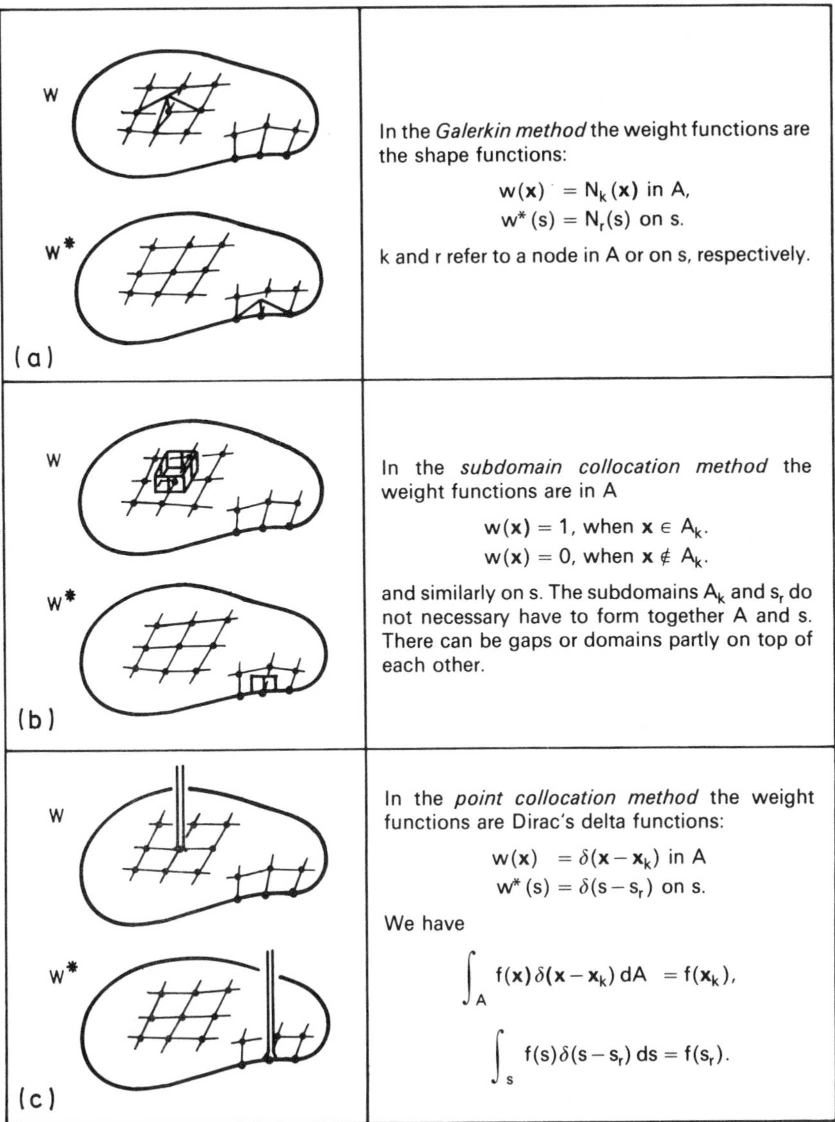

w
w*
(a)
In the *Galerkin method* the weight functions are the shape functions:
$w(\mathbf{x}) = N_k(\mathbf{x})$ in A,
$w^*(s) = N_r(s)$ on s.
k and r refer to a node in A or on s, respectively.
w
w*
(b)
In the *subdomain collocation method* the weight functions are in A
$w(\mathbf{x}) = 1$, when $\mathbf{x} \in A_k$.
$w(\mathbf{x}) = 0$, when $\mathbf{x} \notin A_k$.
and similarly on s. The subdomains A_k and s_r do not necessary have to form together A and s. There can be gaps or domains partly on top of each other.
w
w*
(c)
In the *point collocation method* the weight functions are Dirac's delta functions:
$w(\mathbf{x}) = \delta(\mathbf{x} - \mathbf{x}_k)$ in A
$w^*(s) = \delta(s - s_r)$ on s.
We have
$\int_A f(\mathbf{x})\delta(\mathbf{x} - \mathbf{x}_k)\, dA = f(\mathbf{x}_k)$,
$\int_s f(s)\delta(s - s_r)\, ds = f(s_r)$.

Fig. 2. Different versions of the weighted residual formulation

subdomain and by the point collocation method if one collocation point anywhere in each element is employed. By sketching a typical mesh and by counting the number of equations (11) and by comparing it with the total number of nodal variables it is, however, seen that there are very few if any equations left for the representation of the boundary conditions. Thus a compromise must be made and one cannot hope to satisfy all the equations (11).

A possibility to obtain the right number of discrete equations is to try to associate always two equations with each node. Fig. 2 has been drawn in such a spirit.

There are still difficulties. Let us consider say the Galerkin method. For an inner node k the choice is obvious: we obtain the two discrete equations from the first two equations (10) by taking $w_1 = w_2 = N_k(\mathbf{x})$. For a boundary node r say on s_1 the choice is not so clear. We have two field equations and one boundary condition available. The boundary condition must certainly be used. Thus one discrete equation is obtained from the third equation (10) with $w_1^* = N_r$, where $N_r(s)$ is the value of the shape function N_r at the boundary. How can we get the second discrete equation? Perhaps a linear combination of the first two equations (10) with $w_1 = w_2 = N_r(\mathbf{x})$ and suitable weighting would do.

Numerical tests showed that the best procedure was as follows. On boundary part s_1 (s_2) only the second (the first) equation (10) is employed. A possible explanation for this comes through the formulae

$$\int_A \left(\frac{\partial u}{\partial x} + \frac{\partial v}{\partial y}\right) \mathrm{d}A = \int_s (n_x u + n_y v)\,\mathrm{d}s = \int_s u'\,\mathrm{d}s = 0, \tag{12}$$

$$\int_A \left(\frac{\partial v}{\partial x} - \frac{\partial u}{\partial y}\right) \mathrm{d}A = \int_s (-n_y u + n_x v)\,\mathrm{d}s = \int_s v'\,\mathrm{d}s = 0, \tag{13}$$

which are obtained by Green's theorem. These are the global mass conservation and zero circulation conditions, respectively. So it might be concluded that the first (the second) boundary condition already gives a contribution towards the satisfaction of the first (the second) field equation and it is thus fair to employ the second (the first) field equation at a boundary node for discretization.

3.2 Boundary conditions

In the general form (10) of the weighted residual formulation the boundary conditions have no special status as they are taken into account in principle similarly to the field equations. The discrete equations are nevertheless often obtained in a two-phase process for computational convenience. In the first phase only the field equations are considered and the necessary number of discrete equations are derived. In the linear case (as we have here) they shall look like

$$\mathbf{Ka} + \mathbf{P} = \mathbf{0} \tag{14}$$

where $\mathbf{K}$ and $\mathbf{P}$ are constant with respect to $\mathbf{a}$. In the second phase certain of these equations are replaced by discrete equations due to the boundary conditions.

Further, it is quite obvious that there is no special need to be consistent in the selection of the weight functions in the sense that all of them should be taken according to the same version in an application. On the contrary, considerable simplification is achieved if the boundary conditions are always discretized by point collocation and by taking the boundary nodes as collocation points (see Fig. 2(c)). Thus we would obtain the discrete counterparts of equations (3) and (4) as

$$n_x u_r + n_y v_r - \bar{u}'_r = 0, \quad \text{node } r \text{ is on } s_1 \tag{15}$$

$$-n_y u_r + n_x v_r - \bar{v}'_r = 0, \quad \text{node } r \text{ is on } s_2. \tag{16}$$

So in a discrete equation, the nodal values of only one node appear. However, still further simplification is achieved, if we take the components u' and v' as our nodal variables instead of u and v. Then we have the equations

$$u'_r - \bar{u}'_r = 0, \quad \text{node } r \text{ is on } s_1 \tag{17}$$

$$v'_r - \bar{v}'_r = 0, \quad \text{node } r \text{ is on } s_2. \tag{18}$$

Now each discrete equation contains only one unknown nodal variable, which proves to be useful for computer applications. The well-known large number artifice, described for instance by Zienkiewicz (1977, p. 11), can be conveniently used.

There is a certain price to be paid for this advantage. The original list $\mathbf{a}$ of the nodal variables consists of the components u and v at the nodes. The new list $\mathbf{a}'$ consists of components u and v at inner nodes but has components u' and v' at the boundary nodes. We can write the relationship

$$\mathbf{a} = \mathbf{L}\mathbf{a}', \tag{19}$$

where the transformation matrix $\mathbf{L}$ has a simply structure. Substitution of expression (19) into equations (14) gives

$$\mathbf{K}'\mathbf{a}' + \mathbf{P} = \mathbf{0} \tag{20}$$

where $\mathbf{K}' = \mathbf{K}\mathbf{L}$. In practice the necessary transformations are performed at the element level. As $\mathbf{K}$ is usually non-symmetrical in the weighted residual formulation there is no need for premultiplication with $\mathbf{L}^T$ in an effort to achieve symmetry. Another way of arriving at equations (20) is to substitute expression (19) into approximation (7) and to derive the equations directly starting from $\hat{\mathbf{u}} = \mathbf{N}'\mathbf{a}'$, where $\mathbf{N}' = \mathbf{N}\mathbf{L}$.

3.3 Numerical results

Some results are given here in connection with the problem described in Fig. 3. It is the case of flow around a circular cylinder (radius $= a$) in an infinite

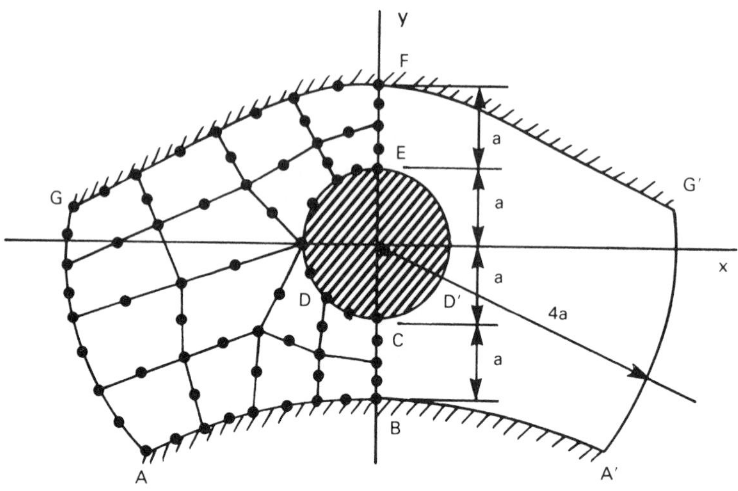

Fig. 3. Computational domain and the mesh

domain with a free stream velocity q_0 in the x-axis direction. The computational domain ABA′G′FG has been selected so that boundaries ABA′ and GFG′ are the streamlines from the known exact solution through points B and F, respectively. Thus these boundaries can as well be considered as solid walls. Boundaries AG and A′G′ are taken as arcs of a circle having its centre at the origin. The stagnation points are at D and D′ corresponding to phase angles $-135°$ and $-45°$, respectively. The value of the circulation around the cylinder is $\Gamma = -2\sqrt{2}\pi q_0 a$. For symmetry reasons only domain ABCDEFG is discretized. A $4 \times 4 = 16$ element mesh with 69 nodes is employed. The elements are eight-noded serendipity ones.

Boundaries AB, CDE, FG, and GA are of type s_1. Velocity component u' differs from zero only on part GA where its nodal values were calculated from the exact solution. Boundaries BC and EF are of type s_2 where from symmetry reasons $v' = (v) = 0$. At the stagnation point D, in addition, the value of the tangential velocity component v' is put to zero.

Calculations were performed with the Galerkin, the subdomain collocation, and the point collocation method for the field equations. The boundary conditions were discretized using formulae (17) and (18). Similarly, at a boundary node, the procedure described previously for the selection of the other discrete equation was employed.

The weighting for the field equations was done in the spirit of Fig. 2. However, in subdomain collocation the subdomain associated with each node was taken to consist of all elements having this node common. This makes the necessary integrations much easier than in the case shown in Fig. 2(b). It is also to be noted, that the residual in point collocation at each node has a different value when calculated from different elements meeting at the node. Actually the mean value of the different residuals at a node was used. Or equivalently by multiplying this equation by the number of elements meeting at the node, we see that the final equation can be obtained by just adding the different residuals together to form the total residual which is then equated to zero.

Fig. 4 shows some results from the calculations, Pramila and Salonen (1979). No solution was obtained with subdomain collocation as the system matrix proved to be singular. Similarly, the results by the Galerkin and the point collocation method do not indicate very convincing performance.

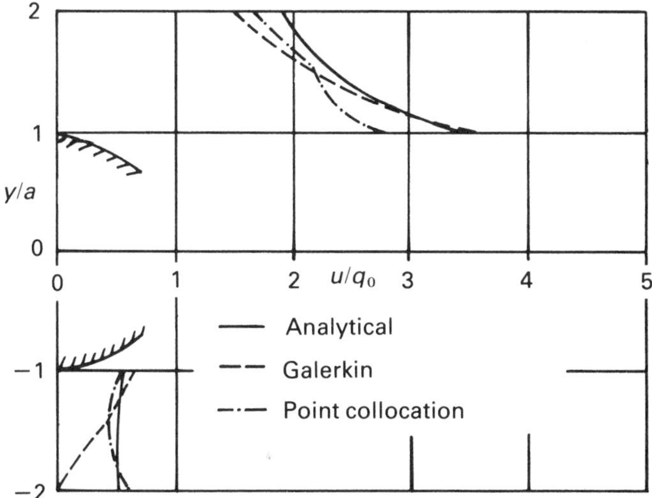

Fig. 4. Horizontal velocity u across sections BC and EF

One can consider this behaviour as demonstrating the unsuitability of our example formulations for teaching purposes or perhaps in the end the other way round. The example gives a lesson by showing us that seemingly logical reasoning can lead to a poorly behaving numerical model.

By applying formulae (12) and (13) for a subdomain associated with an inner node for any mesh we see immediately that the nodal variables of the node in question totally disappear from the two discrete equations. To have a system matrix with many zeros on the diagonal (the non-zero diagonals appear only in the equations corresponding to boundary nodes) is not a good sign for numerical performance in a boundary value problem. This is confirmed by the calculations. As the Galerkin and the point collocation methods resemble much the subdomain collocation method it is then not surprising that also the former ones show rather erratic behaviour.

Further insight on the delicacy of numerical modelling can be obtained from the extremely simple special case shown in Fig. 5(a). It represents uniform flow in the x-direction with $u = q_0$. We take $v = 0$ and consider for a while u as an unknown function of x: $u = u(x)$. The problem left from equations (1) to (4) is

$$\left.\begin{aligned} &\frac{du}{dx} = 0, \quad 0 < x < l \\ &u(0) = q_0, \quad u(l) = q_0. \end{aligned}\right\} \tag{21}$$

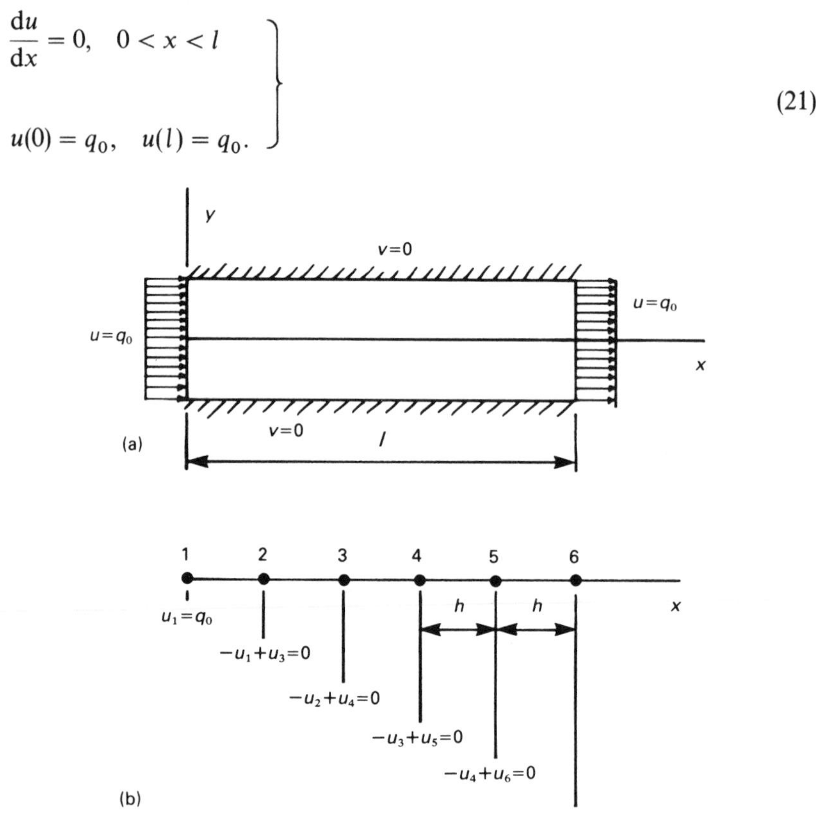

Fig. 5. (a) Uniform flow in the x-axis direction. (b) A finite element mesh and the corresponding discrete equations

This is a wrongly posed situation. Actually we have now a propagation problem and only one of the boundary conditions should be prescribed freely (even if the solution $u = q_0$ happens to satisfy also the other one). The discrete equations obtained by subdomain collocation for a mesh of five equal length two-noded line elements are shown in Fig. 5(b). Exactly equivalent equations are found by the two other versions of the weighted residual method here. It is again seen that the equations have zeros on the diagonal. The set is of such a form that it has a unique solution when the number of elements is odd (as here). When the number of elements is even the system matrix proves to be singular and the computer gives no solution. In the former case the model can be considered as consisting of two separate propagation problems as can be realized by examining the set in Fig. 5(b).

It seems that there exists very little literature on boundary value problems with odd-ordered differential equations. How could we achieve here better numerical behaviour? One possible way could consist of transforming equations (1) to (4) into a more manageable form.

In problem (21) we can for instance take a new variable $u^* = u \exp(-cx)$ or $u = u^* \exp(cx)$ where c is a constant. Instead of (21) we now have a problem

$$\left.\begin{aligned} &\frac{du^*}{dx} + cu^* = 0, \quad 0 < x < l \\ &u^*(0) = q_0, \quad u^*(l) = q_0 \exp(-cl) \end{aligned}\right\} \tag{22}$$

Discretization of differential equation (22) clearly gives a system matrix with non-zero elements on the diagonal. This, perhaps naive, reasoning for hoping for better behaviour was behind the special selection of the form of u^*. Numerical tests showed indeed that a unique solution is now obtained also in the case where the number of elements is even, if $c \neq 0$. The accuracy of the numerical solution finally deteriorates when the value of c grows. This is because the solution $u^* = q_0 \exp(-cx)$ cannot be represented exactly by linear elements. However, for moderate values of c the error is insignificant. For instance with a mesh of four elements and the value $c = 0.1/l$, the maximum error for u with all the three versions was about 0.1 ‰ or under.

On the basis of these results for the one-dimensional example one can speculate that a formulation with new variables say $u^* = u \exp[-c(x+y)]$ and $v^* = v \exp[-c(x+y)]$ could perhaps work in the general case. We have not yet had time to try this.

4. VARIATIONAL FORMULATION

4.1 General

Equations (1) to (4) do not form a self-adjoint system and one cannot devise a functional $\Pi(\mathbf{u})$ which would give equations (1) and (2) directly as its Euler equations. However, we can demonstrate all the important steps in a finite element variational formulation by using the least squares method and by taking expression

$$\Pi(\mathbf{u}) = \frac{1}{2}\int_A \alpha_1 A_1^2 \, dA + \frac{1}{2}\int_A \alpha_2 A_2^2 \, dA + \frac{1}{2}\int_{S_1} \beta_1 B_1^2 \, ds + \frac{1}{2}\int_{S_2} \beta_2 B_2^2 \, ds \tag{23}$$

as our functional. Multipliers 1/2 are not essential and have been taken for convenience. Quantities $\alpha_1(\mathbf{x})$, $\alpha_2(\mathbf{x})$, $\beta_1(s)$ and $\beta_2(s)$ are given positive functions by which we can put different emphasis in expression (23) on each left-hand side in equations (6). The dimensions of α and β must be selected so that Π is dimensionally homogeneous. The exact solution gives a minimum value (zero) to Π.

Substitution of approximation (7) into functional (23) transforms it into a function

$$\hat{\Pi} = \Pi(\hat{\mathbf{u}}) = \frac{1}{2}\int_A \alpha_1 \hat{A}_1^2 \, dA + \frac{1}{2}\int_A \alpha_2 \hat{A}_2^2 \, dA + \\ + \frac{1}{2}\int_{S_1} \beta_1 \hat{B}_1^2 \, ds + \frac{1}{2}\int_{S_2} \beta_2 \hat{B}_2^2 \, ds. \tag{24}$$

When we consider the integrations to be performed, we realize that $\hat{\Pi}$ is a function of the variables $\mathbf{a}$: $\hat{\Pi} = \hat{\Pi}(\mathbf{a})$. The natural counterpart of the condition that functional $\Pi(\mathbf{u})$ should be stationary with respect to $\mathbf{u}$ is that function $\hat{\Pi}(\mathbf{a})$ should be stationary with respect to $\mathbf{a}$. This gives the system equations

$$\frac{\partial \hat{\Pi}}{\partial a_i} = 0, \qquad i = 1, 2, \ldots, n. \tag{25}$$

Thus in a variational formulation the procedure to find the discrete equations is completely predetermined after the functional is given and an approximation (7) is taken. There are no more decisions to be made by the user.

The variational formulation has the well-known advantage that in a linear problem (as we have here) equations (25) written in matrix notation (14) have a symmetric coefficient matrix $\mathbf{K}$. This is seen as follows. The jth coefficient in the ith equation is clearly $K_{ji} \equiv \partial/\partial a_j (\partial \hat{\Pi}/\partial a_i)$. Similarly, the ith coefficient in the jth equation is $K_{ij} \equiv \partial/\partial a_i (\partial \hat{\Pi}/\partial a_j)$. Since the value of the second derivative does not depend on the order of taking the derivatives we have $K_{ji} = K_{ij}$.

As the least squares method is not a variational principle in the classical sense, we nevertheless have here the problem of selecting suitable values for α and β. The normal procedure is indicated schematically in Fig. 6(a). It is interesting to note that when numerical integration is used to perform the integrations in equations (25) or in expression (24), this can be clearly interpreted as shown in Fig. 6(b). The coefficients c are the products of the value of the original α or β at the integration point with the corresponding weight in the integration formula.

A further interpretation and possibility arises. We first form an overdetermined set of discrete equations by any version of the method of weighted residuals. This set is then solved by using the least squares procedure as is usual with overdetermined systems. Thus an expression

$$\hat{\Pi}_d(\mathbf{a}) = \sum \gamma_j F_j^2(\mathbf{a}) \tag{26}$$

is formed, where the γ_j are positive coefficients (usually $= 1$) and the F_j are

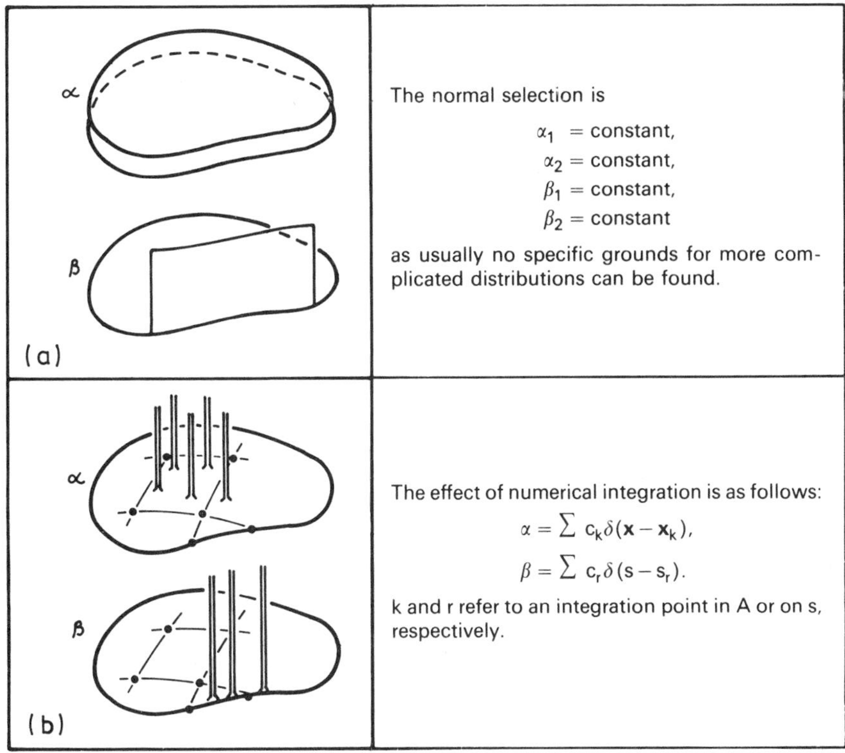

Fig. 6. Quantities α and β in the least squares method

the left-hand sides of the discrete equations. The system is obtained similarly as shown by equations (25). In point collocation, if the collocation points are selected to lie at the integration points used for expression (24) and if suitable values of γ_j are employed, we are back at the result of Fig. 6(b).

4.2 Boundary conditions

We see that the value of one of the quantities α_1, α_2; β_1, or β_2 can be taken arbitrarily (for instance $\alpha_1 = 1$) as it is only the ratios α_2/α_1, β_1/α_1, and β_2/α_1 which are essential. (The change of α_1 just means that all equations (25) are multiplied by a constant.) Further, as the field equations and the boundary conditions are here rather similar, respectively, in nature, it seems reasonable to put $\alpha_1 = \alpha_2 = 1$ and $\beta_1 = \beta_2 = \beta$. Thus expression (24) obtains the form

$$\hat{\Pi} = \tfrac{1}{2}\int_A (\hat{A}_1^2 + \hat{A}_2^2)\,\mathrm{d}A + \tfrac{1}{2}\int_{s_1} \beta\hat{B}_1^2\,\mathrm{d}s + \tfrac{1}{2}\int_{s_2} \beta\hat{B}_2^2\,\mathrm{d}s. \tag{27}$$

The problem then left is the selection of a suitable value for β.

We can consider expression (27) as penalty formulation. The area integral can be thought to represent the original functional and the line integrals represent the penalty terms whose purpose is to enforce the satisfaction of the boundary conditions which are called now constraints according to the terminology of the penalty method.

The typical behaviour of the penalty method is detected. When the value of the penalty number β is low, the constraints are poorly satisfied and the whole solution is inaccurate. When the value of β is increased the results get better but finally no solution is obtained due to ill-conditioning.

The reason behind the ill-conditioning can be explained very simply, if we consider a case where we calculate the line integrals in (27) by the trapezoidal integration rule with the boundary nodes as integration points. The line integrals are replaced by

$$\tfrac{1}{2}\sum \gamma_r(n_x u_r + n_y v_r - \bar{u}'_r)^2 + \tfrac{1}{2}\sum \gamma_r(-n_y u_r + n_x v_r - \bar{v}'_r)^2, \tag{28}$$

where the first (the second) sum is over the nodes on s_1 (on s_2) and the γs are of the order of magnitude $h\beta$, where h is a typical element side length. Let us consider the system equations $\partial\hat{\Pi}/\partial u_r = 0, \partial\hat{\Pi}/\partial v_r = 0$ associated with a node r say on s_1. When a finite word length is employed, the contributions from the area integral in expression (27) disappear in comparison with the penalty terms when the value of β (and thus of γ) is very high and the equations appear in practice in the forms

$$\left.\begin{aligned} n_x\gamma_r(n_x u_r + n_y u_r - \bar{u}'_r) &= 0, \\ n_y\gamma_r(n_x u_r + n_y u_r - \bar{u}'_r) &= 0. \end{aligned}\right\} \tag{29}$$

These equations are clearly linearly dependent and thus the whole set is finally singular.

In more complicated cases the final situation with the penalty method is similar if even one constraint contains more than one nodal variable. Linear dependence is created when the system equations are formed by the process of partial differentiation with respect to the nodal variables.

The suitable range of β must be determined by numerical experiments. The upper limit depends on the word length. The numerical results with a word length of 7 . . . 8 significant digits gave a useful range of $h\beta \approx 10^2 \ldots 10^4$.

It is quite remarkable that we can get wholly rid of the difficulty of selecting the suitable value of β by using u'_r and v'_r as our nodal variables at the boundary nodes. Instead of expression (28) there is now obtained

$$\tfrac{1}{2}\sum \gamma_r(u'_r - \bar{u}'_r)^2 + \tfrac{1}{2}\sum \gamma_r(v'_r - \bar{v}'_r)^2. \tag{30}$$

If we examine again a node r on say s_1, the system equation $\partial\hat{\Pi}/\partial u'_r = 0$ will appear as

$$\gamma u'_r - \gamma\bar{u}'_r = 0 \tag{31}$$

when the value of β is very high. However, no contribution from the penalty terms appear in the second equation $\partial\hat{\Pi}/\partial v'_r = 0$. Thus no linear dependence is created and β can be extremely high without any numerical difficulties.

The discrete equations can be again formed by a two-phase process. Firstly a set $\mathbf{K\,a}+\mathbf{P}=\mathbf{0}$ is obtained from expression (27) without the line integral terms. Thus the boundary conditions are not yet represented at all. Then relationship (19) is employed and a set of the form (20) is obtained. Premultiplication with $\mathbf{L}^{\mathrm{T}}$ then gives a set

$$\mathbf{K}^*\mathbf{a}'+\mathbf{P}^*=\mathbf{0} \tag{32}$$

where the system matrix $\mathbf{K}^*=\mathbf{L}^{\mathrm{T}}\mathbf{K}\mathbf{L}$ is now symmetric and where $\mathbf{P}^*=\mathbf{L}^{\mathrm{T}}\mathbf{P}$. In practice these transformations are performed at the element level. Another way of arriving at equations (32) is to use approximation $\hat{\mathbf{u}}=\mathbf{N}'\mathbf{a}$, where $\mathbf{N}'=\mathbf{N}\mathbf{L}$, directly in expression (27). Secondly those equations which correspond to a given nodal variable say u'_r are replaced by equations like (31). This is in fact achieved most conveniently by the large number artifice. It is further seen that the symmetry of the system matrix is preserved in this operation.

4.3 Modified least squares

Numerical results showed that extremely fine meshes are needed to obtain a satisfactory accuracy with the conventional least squares method. For instance the global conditions (12) and (13) are usually greatly in error with moderate meshes and the formulation seems to be useless in practice.

The reason for this behaviour can be understood by looking at the Euler equations of the least squares functional. For simplicity we consider functional (23) with $\alpha_1=\alpha_2=1$ and without the line integrals. This means that the admissible functions u and v are assumed here to satisfy conditions (3) and (4), i.e. they are the essential boundary conditions. By taking the first variation of Π and equating the result to zero we obtain the Euler field equations

$$\left.\begin{aligned}\frac{\partial^2 u}{\partial x^2}+\frac{\partial^2 u}{\partial y^2}&=0,\\ \frac{\partial^2 v}{\partial x^2}+\frac{\partial^2 v}{\partial y^2}&=0.\end{aligned}\right\} \tag{33}$$

The natural boundary conditions are found to be field equation (2) on s_1 and field equation (1) on s_2.

These results are typical of the least squares method. The Euler equations and the natural boundary conditions do not correspond to the original problem statement as is the case with a real classical variational formulation. The new field equations (33) are obtained by differentiation and combination from the original field equations (1) and (2). Thus each u and v satisfying the original equations, also automatically satisfy the new ones. Unfortunately, however, this reasoning does not work in the other direction. Similarly the discrete equations for inner nodes will be actually approximations to the new field equations and not to the original ones.

Let us for the sake of argument consider for a while a case where the original field equations (1) and (2) would have had say constant non-zero right-hand sides. This would mean a flow with a constant distributed volume source and a constant vorticity. Equations (12) and (13) would no more be valid. It is easy to show that the corresponding least squares functional would again give the same Euler field equations (33) as before. So we can say that the least squares method 'does not know' which equations it is trying to solve. The information about the zero right-hand sides comes only through the natural boundary conditions which consist of the original field equations. As the natural boundary conditions are satisfied just approximately in the variational finite element method, it is conceivable that large errors can appear.

One possibility of increasing the accuracy is feeding more information about the original field equations into the formulation. Thus one can for instance use any version of the weighted residual method to obtain a suitable number of discrete equations. These can then be considered as constraints on the original least squares functional Π and can be taken into account by the Lagrange multiplier method.

We have experimented with the subdomain collocation method. Thus we obtain the constraints

$$\left.\begin{aligned} \int_{A_k} A_1\,\mathrm{d}A &\equiv \int_{A_k}\left(\frac{\partial u}{\partial x}+\frac{\partial v}{\partial y}\right)\mathrm{d}A = 0, \\ \int_{A_k} A_2\,\mathrm{d}A &\equiv \int_{A_k}\left(\frac{\partial v}{\partial x}-\frac{\partial u}{\partial y}\right)\mathrm{d}A = 0. \end{aligned}\right\}\quad k = 1, 2, \ldots, m \tag{34}$$

Notation A_k refers to the kth subdomain and m is the total number of subdomains. (Notations A_1 and A_2 appear here in two meanings.)

We now have a modified functional

$$\Pi_{\mathrm{L}}(\mathbf{u},\boldsymbol{\lambda}) = \Pi(\mathbf{u}) + \sum_{k=1}^{m}\left(\lambda_{2k-1}\int_{A_k} A_1\,\mathrm{d}A + \lambda_{2k}\int_{A_k} A_2\,\mathrm{d}A\right), \tag{35}$$

where the λs are the Lagrange multipliers. It is to be noted that $u(\mathbf{x})$ and $v(\mathbf{x})$ are unknown functions but the λs are unknown constants.

Substitution of approximation (7) into functional $\Pi_{\mathrm{L}}(\mathbf{u},\boldsymbol{\lambda})$ transforms it into a function $\hat{\Pi}_{\mathrm{L}}(\mathbf{a},\boldsymbol{\lambda})$. The system equations are thus

$$\left.\begin{aligned} \frac{\partial \hat{\Pi}_{\mathrm{L}}}{\partial a_i} &= 0, \qquad i = 1, 2, \ldots, n \\ \frac{\partial \hat{\Pi}_{\mathrm{L}}}{\partial \lambda_j} &= 0. \qquad j = 1, 2, \ldots, 2m \end{aligned}\right\} \tag{36}$$

This set still has a symmetrical coefficient matrix.

4.4 Numerical results

Some results for the example problem of Fig. 3 by the modified least squares method are given in Table 1 (Salonen, Lehtonen, Pramila, 1981). Quantity q is the magnitude of velocity at the node ($x = -1.75\,a, y = -0.25\,a$) marked with

Table 1. Percentage error of q and Γ against the number of subdomains m

m	$\frac{q-q_a}{q_a}\cdot 100$	$\frac{\Gamma-\Gamma_a}{\Gamma_a}\cdot 100$
0	−22.45	−30.52
1 	−8.80	−11.62
2 	−4.71	−8.82
2 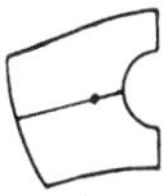	−6.69	−10.55
4 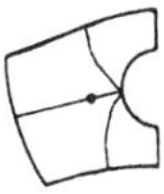	−2.59	−6.72
8 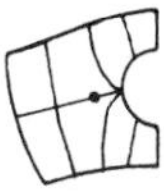	−1.63	−5.04
8 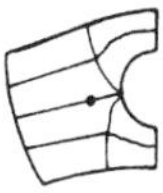	−1.87	−5.80
16 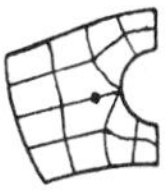	−0.53	−3.83

a dot in the insets and Γ is the circulation around the cylinder. Subscript 'a' refers to the analytical values ($q_a \approx 0.9107\, q_0$, $\Gamma_a = -2\sqrt{2\pi}\, q_0 a \approx -8.886\, q_0 a$). The table shows the effect of the number and the way of selection of the subdomains on the accuracy. The case $m = 0$ corresponds to the standard least squares method. It is interesting to note that the use of only one subdomain (the whole domain, which means only two additional unknowns) already increases the accuracy substantially. The best results are obtained by taking each element as a subdomain ($m = 16$). The velocity profiles are found to be quite smooth and accurate and no erratic behaviour is detected.

Also in practice it is most convenient to have each element as a subdomain as no additional data is thus needed to give information about the subdomains. Expression (35) obtains then the form

$$\Pi_L(\mathbf{u}, \boldsymbol{\lambda}) = \Pi(\mathbf{u}) + \sum_e \left(\lambda_1^e \int_{A^e} A_1 \, \mathrm{d}A + \lambda_2^e \int_{A^e} A_2 \, \mathrm{d}A \right), \tag{37}$$

where the meaning of the notation is self-evident.

As the constraints are now local in the sense that only the nodal variables of one element at a time appear in each constraint, we can also alternatively apply the penalty method instead of the Lagrange multiplier method without destroying the sparse and banded nature of the coefficient matrix. Thus we can take the expression

$$\Pi_P(\mathbf{u}) = \Pi(\mathbf{u}) + \frac{\alpha}{2} \sum_e \left[\left(\int_{A^e} A_1 \, \mathrm{d}A \right)^2 + \left(\int_{A^e} A_1 \, \mathrm{d}A \right)^2 \right] \tag{38}$$

as the basis. Observe the position of the exponent 2 in the penalty terms. After approximation (7) we obtain the discrete equations from $\partial \hat{\Pi}_P / \partial a_i = 0$, $i = 1, 2, \ldots, n$. This procedure has the advantage that the number of unknowns is not increased and the drawback that the suitable value for the penalty number α must be determined by numerical experiments. In the cases we have tried this far the penalty version has worked very well.

We have not tried to use Galerkin or point collocation for creating constraints. Subdomain collocation is here the physically most appealing version as then the important global mass conservation and zero circulation conditions around any reducible loop consisting of element edges can be seen to be exactly (barring errors from numerical integration and roundoff) satisfied if each element is used as a subdomain.

It should finally be noted that the reported accuracy with the modified least squares method taking each element as a subdomain was achieved with the eight-noded serendipity element. Little reflection reveals that for instance with the three-noded triangular element this procedure would not succeed. Somewhat larger subdomains should then be used.

5. CONCLUSIONS

The formulations discussed in the article show that astonishingly many features of numerical modelling in the finite element method can be explained in a simple way by employing the ideal fluid flow as an example problem.

ACKNOWLEDGEMENT

We thank R. Salminen M.Sc. for his help in some calculations.

REFERENCES

Pramila, A., Salonen, E.-M., 1979, Weighted Residual u, v Finite Element Formulations for Ideal Fluid Flow, in *Proc. International Conference on Computer Applications in Civil Engineering*, October 23–25, 1979. University of Roorkee, Nem Chand & Bros, p. V-1.

Salonen, E.-M., Lehtonen, P., Pramila, A., 1981, Further u, v-Formulations for Ideal Fluid Flow, in Proc. *Second International Conference on Numerical Methods in Laminar and Turbulent Flow*, July 13–16, 1981. Ed. C. Taylor, B. A. Schrefler, Pineridge Press, p. 1185.

Zienkiewicz, O. C., 1977, *The Finite Element Method*, 3rd edn, McGraw-Hill.

A Control Model of the Severn Barrage

S. C. Ryrie
Bristol Polytechnic

SUMMARY

A simple dynamic model is presented of the operation of a tidal power scheme, as an example of a model which involves only simple mathematics, but which has obvious and realistic practical significance. The aim of the model is to determine how the total available energy output depends on the control of the flow through the barrage, and to find that control which gives the greatest energy output, subject to typical practical constraints.

1. INTRODUCTION

It is often difficult, when teaching students how mathematics can be applied to real life, to find examples of models which realistically describe the main features of the real world, and yet whose mathematical techniques remain unobscured by details which can often seem far removed from the real world. My purpose here is to describe a mathematical model of a physical system, whose main features are readily understandable, but which has obvious practical significance and whose mathematical structure is clear.

A further problem in teaching mathematical modelling arises from the fact that, while the real world often poses open-ended problems, with no clear-cut 'correct' answer, many examples of models are, of necessity, posed as closed problems which can easily seem artificial. I hope to show that, within the model I shall describe, there is scope for posing questions to which there may be no clear-cut answer.

I shall, therefore, show the use of a model of a tidal energy scheme through an example which requires only simple mathematical techniques; to show that it may also be used as a simple example of the application of optimal control theory, and that constraints arising from the real world can lead to more complex and less clear-cut problems, for whose analysis mathematical techniques do exist.

2. THE SEVERN BARRAGE

The Severn Estuary, in the south-west of Britain, has a tidal range amongst the highest in the world, and has often been suggested as a site for an energy-extracting barrage. Recent interest in such a scheme began in the 1970s, following the large increases in world energy prices. There is now a definite proposed scheme, whose technical feasibility is established. Fig. 1 shows the likely location of a barrage. A report to the Department of Energy (1981) fully describes the likely scheme.

Energy is extracted by a barrage by allowing the tidal flow to pass through turbines which drive generators. The instantaneous rate of flow through the turbines may be controlled by an operator; it is clearly desirable that this control should be such as to extract the maximum possible energy from the barrage. We are concerned with the problem of determining this control.

3. MODEL OF THE SYSTEM

Fig. 2 shows the system model which we shall use. We assume that the water level $x_0(t)$ outside the barrage is a given function of time, and is unchanged by the operation of the barrage. We shall use

$$x_0(t) = \tfrac{1}{2}R \cos wt \tag{1}$$

where R is the tidal range, and w is the tidal frequency; the tidal period, $T = 2\pi/w$, is normally about 12.4 hours.

We assume that the water surface inside the barrage always remains flat, and that its depth $x(t)$ is related to the discharge $Q(t)$ through the turbines by the equation.

$$\frac{\mathrm{d}x}{\mathrm{d}t} = -\frac{Q(t)}{A} \tag{2}$$

where A is the surface area of the basin enclosed by the barrage; for the proposed Severn Barrage, this area is about 500 km^2. The power available for the turbines is

$$p = \rho g h Q \tag{3}$$

where h is the head across the barrage, i.e.

$$h = x - x_0$$

ρ is the density of sea-water, and g is acceleration due to gravity.

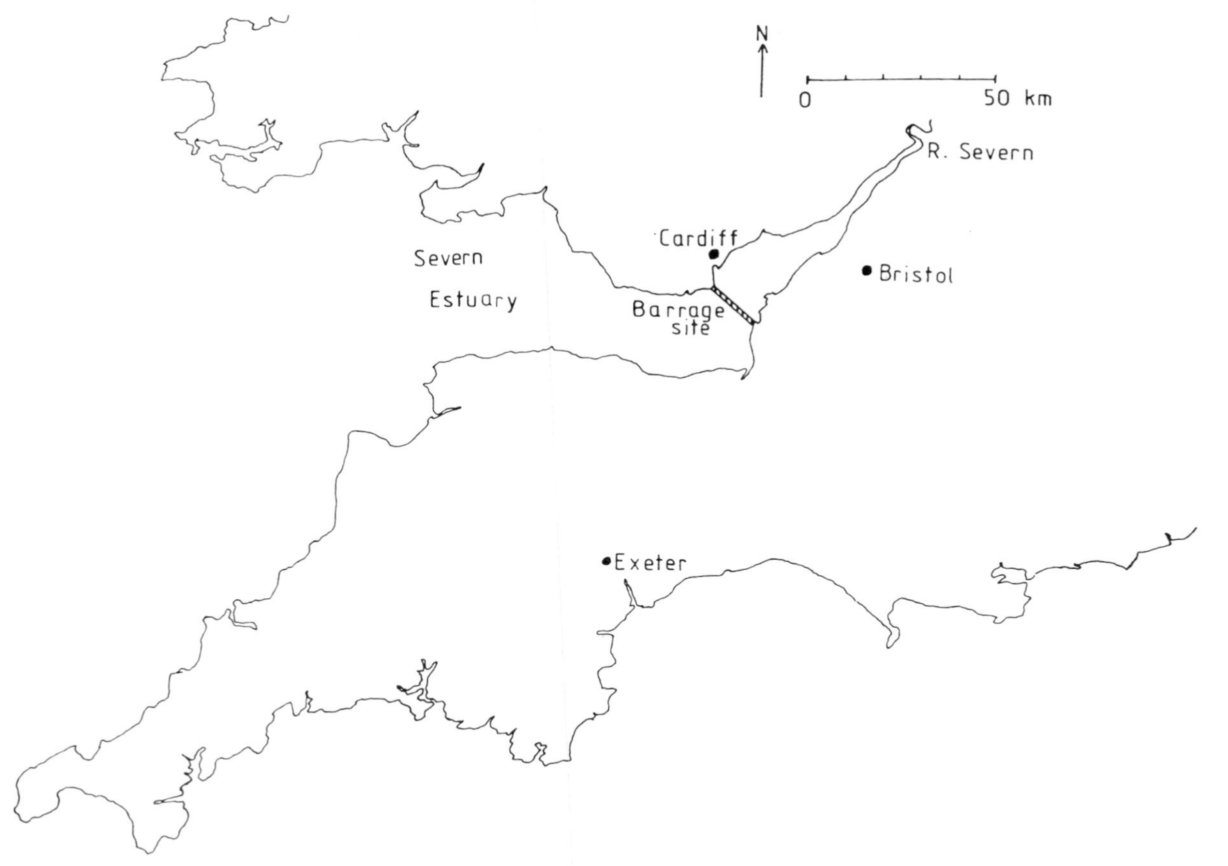

Fig. 1. Location of the proposed Severn Barrage

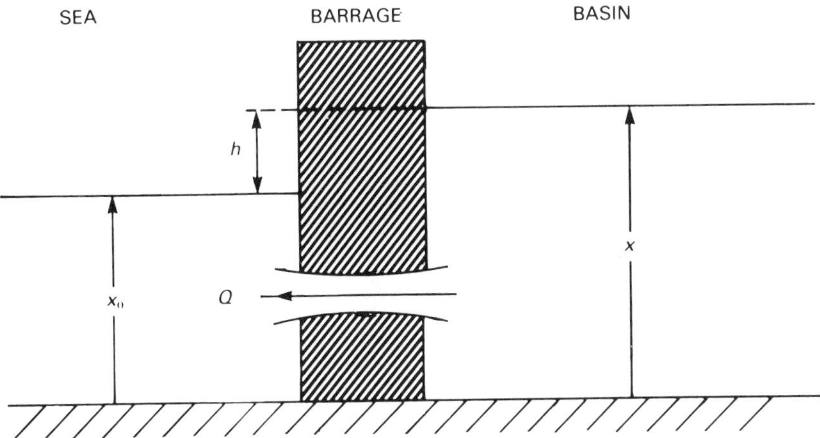

Fig. 2. Sketch showing variables used in the model of the operation of the barrage

This is the power which would be achieved if turbines of 100% efficiency were used.

The total energy E generated during one tidal period T is

$$E = \int_0^T p \, dt$$

so

$$\frac{E}{\rho g} = \int_0^T hQ \, dt = \int_0^T (x - x_0) Q \, dt \tag{4}$$

Our aim is therefore to choose Q to maximize this integral, in which Q is related to x through (2).

At this point, the presence of physical constraints, and how they are modelled, is important. There are various types of constraint on the operation of a barrage, of which the most obvious are the most easily modelled. Typically, we might have some or all of:

$|x| < \frac{1}{2}R$ (a limit on the water level, to prevent excessive flooding or draining in the basin) (5)

$|Q| < Q_m$ = constant (a limit imposed by the operation and number of turbines installed) (6)

$Qh > 0$ (i.e. $p > 0$, otherwise pumping against the head may be needed). (7)

Such constraints as these are, however, not to be seen as immutable. If, for instance, one imposes (6) as a constraint, one may then sensibly ask: what would be the effect of increasing Q_m by installing more turbines? Would this be justified by the increase in output? Some knowledge of the costs of turbines and the marginal value of electricity would then be required.

We shall show as an example, how (4) may simply be maximized under constraint (6) only. Constraint (5) may similarly be treated, but (7) is somewhat less tractable.

We have, from (4),

$$\frac{E}{\rho g} = -A\int_0^T (x-x_0)\dot{x}\,\mathrm{d}t$$

$$= -\tfrac{1}{2}A\,[x^2]_0^T + A\int_0^T x_0(t)\dot{x}(t)\,\mathrm{d}t$$

$$= -\int_0^T x_0(t)Q(t)\,\mathrm{d}t \tag{8}$$

assuming that $x(t)$ will be periodic in time. This integral is clearly maximized, subject to $|Q| < Q_m$, by choosing

$$Q(t) = \begin{cases} Q_m & \text{if } x_0 < 0 \\ -Q_m & \text{if } x_0 > 0 \end{cases} \tag{9}$$

so that the turbines always operate at their limit, either forwards or in reverse. By using (9) in (8), we find the maximum available energy over one tidal cycle to be

$$E = 2\rho g\,\frac{Q_m R}{w}$$

Using approximate figures typical of the proposed Severn Barrage scheme:

$$Q_m = 10^5\ \mathrm{m^3/s}$$
$$R = 8\ \mathrm{m},$$

this gives, approximately,

$$E = 30\ \mathrm{GWh}.$$

The corresponding average output found by the Department of Energy (1981) is about 20 GWh.

Fig. 3 shows the corresponding pattern of variation with time, over one cycle of 12.4 hours, of water levels inside and outside the estuary.

4. ALTERNATIVE FORMULATION

The same problem can also be posed in terms of formal optimal control. To maximize

$$\rho g\int_0^T (x-x_0)\,Q(t)\,\mathrm{d}t$$

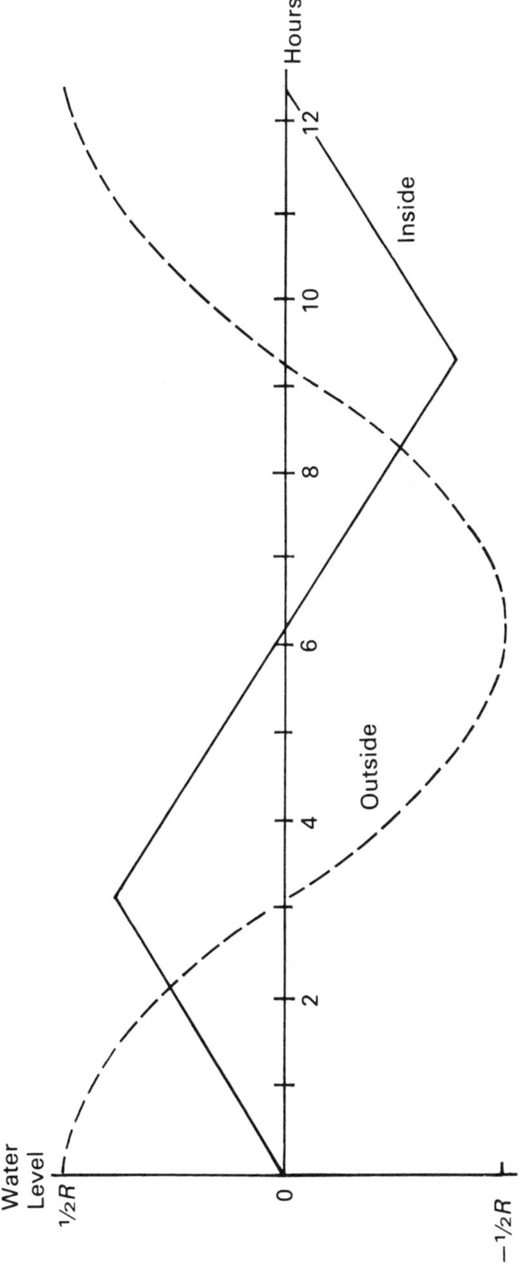

Fig. 3. Water levels inside and outside the barrage during one tidal cycle of 12.4 hours. Limit on discharge is $Q_m = \frac{1}{4}RAw$

where $\dot{x} = -Q/A$ and subject to $|Q| < Q_m$, we use a Hamiltonian:

$$H = \rho g(x - x_0)Q + \lambda(-Q/A)$$

which, using Pontryagin's Maximum Principle, should be maximized with respect to Q.

Clearly, H is maximized by taking

$$Q = \begin{cases} Q_m & \text{if} \quad \rho g(x - x_0) - \lambda/A \geqslant 0 \\ -Q_m & \text{if} \quad \rho g(x - x_0) - \lambda/A < 0 \end{cases} \tag{10}$$

The adjoint equation is

$$\dot{\lambda} = -\frac{\partial H}{\partial X} = -\rho g Q$$

Since also $\dot{x} = -Q/A$, we see that the solution for λ may be written as:

$$\dot{\lambda}(t) = \rho g A x(t) + \alpha \tag{11}$$

for some constant α.

Remembering that $x(t)$ must be periodic, so that

$$\int_0^T Q(t)\,\mathrm{d}t = 0$$

we see, by considering (10), that $\alpha = 0$, so that (11) is just

$$\lambda = \rho g A x \tag{12}$$

(9) may now be retrieved from (10).

In addition to providing a simple yet realistic problem in optimal control, whose solution is also available by an alternative methods, this example shows, from equation (12), the physical interpretation of the adjoint variable $\lambda(t)$.

The usual interpretation of $\lambda(t)$, in the present context, is that it represents the rate of change, with respect to a change in x, of the total available energy which can be extracted in the remaining time, from t to T; (12) indeed shows that λ is the rate of change with x of the total potential energy of the water in the basin, a result which satisfies one's physical common-sense.

5. POSSIBLE EXTENSIONS

There are many further questions which can arise from the simple model used here. One might, for instance, model the fact that the efficiency of real turbines is less than 100 %, and varies with both head and discharge: given that this is so, what is then the optimum design head and design discharge for the turbines? Use of the constraint (7) would allow an analysis of whether or not the cost of installing pumps would be justified. A reading of recent research (for instance,

Proctor (1981)) shows that the assumptions made at the start of section 3 are unlikely in practice to be valid.

I am grateful to Dr. D. T. Bickley of Bristol University, for causing my interest in the modelling of tidal power schemes.

REFERENCES

Department of Energy, 1981, *Report of the Severn Barrage Committee*, Vol. 1. H.M.S.O.

Proctor, R., 1981, Mathematical Modelling of Tidal Power Schemes in the Bristol Channel. Proc. 2nd Internat. Symp. on Wave and Tidal Energy, Cambridge, U.K., Organized by BHRA Fluid Engineering, Cranfield, U.K.

Automata and Switching Theory in Teaching Discrete Mathematics

P. Siwak,
Technical University of Poznan

SUMMARY

There is a very important problem in teaching mathematics: how to expose some general and abstract notions not only clearly and correctly but in a deeply intelligible manner as well. There is an idea leading to this end (see, for example, Cohors-Fresenberg, 1977), that is, the idea to present all mathematical formalisms included in the course for students together with a variety of devices which could help to convince them about the significance and practical usefulness of a theoretical approach. The realization of this idea is especially possible in teaching automata and switching theory and probably in all related branches where discrete modelling is emphasized.

In this chapter a classical interpretation of synchronous realizing net of automata is discussed and some realizing net which has the ability to control the spreading of clock pulses is proposed. The problems concerning a flow of information in such nets are presented and some devices are shown to be useful in stimulating students' interest and in intensifying their education.

1. INTRODUCTION

The model of finite automaton was formalized and developed almost 30 years ago (with one of the first monographs by Hartmanis, Stearns (1966)), when its virtues in both the theory and the applications were realized. Formal simplicity and immediate implications in the engineering methods of synthesis and analysis made this model fundamental in discrete processes and systems for many years. The model has proved to be especially well suited to logical design with small and medium scale integration digital elements.

Because of the rapid progress in VLSI technology practising engineers are rather more interested in other tools to-day. Nevertheless, the automata theory still remains an important component of their education. This discipline, because of its clearly defined notions, well-determined sphere, and connections with many fields of mathematics, is a valuable pedagogical tool in teaching whenever there are the elements from discrete mathematics. Since it is rather easy and not expensive to design and build various types of automata-like devices, they can be employed to help students to understand the formalisms and their practical significance. They undoubtedly permit to win the students for theory, which of course intensifies the endeavour of the teacher during the educational process.

Teaching for many years automata and switching theory in courses given at TU of Poznań for students of electrical engineering, it came home to me that traditional lectures with chalk and blackboard should be supported by simultaneously given elements from the laboratory. Hearing, for example, about the formal definition of an automaton with its input alphabet, states, transitions and outputs, each listener should have a possibility to 'touch' these input letters by individually changing certain physical situations at the input side of the object considered as an automaton, or to observe that output and input letters are not related by a function because of the influence of internal states. Some special devices were designed and constructed for students then, to stimulate their interest and imagination and to support their intuition to understand formal notions.

In this chapter we present the notion of realizing automaton as a fundamental model for automaton-like objects not only from digital techniques but biology, psychology and many other domains, as well. The realizing automaton is understood here as a net of automata A_i which can cooperate either synchronously as in the classical models from Hartmanis, Stearns (1966) or asynchronously as in the proposed model of Siwak (1977). On the basis of the realizing net of component automata we can ask questions which are in their nature from discrete mathematics and which can be interpreted with the help of simple sequential switching structures. We will concentrate on these aspects of modelling which treat information flow problems, e.g. what is the least amount of information about the state of the net which is needed by component automaton A_i if it has to represent the behaviour of a given automaton B in some way (with a given state assignment) in a certain realizing net A. Realizations with the flow of clock pulses gated show the role of partially preserved morphisms. We present their application in the sequences extrapolating machine.

Many problems from theory and practice lead to the analysis of languages. We propose a universal acceptor RLA which can facilitate illustrating some properties of regular languages.

2. INTERPRETATION OF SYNCHRONOUS REALIZING NET

The model of an automaton without outputs will be mainly considered. Let $A = (S, \Sigma, \delta)$ be a finite state automaton, where S is a finite state set, Σ is a nonempty finite input set and $\delta : S \times \Sigma \to S$ is a transition function. Any pair (s, t) of states such that $\delta(s, \sigma) = t$ for some $\sigma \in \Sigma$ is said to be the transition which starts from s.

For a given $A = (S, \Sigma, \delta)$ an automaton $A' = (S', \Sigma, \delta')$, with $S' \subseteq S$ and a restriction δ' of δ to $S' \times \Sigma$ which is into S', is called the subautomaton of A.

An automaton $A = (S, X, \delta)$ is said to be a realization of $B = (S_B, X, \delta_B)$ if there is a subautomaton $A' = (S'\ X,\ \delta')$ of A and a one-to-one assignment $\alpha: S_B \to S'$ such that

$$\delta'(\alpha(s), x) = \alpha(\delta_B(s, x))$$

holds for all $s \in S_B$ and $x \in X$.

The homomorphic realization of automaton B is sometimes considered to be such automaton A that state assignment α is the function onto disjoint subsets of S'. The function $h = \alpha^{-1}$ fulfils in this case

$$h(\delta'(s', x)) = \delta_B(h(s'), x)$$

for all $s' \in S'$ and $x \in X$.

We assume then the input set to be unassigned. Let us take a look how the realizing automaton is interpreted.

Consider a set of automata $A_i = (S_i, \Sigma_i, \delta_i)$; $i \in \{1, 2, \ldots, n\} = N$. Let automaton $A = (S, X, \delta)$ be such that $S = S_1 \times S_2 \times \ldots \times S_n$ and $\delta(s, x) = \delta((s_1, s_2, \ldots, s_n), x)$

$$= (\delta_1(s_1, f_1(s, x)), \ldots, \delta_n(s_n, f_n(s, x))),$$

where $f_i : S \times X \to \Sigma_i$ are the excitation functions of automata A_i. It is seen that such automaton A forms a net of component automata A_i. Such a net precisely is in fact meant each time when the realizing automaton A for a given realized B is considered. The state assignment α: $S_B \to S'$, with $\alpha(s) = (s_1, s_2, \ldots, s_n)$ for $s \in S_B$, can now be expressed by partial assignments $\alpha_i(s) = s_i$ or by partitions $\tau_i = \{b_1^{(i)}, \ldots, b_k^{(i)}\}$ on S_B, which are such that $s \equiv t(\tau_i) \Leftrightarrow s_i = t_i$.

The clock pulses are not considered explicitly in this realizing net, but note that all component automata are assumed to operate in parallel, being able to change their states precisely at the moment when clock pulse CL comes.

In Fig. 1 an ith node of the net with automaton A_i is shown. And here is the characteristic feature of commonly used interpretation of this synchronous realizing net, which we would like to emphasize.

The spreading of inputs x among the component automata of the net depends on its state, while clock pulses on the contrary are conducted immediately to each one A_i.

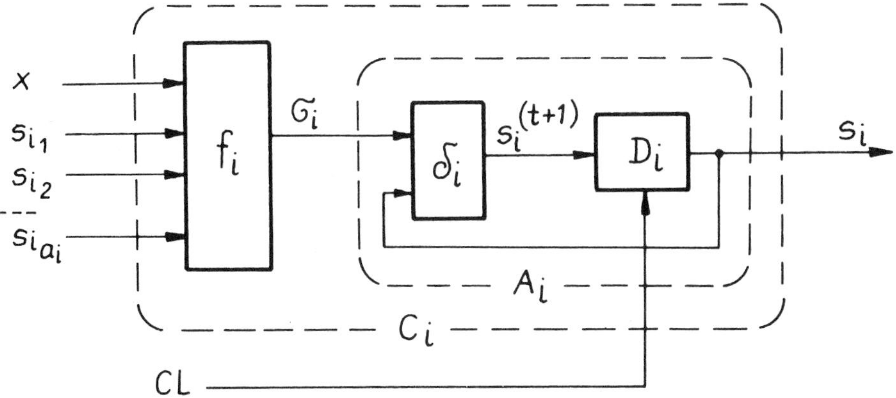

Fig. 1. Component automaton A_i of the synchronous realizing net

2.1 Reduction of dependencies

From Fig. 1 we have:

$$\sigma_i = f_i(y_{i_1}, y_{i_2}, \ldots, y_{i_{a_i}}, x) \quad \text{for all } i \in N,$$

where $y_i \in S_i$. This means that the excitation function of each component automaton A_i determines its set of neighbours, i.e. the set $N_i = \{i_1, i_2, \ldots, i_{a_i}\}$ of all those automata from the net which influence input σ_i. Therefore we can define a directed graph $G = (N, I)$, where N is its set of vertices and the set of edges $I \subseteq N^2$ is given by $(j, i) \in I \Leftrightarrow j \in N_i$. This graph is referred to as the structure of the realizing automaton. When the flow of only internal information is considered then card $(N_i) = a_i$ is just the indegree of a node i. The sequence $(a_1, a_2, \ldots, a_n)$ was introduced by Weiner, Smith, (1967), as the profile of dependencies. The reduction of dependencies (in excitation functions) can be characterized by the number $R(\alpha) = \Sigma_{i=1}^{n} a_i$.

Looking for state assignment α which implies the minimal possible value for $R(\alpha)$ is known as state assignment problem, Hartmanis, Stearns (1966), but for arbitrary component automata it still remains open.

Any net of automata A_i may be physically demonstrated for students when this problem is explained, to show that reduction of dependencies means just the reduction of the amount of information which has to be interchanged between the nodes of this net. The device from Fig. 2 with PLAs (Programmable Logical Arrays) may be used. Clear visualization of the states of all component automata is, however, desired.

Let us illustrate the above-mentioned notions here.

Example 1. Consider automaton $B = (\{1, 2, 3, 4, 5\}, \{00, 01, 10\}, \delta_B)$ with transition function δ_B given by transition graph in Fig. 3. Choose the delay type

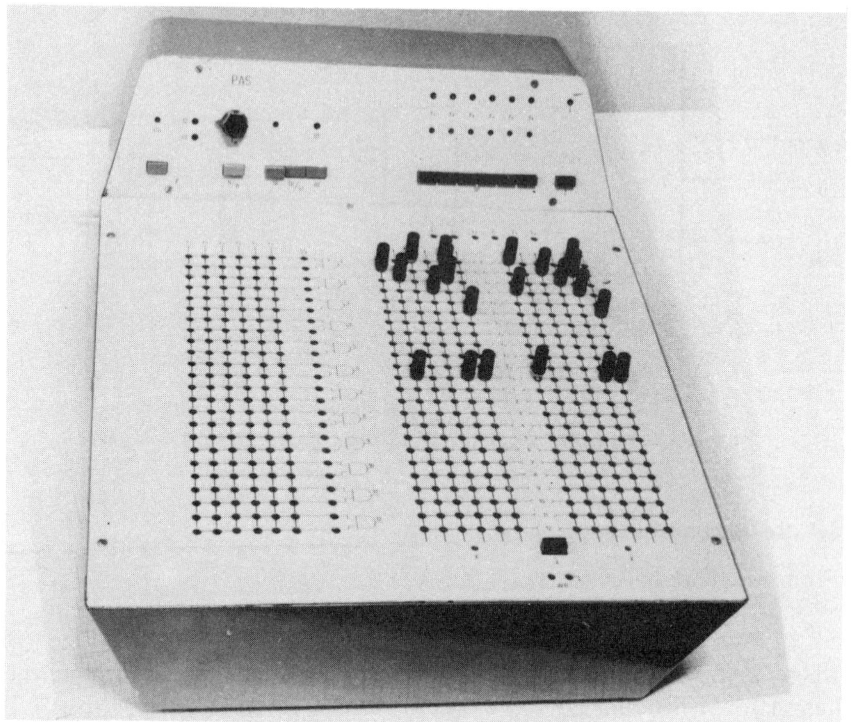

Fig. 2. Realizing device with a possibility of changing the contents of PLA

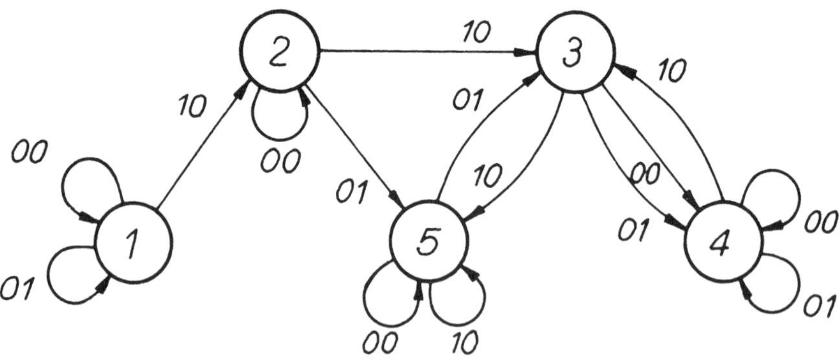

Fig. 3. Transition graph for Example 1

component automata $A_i = (\{0, 1\}, \{0, 1\}, \delta_i)$ with $\delta_i(s, \sigma) = \sigma$ for all $s \in S_i = \{0, 1\}$; $i = 1, 2, 3$. Compare two different state assignments α_{I} and α_{II} given by partitions:

$$\mathrm{I}: y_1 \triangleq \{\overline{1,2,3,4};\ \overline{5}\},\quad y_2 \triangleq \{\overline{1,2,5};\ \overline{3,4}\},\quad y_3 \triangleq \{\overline{1,3,5};\ \overline{2,4}\} \quad \text{or}$$

$$\mathrm{II}: y_1 \triangleq \{\overline{1,4};\ \overline{2,3,5}\},\quad y_2 \triangleq \{\overline{1,3,4};\ \overline{2,5}\},\quad y_3 \triangleq \{\overline{1,3,5};\ \overline{2,4}\}.$$

These assignments imply the following profiles of dependencies:
I: (3, 3, 3) and II: (1, 2, 3) and two structures G_{I} and G_{II} shown in Fig. 4.

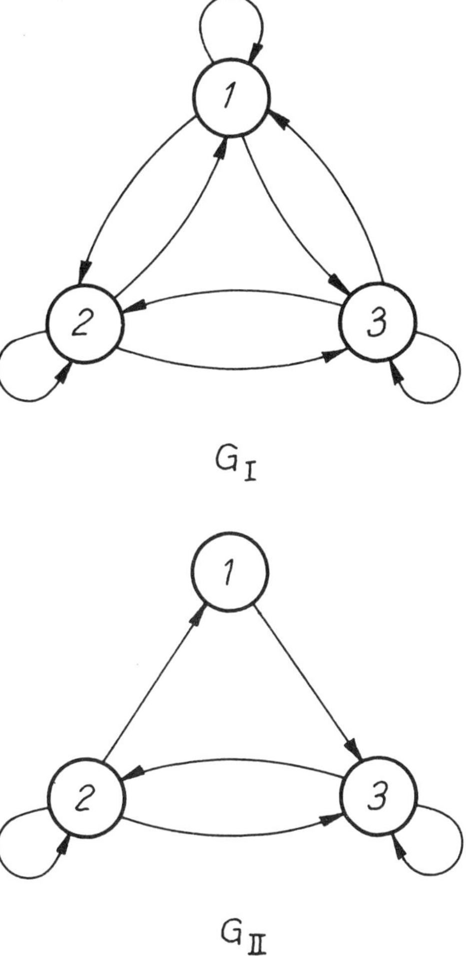

Fig. 4. Two structures of realization of automaton B with state assignment α_{I} and α_{II}

2.2 Input information for component automata

The model of realizing net permits to evaluate any automaton A_i whether or not it can be employed with partial assignment α_i as the component automaton in certain realization of a given automaton B. We can even determine the least amount of information which must be attainable from any net to ensure that

$$\alpha_i(\delta_B(s, x)) = \delta_i(\alpha_i(s), f_i(s, x))$$

will hold for all $s \in S_B$ and $x \in X$.

For delay type component automata (see the automaton D_i in Fig. 1) this amount of information is determined by maximal partition π_i which can be mapped by all functions $f_x : S_B \to S_B$ into τ_i, where $f_x(s) = \delta_B(s, x)$ for all $s \in S_B$; $x \in X$. Such a partition, denoted by $M(\tau_i)$, was introduced by Hartmanis, Stearns (1966).

We propose here an idea how this input information can be defined and determined for arbitrary component automaton, i.e., one being 'between' automata C_i and A_i in Fig. 1.

Let the states of an automaton $B = (S_B, X, \delta_B)$ be assigned to $A_i = (S_i, \Sigma_i, \delta_i)$ by partial assignment α_i. Then any maximal partition π on the set S_B such that

$$s \equiv t(\pi) \Leftrightarrow (\forall x \in X)(\exists \sigma_i \in \Sigma_i)\ [\delta_i(\alpha_i(s), \sigma_i) = \alpha_i(\delta_B(s, x))]^\wedge$$
$$[\delta_i(\alpha_i(t), \sigma_i) = \alpha_i(\delta_B(t, x))],$$

will be called input information for automaton A_i, and denoted by $\pi \in M_{A_i}(\tau_i)$, where τ_i is determined by α_i.

Let $X_{u,v} = \{\sigma \in \Sigma \mid \delta_B(u, \sigma) = v\}$. We have then:

Theorem 1. Let α_i: $S_B \to S_i$(or τ_i) be a partial assignment of S_B into S_i for $A_i = (S_i, \Sigma_i, \delta_i)$ and $B = (S_B, X, \delta_B)$. Then $s \equiv t(\pi)$, where $\pi \in M_{A_i}(\tau_i)$, if and only if

$$X_{\alpha_i(s), \alpha_i(\delta_B(s, x))} \cap X_{\alpha_i(t), \alpha_i(\delta_B(t, x))} = Q_{s,t}(x) \neq \varnothing.$$

Proof. Is omitted.

If $Q_{s,t}(x)$ is nonempty for all $x \in X$ then we say that input control conditions in A_i are not inconsistent for all transitions starting from s or from t. Determination of $M_{A_i}(\tau_i)$ from the graph of inconsistencies leads us to disjoint decomposition of graphs on maximal cliques.

Example 2. Let us determine input information for JK flip-flop in case it were to realize the automaton B given in Example 1, with state assignment $\tau = \{b_0, b_1\} = \{\overline{1, 4, 5}; \overline{2, 3}\}$. The table of input control conditions is:

δ_B	1 31	2 42	3 54	4 24	5 53	inconsistencies
$\tau = \{\overline{1, 4, 5};\ \overline{2, 3}\}$	$J\bar{J}$	$K\bar{K}$	KK	$J\bar{J}$	$\bar{J}J$	$1 \not\equiv 5, 2 \not\equiv 3, 4 \not\equiv 5$

where $J = \{10,\ 11\}$, $\bar{J} = \{00,\ 01\}$, $K = \{01,\ 11\}$ and $\bar{K} = \{00, 10\}$. Decomposing the graph of inconsistencies on disjoint maximal cliques we obtain:

$$M_{JK}(\tau) = \{\{\overline{1,2,4}; \overline{3,5}\}, \{\overline{1,3,4}; \overline{2,5}\}, \{\overline{1,2}; \overline{3,4}; \bar{5}\}, \{\overline{1,3}; \overline{2,4}; \bar{5}\}\}.$$

From cardinality of elements $\pi_{JK} \in M_{JK}(\tau)$ one can deduce that indegree a_{JK} can not be less than 1 in any realizing net containing this JK component.

2.3 Realizability with constraints

The evaluation as above when executed for such automaton A_1 which has to be the only component of a realizing automaton A, permits us to recognize the realizability of a given automaton B with the help of considered A_1. This leads to many subtle relations between digraphs, studied by Geller (1974), and Siwak (1976) for example. We will not discuss the details here, but only illustrate a state splitting in given B. This splitting may be necessary when there are some constraints on possible assignments. In such a case we have the homomorphic realization of B.

Example 3. We will implement the program (automaton B) determined by a flow-chart in Fig. 5 using a microprogrammable automaton A_1 shown in Fig. 6. Note that the sequential structure with a state buffer and ROM guarantees the realizability of any automaton B. But here the next state codes $(Y_1, Y_2\ Y_3, W)$ can be modified only by changing the value of $W \in \{0, 1\}$. To overcome this constraint we have to consider the graph of relation: 'common predecessing state' and decompose it on maximum two state cliques. This is shown in Fig. 7.
Now the contents of ROM can be determined, e.g. as in Table 1.
Note that in pairs (1′, 4′), (1″, 6′), (3, 5) and (4″, 6″) the codes differ only on l.s.b. position.

In presenting the realizability problems to students the typical, widely used digital circuits of MSI, like counters or shift registers, are quite illustrative.

3. REALIZATIONS WHERE GATING OF CLOCK PULSES IS ALLOWED

Let for automata $A_i = (S_i, X, \delta_i)$, $i \in N = \{1, 2, \ldots, n\}$ the function f: $S \to 2^N$ be given, where $S = S_1 \times S_2 \times \ldots \times S_n$. By f-net of automata A_i we mean automaton $A = (S, X, \beta)$ such that for $s = (s_1, s_2, \ldots, s_n)$ and $t = (t_1, t_2, \ldots, t_n)$ from S and $x \in X$, if $\beta(s, x) = t$, then

either $t_i = s_i$ if $i \in f(s)$,
or $t_i = \delta_i(s_i, x)$ if $i \notin f(s)$.

The function f can be also expressed with the help of n-tuples of such functions g_i: $S \times \{CL, \overline{CL}\} \to \{CL_i, \overline{CL_i}\}$ so that $g_i(s, CL) = CL_i \Leftrightarrow i \in f(s)$. These func-

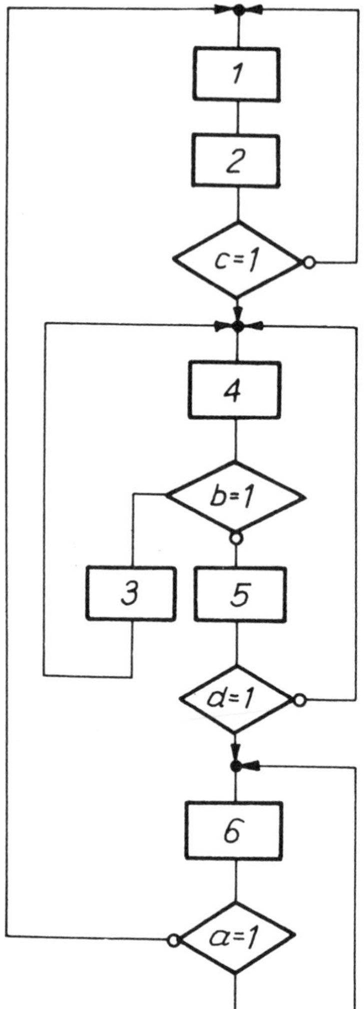

Fig. 5. Flow-chart realized in Example 3.

tions play the same role in automaton $A = (S, X, \beta)$ as functions f_i play in realizing net $A = (S, X, \delta)$. In Fig. 8 we compare the node of the proposed model with the node of the previous model.

In automaton $A = (S, X, \beta)$ the component automata A_i are operating asynchronously in the sense that individual component automata are allowed to change their states at certain states of A and not at others. It can be seen now that in the proposed model the spreading of clock pulses is gated and rests

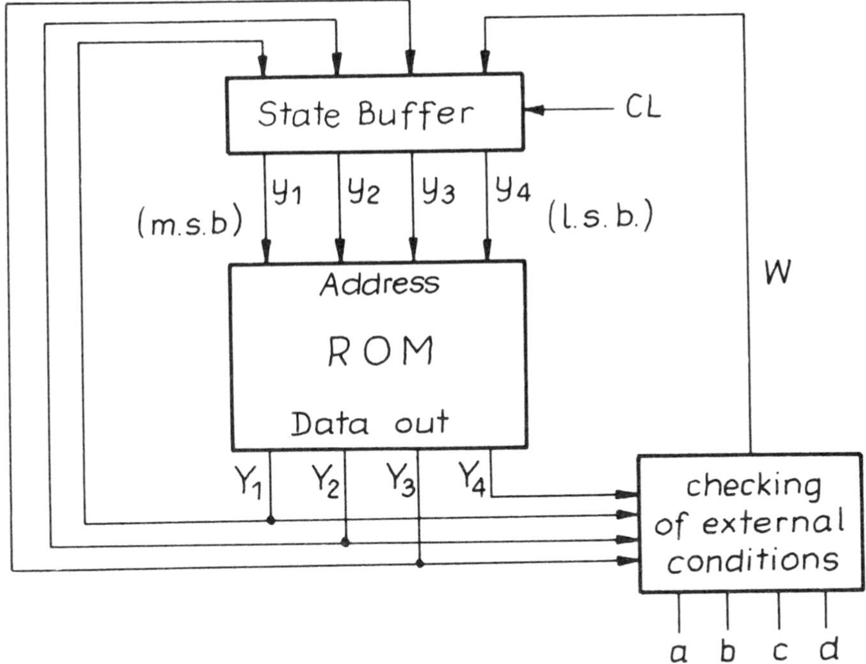

Fig. 6. Microprogrammable automaton A_1

under the control of neighbouring automata from N_i, while the inputs are conducted directly to each one A_i.

The concept of the f-net, Siwak (1977), seems to promise some applications for Cartesian composition of automata introduced and investigated in algebraic automata theory by Dörfler (1978).

Let $A = (S, X, \delta)$ be an arbitrary automaton and let $T \subset S$. A partition π on the set S is said to be T-valid with respect to A if the following conditions hold:

(1) $\delta(s,x) \equiv s(\pi)$ for all $x \in X$ and $s \in T$,

(2) if $s, t \notin T$ then $s \equiv t(\pi)$ implies $\delta(s, x) \equiv \delta(t, x)(\pi)$ for all $x \in X$.

The partition π is preserved by all functions f_x, except transitions which are started from T. Thus partition π has SP (substitution property) for A if it is T-valid with respect to A and T is empty. Nevertheless, we can define some quotient automaton in the general case.

For a partition π on S which is T-valid with respect to $A = (S, X, \delta)$ let the (T, π)-image of A be such an automaton $A/(T, \pi) = (\pi, X, \overline{\delta})$ that $\overline{\delta}\,(b, x) = b'$ if $\delta((b \setminus \mathrm{T}), \mathrm{x}) \subseteq b'$, where $b, b' \in \pi$.

We can now state the following.

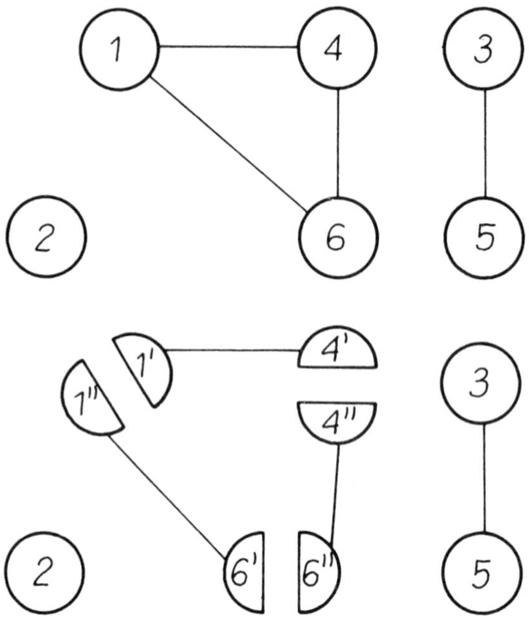

Fig. 7. Graph of the relation 'common predecessing state' and state splitting to obtain maximum two state cliques

Table 1. Content of ROM when flow-chart from Fig. 5 is realized by the automaton A_1 from Fig. 6.

s	Address (in dec.) $\alpha(s)$	Data $Y_1 Y_2 Y_3$	W
1′	0	1 0 0	0
4′	1	0 1 0	0, 1
1″	2	1 0 0	0
6′	3	0 0 1	0, 1
3	4	0 0 0	1
5	5	0 1 1	0, 1
4″	6	0 1 0	0, 1
6″	7	0 0 1	0, 1
2	8	0 0 0	0, 1

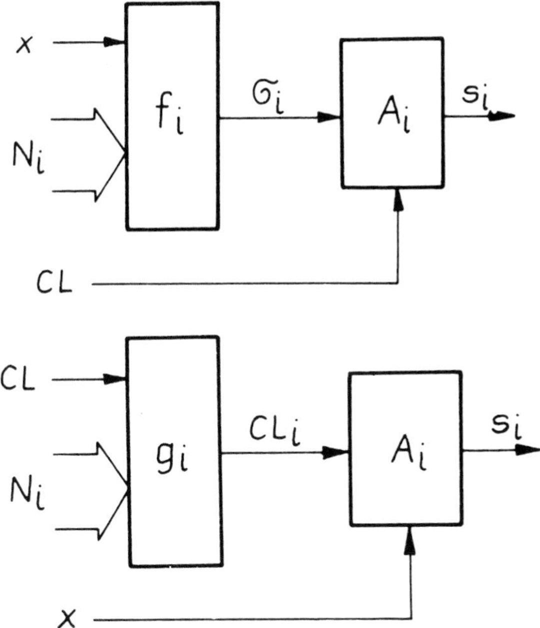

Fig. 8. Nodes containing an automaton A_i; for a synchronous net and for an asynchronous one

Theorem 2. An automaton $B = (S_B, X, \delta_B)$ is realizable by an *f*-net if and only if there exist *n* partitions π_i on the set S_B, which are T_i-valid with respect to *A* for some $T_i \subset S$, and such that $\pi_1 . \pi_2 . \ldots . \pi_n = 0$.

Proof. Is omitted.

3.1 Applications of f-nets in modelling

On the basis of the above realizability theorem we are able to synthesize and to analyse any discrete process *B* with its elements cooperating asynchronously.

Procedure of synthesis

1. Computing all partitions π_i which are T_i-valid with respect to a given *B*.
2. Choosing the components $A_i = B/(T_i, \pi_i)$ for partitions π_i which are T_i-valid with respect to *B*, and such that

$$\pi_1 . \pi_2 . \ldots . \pi_n = 0.$$

3. Assigning the states by α: $S_B \rightarrow S_1 \times S_2 \times \ldots \times S_n$, where

$$\alpha(u) = (b^{(1)}(u), b^{(2)}(u), \ldots, b^{(n)}(u)) \quad \text{for } u \in S_B.$$

4. Determining the function $f: S \rightarrow 2^N$;

$$i \in f(\alpha(u)) \Leftrightarrow u \in T_i.$$

Procedure of analysis

1. Determining the partitions π_i: $u \equiv v(\pi_i) \Leftrightarrow u_i = v_i$, where $\alpha_i(u) = u_i$ and $\alpha_i(v) = v_i$.
2. Defining the sets of validity: $T_i = \{u \in S_B | i \in f(\alpha(u))\}$.

The flow of information can be easily demonstrated in *f*-nets. The appropriate simulating device with universal, programmable component automata and with arbitrarily choosen functions g_i gating the clock pulses, was used in our Institute. It is shown in Fig. 9.

The following example will additionally illustrate the idea.

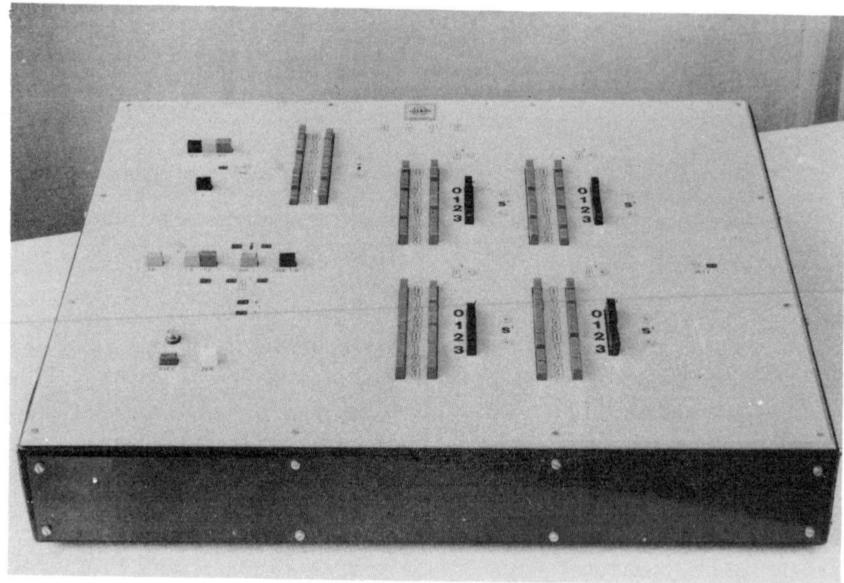

Fig. 9. *f*-nets simulating device

Example 4. Let automaton $B = (S_B, X, \delta_B)$ be given as in Fig. 10. There are shown also some (T, π)-images of B. If the first three partitions are applied (note that $\pi_1 . \pi_2 . \pi_3 = 0$) then the realizing *f*-net can be easily foreseen. The flow of information in the *f*-net, and the *f*-net itself are shown in Fig. 11 and in Fig. 12. The gating functions g_i are such that for

$$i = 1, 2, 3: i \in f((b^{(1)}, b^{(2)}, b^{(3)})) \Leftrightarrow b^{(1)} \cap b^{(2)} \cap b^{(3)} \in T_i.$$

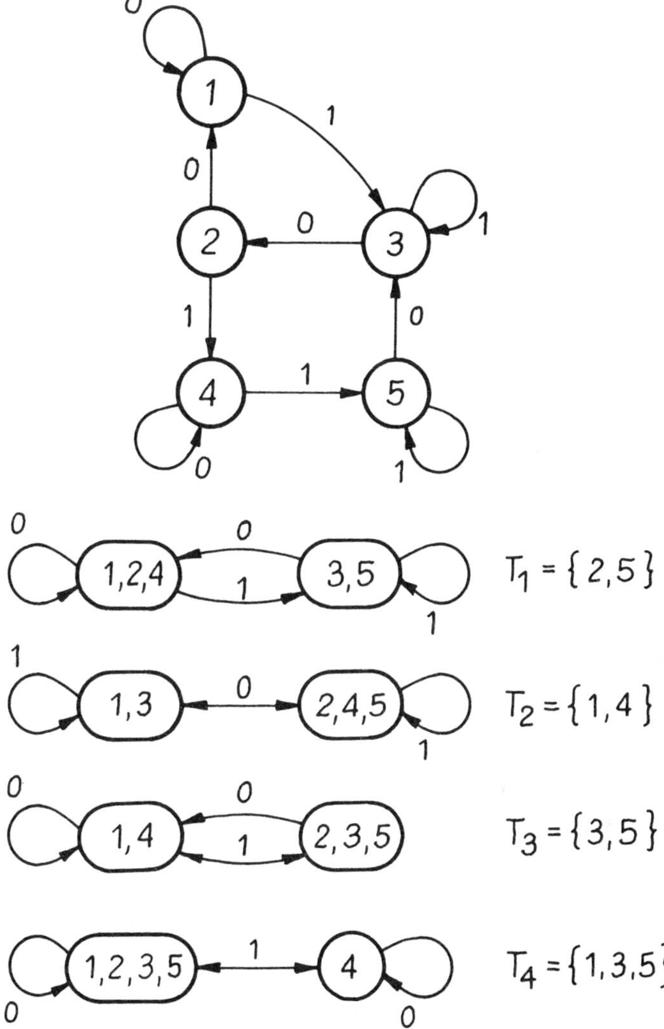

Fig. 10. Automaton B being realized by an f-net and some partitions which are T_i-valid with respect to B

3.2 The sequence extrapolating structures

It was shown (Siwak, 1981) that the f-nets can be effectively employed in machines which extrapolate the input sequences. These machines can be interpreted as penny-matching game-playing machines or learning automata. They are usually looking for correlations between the opponents' choices and the pattern of moves of both players for the previous games.

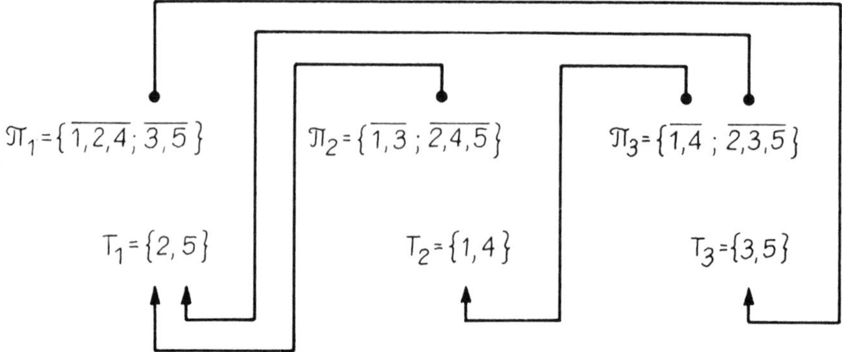

Fig. 11. Schema of dependencies between coding partitions and their sets of validity

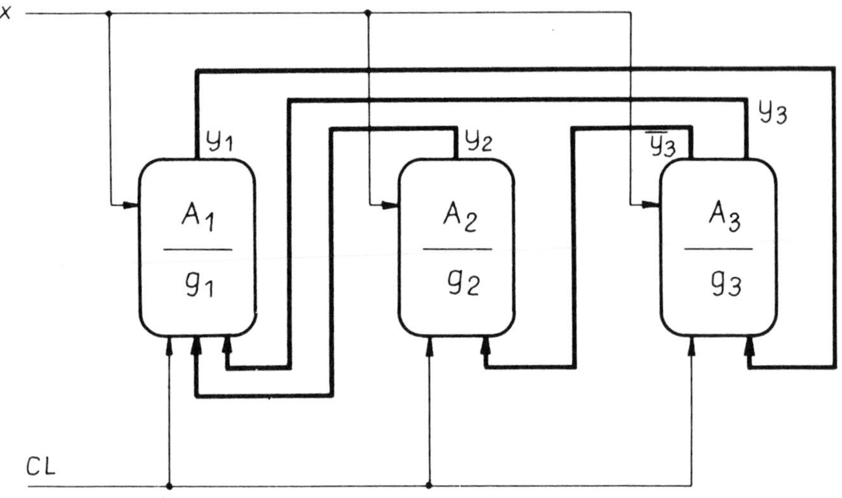

Fig. 12. *f*-net realizing a given automaton *B*

In Fig. 13 a binary sequence extrapolating device (BSE) is shown. BSE is an effective demonstration device for students in that the play is simple (two pushbuttons to choose) and the adaptive behaviour readily perceived. In Fig. 14 the structure of the BSE is shown. Note that at each moment only one component automaton A_i is allowed to change its state. The front machine (shift register has been chosen here) operates all the time and addresses this one allowed. One may find this structure to be a useful tool in studying the influence of reinforcement schema on the rate of learning or some other factors, because

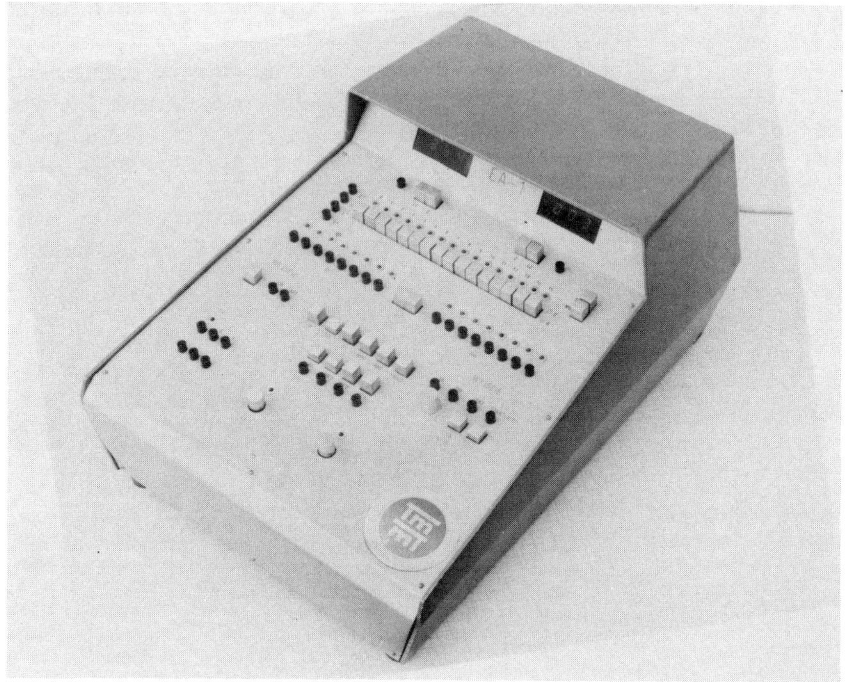

Fig. 13. Binary sequence extrapolating automaton device

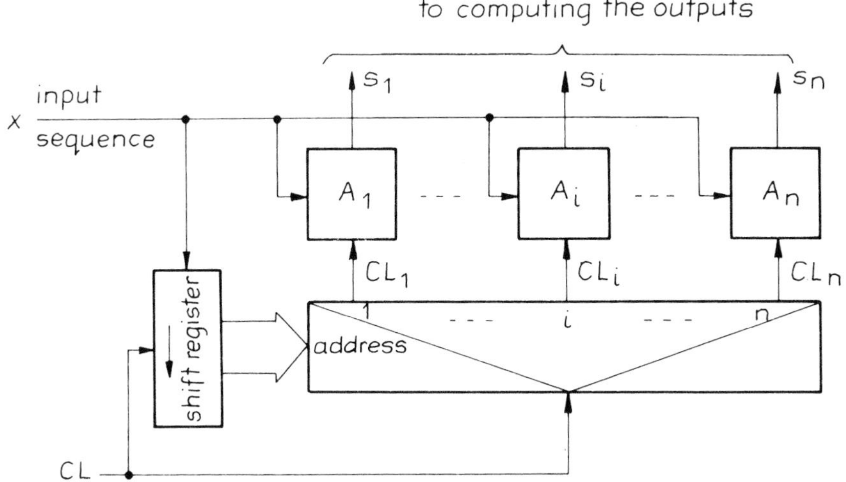

Fig. 14. *f*-net-like structure of the BSE

there is the possibility of choosing individually the component automata A_i from a given set.

In Fig. 15 the transition graph is given for BSE with four component automata A_i of delay type. Comparing outputs from vertices with the labels of ingoing edges one can estimate the efficiency of extrapolation procedure for any input string and any starting state.

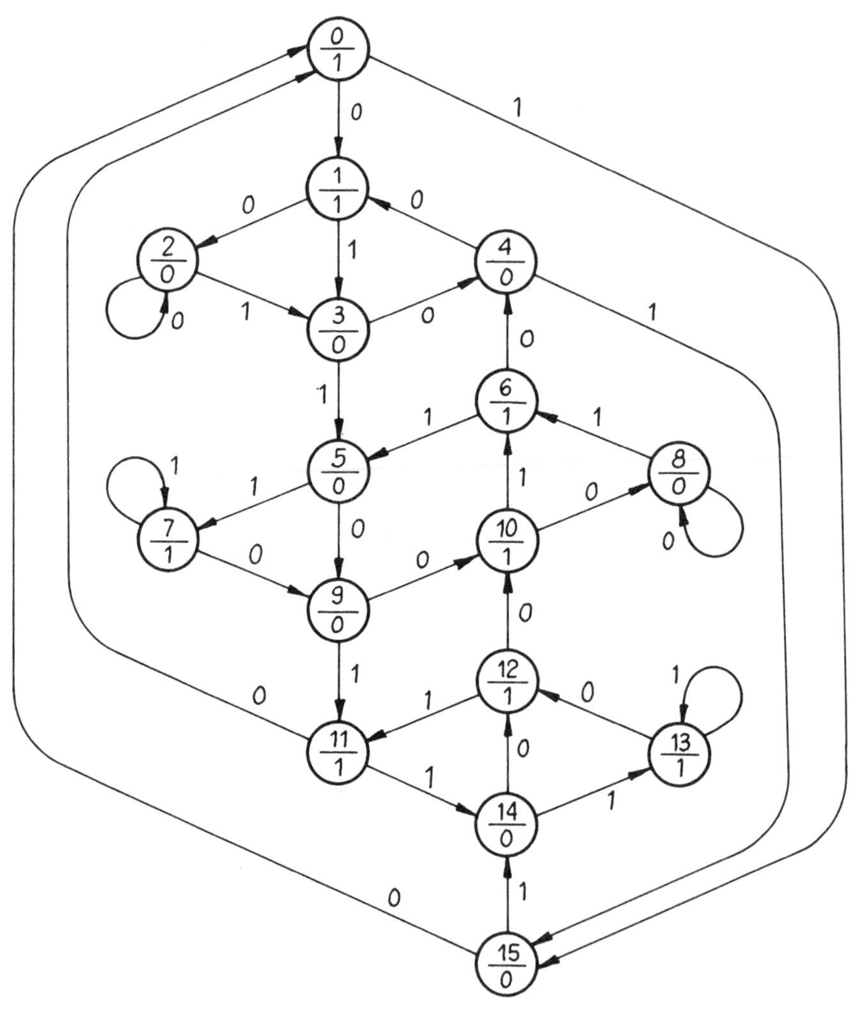

Fig. 15. Some binary sequence extrapolating automaton

Fig. 16. Binary images processing cellular machine

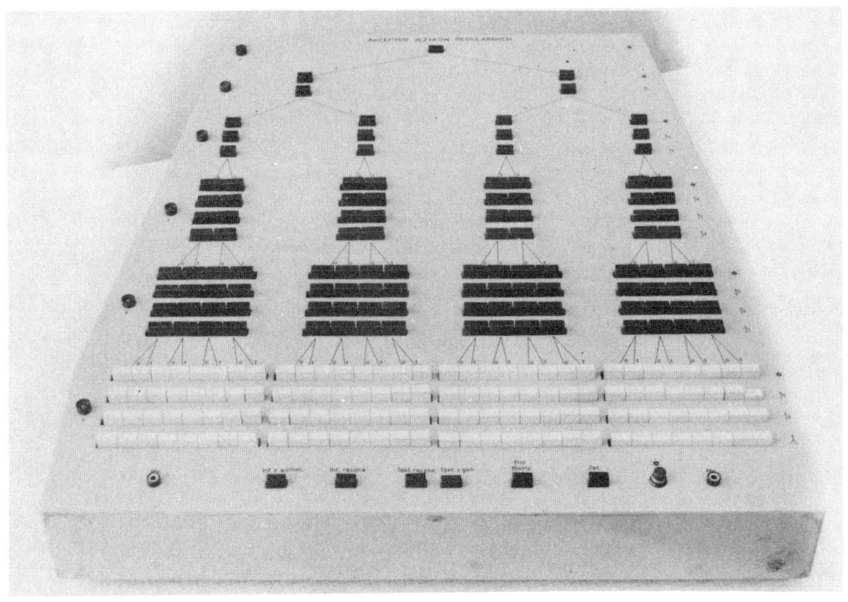

Fig. 17. RLA device

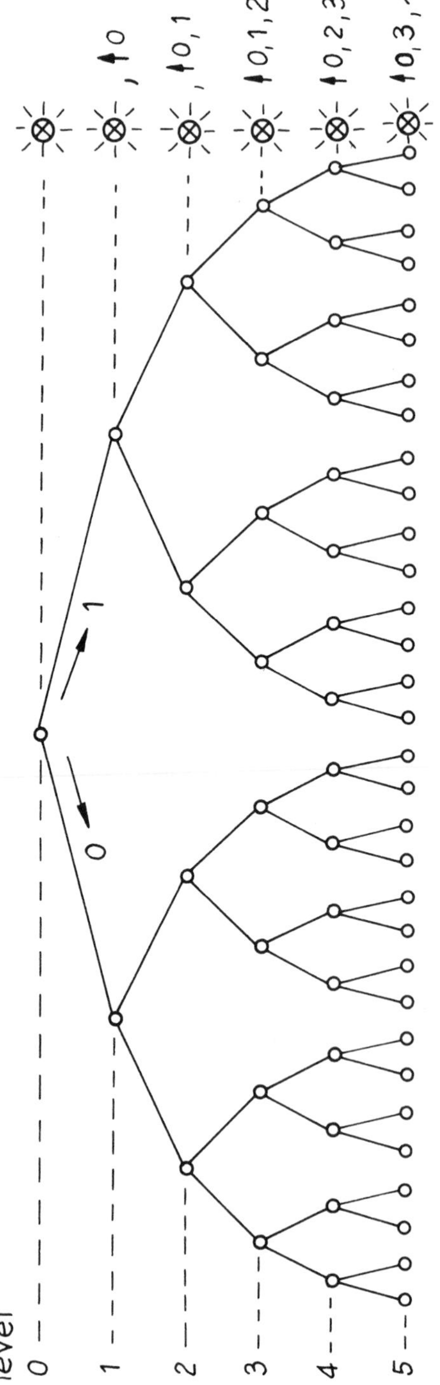

Fig. 18. Universal tree accessible at the front of RLA

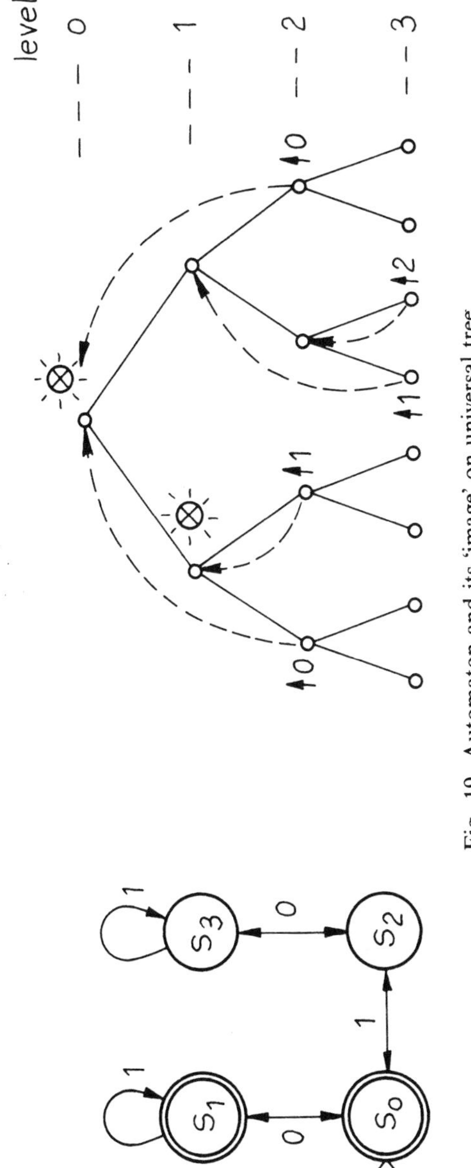

Fig. 19. Automaton and its 'image' on universal tree

4. ACCEPTOR OF REGULAR LANGUAGES

The domain of regular languages seems to have not too much in common with the problems of flow of information in modelling nets of automata and with their structures, at first glance. But in the author's opinion it is not so. There is and probably there will be the real and fundamental need to express the behaviour of great and regularly structured sets of objects which interchange the information, in some precisely formalized and concisely shaped forms. This need comes mainly from progress in VLSI technology. The attention of computer scientists which is focused on languages, proves the same.

In the process of teaching it may be observed that students have some problems with understanding and with the rate of acquisition of the basic properties of regular languages. In an academic course the automaton as a classifier is mainly discussed, with Nerode theorem and with automata relations to some other systems, e.g. rational numbers (Even, 1964).

The problem of looking for the GOE configurations in cellular networks, especially in Conway's cellular space, considered by Siwak (1983) has also brought us to some properties of regular languages. Instead of details we present a BIP machine (binary images processing device) (see Figure 16). Note also an interesting paper by Takahashi (1976) where the connections between cellular nets and languages are studied.

All these facts leads to corollary that certain universal device accepting regular languages would be helpful and is really needed in many fields.

We have built the RLA (regular languages acceptor) shown in Fig. 17. It can accept some regular languages on the basis of some form of its regular expression. The front of the machine is shown in Fig. 18. It has the form of a universal tree over two-element alphabet, with the sets of pushbuttons located in its nodes. Receiving an input the acceptor changes the node of the tree, showing his new 'position'. The level usually increases provided there are some requests at the new position. They can force jumps to previous levels or can stimulate the state of acceptance at the output of RLA.

Example 5. To illustrate the RLA we have chosen an automaton from McNaughton, Papert (1971), defined by

$$L = (ala)^{*}\mathrm{a}(e+01^{*}) \quad \text{and} \quad a = e+0(00+1)^{*}0.$$

The requests which have to be determined to force RLA to accept this L are shown in Fig. 19 together with a transition graph of appropriate automaton.

REFERENCES

Cohors-Fresenberg E., 1977, *Mathematik mit Kalkülen und machinen*, Vieweg and Sohn GmbH.

Dörfler W., 1978, The Cartesian Composition of Automata, *MST*, **11**, 239–257.

Even S., 1964, Rational Numbers and Regular Events, *IEEE Tr. on El. Comp.*, **13**, 740–741.

Geller D. P., 1972, Realization with Feedback Encoding, TR, The University of Michigan.

Geller D. P., 1974, Walkwise and Admissible Mappings Between Digraphs, *Disc. Math.*, **9**, 375–390.

Hartmanis J., Stearns R. E., 1966, *Algebraic Structure Theory of Sequential Machines*, Prentice-Hall Inc.

McNaughton R., Papert S., 1971, *Counter-free Automata*, The MIT Press.

Siwak P., 1976, Some Extensions and Memory of Automata, *Found. of Control Eng.*, **1**, 235–242.

Siwak P., 1977, Decomposition of Automata onto Components with Gated Clock Signals, *Found. of Control Eng.*, **2**, 205–210.

Siwak P., 1981, Realization of Automata with Controlled Time-Sharing of Components, *Elektryka ZNPP*, **23**, 35–47 (in Polish).

Siwak P., 1983, The Problem of Looking for the GOE Configurations in Cellular Network (submitted for publication).

Takahashi H., 1976, The Maximum Invariant Set of an Automaton System, *Inf. and Control*, **32**, 307–354.

Weiner P., Smith E. J., 1967, Optimization of Reduced Dependencies for Synchronous Sequential Machines, *IEEE Tr. on El. Comp.*, **16**, 835–847.

Can Mathematical Modelling be Taught?

R. R. McLone
Southampton University

When I first heard from the organizers of the prospect of a conference on the teaching of mathematical modelling, my immediate reaction was 'what a good idea', that this was a most opportune time for such a conference coming as it does after a period when there has been an evident growing interest in putting on modelling courses and at the same time a considerable increase in the literature available for the organizers of such courses. Yet the very development of these courses and the rapidity with which many institutions (especially those involved with the training of mathematics teachers) have included them in their study programme made me wonder—just ever so slightly!—whether there were perhaps some fundamental questions being missed—in particular what is it that these courses are setting out to achieve and, indeed, what evidence have we that mathematical modelling *per se* can be 'taught' at all?

The *activity* of applying mathematics is almost as old as mathematics itself. There is a good deal of evidence throughout the ages that many new mathematical ideas have arisen through the attempts of workers to grasp and develop a mathematization of practical problems. One does not need to catalogue the historical perspective of such events to see that this activity, stemming in some instances from taking known mathematics and finding that it can be used to describe a practical situation with resultant insight and understanding, and in other cases from tackling new situations where current mathematical knowledge proved insufficient or inadequate, has been a central feature of the work of those trained in mathematics and whose careers subsequently demand the application of this training. It is also true that the range of application of mathematical techniques has broadened considerably in recent decades and mathematics in some form (by which we include statistics, operational research, numerical analysis and related topics as well as more traditional classical mathematics) is now applied in areas where it would probably have been unheard of thirty years ago if not more recently (for example to biology, social studies, management and business studies, and so on). These two aspects brought into conjunction have increased a general greater awareness of the applicability of mathematics amongst a larger number of potential users, and has increased the expectation that mathematicians—or, more precisely and especially, mathematics graduates—are willing, and able, to use their mathematical expertise in tackling 'real' problems as they arise in their chosen field of employment.

On the other hand there is some established evidence that the very same graduates did not show such ability as might have been expected of them. Traditionally the activity of applying mathematics has been expressed in curriculum terms (in the UK especially) in 'Applied Mathematics'. Yet the courses offered and followed under such a designation generally appear not to develop the practical abilities expected of a graduate. Perhaps we should look more closely at what these abilities are. Many have attempted to describe what a professional applied mathematician does. A fairly straightforward description might be on the lines of 'taking an often apparently ill-defined, complex practical situation and constructing from it a more precisely defined problem to which mathematical expertise can be brought. The defining of that problem is a skill of great importance, but evidently very hard to acquire. The result of such mathematization is a 'solution' which is reinterpreted in terms of the original situation. This description appears, of course, in many places and has been repeated, not always precisely in this form, many times; but I restate it here because I find it helps both to constantly remind myself what it is we are trying to achieve in the courses we devise and also to find some framework by which I can begin to measure the effectiveness, or otherwise, of the curriculum developments being introduced. One particular summary of the above, perhaps illustrated in its simplest form, appears in *Case Studies in Mathematical Modelling* (James and McDonald, 1981). It is set out as shown in Fig. 1. This does not yet suggest what abilities we are seeking however, even less

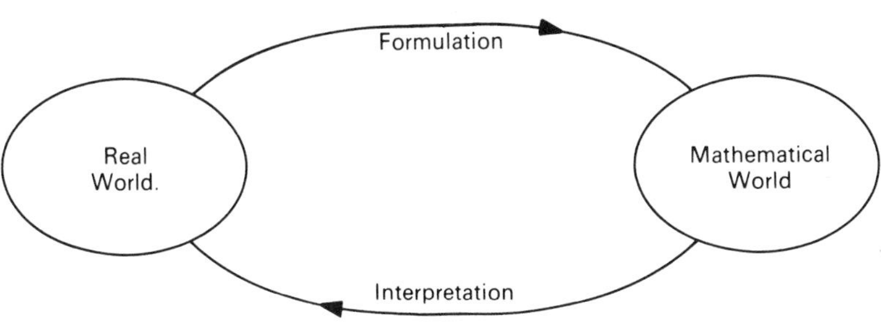

Fig. 1.

how to achieve them. Perhaps we should turn to the traditional teaching of 'applied mathematics', which as part of the mathematics curriculum has a long history. One problem lies in defining just what we mean by 'applied mathematics'. Here I turn for help to Henry Pollak's contribution to *New Trends in Mathematics Teaching*, Volume IV (Chapter XII—The interaction between mathematics and other school subjects (Pollak, 1976)). Here he makes four different definitions, which I shall attempt to summarize:

(1) *Applied Mathematics means classical applied mathematics*, by which he means those classical branches of mathematics such as analysis (including calculus), ordinary and partial differential equations, integral equations, theory of functions, in fact those branches of mathematics most applicable to classical physics although he stresses that no actual connection with physical problems is implied.
(2) *Applied Mathematics means all mathematics that has significant application*, which enlarges (1) more especially by including such topics as set theory and logic, linear algebra, probability, statistics and computing.
(3) *Applied Mathematics means beginning with a situation in some other field or in real life*, making a mathematical interpretation, or model, doing some mathematics on the model and applying the result to the original situation. The 'other field' is by no means restricted to physical science and includes the areas of application mentioned earlier in this talk.
(4) *Applied Mathematics means what people who apply mathematics in their livelihood actually do*. (Like (3), but more so!)

These four definitions may be represented as shown in Fig. 2. Here (2) is seen to overlap (1) rather than include it, as there are many branches of (1) not currently 'applicable' in practical terms. Both (1) and (2), however, are defined

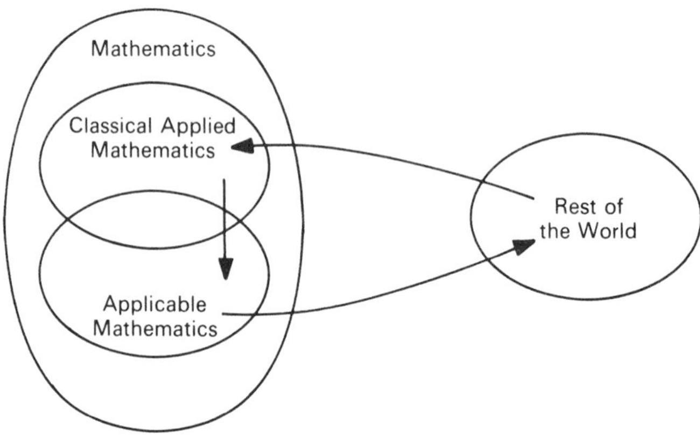

Fig. 2.

within mathematics; it is only in definitions (3) and (4) that we see the interaction between mathematics and the rest of the world. In curriculum terms much of Applied Mathematics is contained in definitions (1) and (2); but one is struck by the similarity in concept between Fig. 1 and Fig. 2, when one extends the definition to (3) or (4). What constitutes the essential change made here? It appears that it is not the particular content of what we call 'Applied Mathematics' which lies at the root of its inability to convey the qualities needed by a practising mathematician, but something more to do with the mode of presentation. 'Applied Mathematics' as we have come to know it is taught as a series of topics with the emphasis on confronting the student with more topic material, just as might be expected in a course on some branch of Pure Mathematics (this was certainly true in 1973 and the intervening ten years have seen little change (McLone, 1973)), to be absorbed, reflected, deflected, ducked—or whatever—by the students as they do in any other traditionally taught unit.

If we accept the premise that it is not necessarily the content but the style of courses in Applied Mathematics which belies the nature of the activity, then it must surely follow that a change to other—perhaps more 'acceptable', 'attractive' in terms of student interest—areas of application, whilst maybe improving somewhat student motivation, nevertheless will not of itself provide a means of extending the range of abilities the student is expected to develop. The activity of mathematizing in, say, geography or biology is not fundamentally different to the activity carried out in more traditional applications; changing the area of application may produce new models, but does not of itself further any possible *teaching* of modelling. A little too often a change in the area to model has been confused with a development of the activity of modelling.

To return then to my original question 'Can mathematical modelling be *taught*?'; or is it perhaps merely *acquired* by the student (through experience, maturity, what you will). As I said at the beginning there has been in this country at least a great interest in 'Mathematical Modelling' as an explicit part of the curriculum in recent years. Many colleges and polytechnics—and a few universities—have introduced units with this specific title, which indicates that there is a belief that there is something which can be identified and conveyed as a specific unit of the course. Here I make a personal confession. I have been running a unit with such a title at Southampton for over ten years. But what does it achieve? More important, how can I make any appraisal of what it achieves? By what criteria does one assess the value and attainment of such units? Certainly *not* by setting standard examinations and judging the students' overall performance on them—a measure generally adopted in traditional courses.

I am conscious that I still have not progressed to consider what are the 'abilities' expected of the graduate who proceeds into employment on the basis of a mathematics degree. There has been some research in this area which points to four central abilities which may be labelled Technical Knowledge, Discovery, Criticism and Communication. These qualities are not the sole province of the mathematician. It is how they are expressed in terms of the needs of the mathematics student which concerns us. Some while ago I offered the following (McLone, 1979) as an attempt at interpreting these central abilities for the mathematics student. I reproduce them again (as Table 1) simply as an indication of the possibilities and not as any final definition—indeed I find myself continually questioning and revising them. But I still find the *process* of great value and I believe it is important for anyone who is desiring to convey in some way the activity of mathematical modelling to establish for themselves an analysis of the qualities they are seeking to foster.

Table 1. Abilities for the mathematics student

Technical knowledge
A. The acquisition of basic techniques and theory.
B. Use of these techniques to solve standard problems.
C. Use of existing techniques in unfamiliar situations.
Discovery
D. Improvisation of new techniques when existing ones are inadequate.
E. Abstraction of a unifying principle from a class of situations.
F. Formulation of a problem in mathematical terms.
Criticism
G. Organization of material (from books, papers, etc.)
H. Assessment of suitability of different mathematical models.
I. Asking of pertinent questions in a mathematical situation leading to criticism of (a) own, (b) others' work.
Communication
J. Communication of ideas, results, in mathematical form.
K. Communication of mathematical results to non-mathematicians.
L. Work effectively in a group.

What many modellers have *said* about the way of conveying the activity of modelling has been analysed by Dick Clements in his article on 'The Development of Methodologies of Mathematical Modelling' (Clements, 1982). He says, '. . . one of the purposes of having a methodology of modelling is to give, for the teacher, a structure to the knowledge and skills he is trying to inculcate into the students'. Such is true for almost any teacher. However, as Dick Clements acknowledges, mathematical modelling is not taught by lecturing on a methodology and setting questions on a tutorial sheet! He goes on, '. . . Rather, the effective teacher will give the students a few modelling problems to tackle in an unstructured way. This will allow the students to arrive at the recognition that there are common features in the way in which they tackle different problem situations and a knowledge of their need for an organised way of tackling unfamiliar problems, i.e. a methodology'.

Well will it? This brings me back to the confessional. Am I 'effective' in the course I run? Certainly my students are placed in unfamiliar situations, and tackle these situations with an apparent lack of structure to the course, in a way quite different from almost all other courses that these students take. On the other hand, as my colleague Keith Hirst has observed in his discussion (Hirst, 1981), of a course run by him called 'Mathematical Workshop', this approach is not unique to modelling as such, but is part of investigational learning wherever it occurs. The aspect which differentiates mathematical modelling is that the problems are based in some way in *the real world* (cf. James, or definitions (3) and (4) of Pollak). Moreover, the students must *recognize them as real*, in their own terms. My own experience suggests that students soon recognize 'a phoney'. Contrived situations, however interesting or unfamiliar, can very quickly get the brush-off when once observed to be such, and since motivation is vital to such a course anything counterproductive in this way must be avoided. To me one of the most important aspects of 'setting the scene', and also most difficult, is to describe the situation impartially and yet (if this is not a contradiction) from the point of view of someone who is involved. In other words the situation must *matter* to someone. In our rush to find 'modelling problems' this is often overlooked. Unimportant or irrelevant problems are just as unmotivating as unreal problems.

Another important feature to me in judging effectiveness is the student's confidence. Already put at risk by the style of the course, it has to be carefully nurtured. In particular students who survive well on the traditional diet of lecture notes and related exercises can find the apparently unprepared (uncooked?) dish placed before them in the form of a modelling situation quite indigestible. It is also often said that it is important to take care over the level of mathematical expertise that might be expected. From my own experience, students for whom the development of many of the skills mentioned in Table 1 is a new experience cannot develop them effectively without a high degree of confidence that the mathematics required will be within their competence. Confidence is a fragile plant and easily withered. One has to walk a tightrope between the apparent void of unstructured problems on the one hand and the hard rocks of difficult mathematics on the other.

However well prepared the choice of problems, if students only experience the activity of mathematical modelling from their own interaction with these problems there is the possibility, and in some cases a high degree of probability, of some misconceived understanding of what the activity means. In this respect I find Clements's assertion not entirely convincing. I find it helpful to introduce into the course a series of talks by 'outside' speakers who describe situations that they have recently grappled or are currently grappling with and the way in which they are mathematizing them. The speaker is encouraged *not* to produce 'finished' work and the students are offered the opportunity to develop an aspect of the problem in an open-ended way. Another feature I would like to mention is the group interaction aspect of the work of most practising mathematicians in the 'real world'. The importance of the communication skills is stressed by many who employ our finished product and this can only be practically experienced within the classroom by working in small groups, to encourage cooperation in a positive sense. There will be problems—even when students are allowed to form their own groups there will still be those who do not interact well, if at all, with the others. Yet this is also part of the preparation for the *reality* of a working experience.

What of the question in the title of my talk? Have I a measure to enable a response? In one respect—no. I have already indicated that an assessment system such as that within which I have to work, designed largely for the development of very different skills and leading to a single representation on a scale of 0–100, can in no sense be regarded as such a measure. So what can I use? Do the students acquire some of the abilities such as are quoted in the list in Table 1? Sometimes I have reason to believe that students *do* exhibit Discovery and Communication skills when they were thought elsewhere not to possess them—but did the course generate them, or did I simply give the students the opportunity to express skills already there? I sometimes suspect the latter, although that also would seem to be a worthwhile experience. On the other hand Criticism is often superficial and one on which I believe we spend too little time.

If effectiveness means that for some students there has been a revival of motivation to use their mathematical expertise and renewed confidence in their own abilities, then with some temerity I would claim a measure of effectiveness. But I would wish to conclude this talk by suggesting some questions which individual teachers of mathematical modelling might consider—and then ask of themselves whether mathematical modelling can be taught.

- —What are the abilities which the course in Mathematical Modelling is intended to develop?
- —How does the course set out to achieve these?
- —How do the students measure up to these abilities—before the course, and afterwards?
- —Are the problems *real*—do they *matter*?
- —Do the students display *confidence* in their modelling ability?

Maybe because I am still working out what it means to develop the activity of applying mathematics in curriculum terms that I am not sure of the answer, but if I have given my students an opportunity to develop for themselves fresh skills and to display them, then may be I have taught mathematical modelling.

REFERENCES

Clements R. R., (1982), The development of methodologies of mathematical modelling, *Teaching Math. and its Applic.*, **1**, 125–131.

Hirst K, E., 1981, Undergraduate investigations in mathematics, *Educ. Stud. Math.*, **12**, 373–387.

James D. J. C. and McDonald J. J., 1981, *Case Studies in Mathematical Modelling*, Stanley Thornes.

McLone R. R., 1973, *The Training of Mathematicians*, Social Science Research Council.

McLone, R. R., 1979, Teaching mathematical modelling, *Bull. I.M.A.*, **15**, 244–246.

Pollak H. O., 1976, The interaction between mathematics and other school subjects, *New Trends in Mathematics Teaching*, Vol. IV, pp. 232–248.

International Conference on The Teaching of Mathematical Modelling

*University of Exeter, July 12–***15, 1983**

ORGANIZING COMMITTEE AND THE PROCEEDINGS EDITED BY

J. S. Berry, *The Open University*
D. N. Burghes, *Exeter University*
I. D. Huntley, *Sheffield City Polytechnic*
D. J. G. James, *Coventry (Lanchester) Polytechnic*
A. O. Moscardini, *Sunderland Polytechnic*

ADVISORY PANEL

LIST OF DELEGATES

An asterisk indicates a member of the organizing committee.

Mr. R. L. Abrines
School of Mathematics
Kingston Polytechnic
Penrhyn Road
Kingston upon Thames KT1 2EF

Professor U. D. Ambrosio
Universidade Estadual de Campinas
IMECC-UNICAMP
Caixa Postall 1170
13.100 Campinas
S.P. Brasil

Professor M. Singh Arora
Department of Mathematics
University of Bahrain
Bahrain

Mr. Ib Axelsen
Sarabjerg 14
8660 Skanderborg
Denmark

Mr. J. E. Baker
Faculty of Mathematics
The Open University
Milton Keynes
MK7 6AA

Mr. J. M. Barclay
21 Strachan Road
Edinburgh EH4 3RH

Professor M. S. Bell
The University of Chicago
The Department of Education
5825 Kimbark Avenue
Chicago Illinois 60637

Dr. J. S. Berry*
Department of Mathematics
The Open University
Walton Hall
Milton Keynes
MK7 6AA

Rolf Biehler
IDM
Universität Bielefeld
Postfach 8640
D-4800 Bielefeld 1 FRG

Professor Martin Billik
Professor of Mathematics
San Jose State University
San Jose
California
U.S.A.

Dr. W. T. F. Blakeley
20 Saxoncourt
Tettenhall
Wolverhampton WV6 8SA

Professor Dr. Werner Blum
Wegmannstrasse 1
D-3500 kassel FRG

Dr. A. Bossavit
1 Av. du Gal de Gaulle
92141 Clamart France

Dr. G. Brandell
Department of Mathematics
University of Lulea
S-95187 Lulea Sweden

Professor D. N. Burghes
School of Education
University of Exeter
St. Luke's
Exeter EX1 2LU

Dr. H. Burkhardt
Shell Centre for Mathematical Education
University of Nottingham
Nottingham NG7 2RD

Dr. D. M. Burley
Dept. of Applied & Computational Mathematics
The University Sheffield
Sheffield S10 2TN

Dr. P. Bungartz
535 Euskitchen

E-Brandstrom Strasse 6
Bonn

Dr. Hans Joachim Burscheid
Universität zu Köln
Seminar für Mathematik und ihre Didaktik
Gronewaldstrasse 2
5000 Kőln 41 FRG

Dr. D. I. Calvert
Mathematics Department
Plymouth Polytechnic
Drake Circus
Plymouth

Mr. Paul Carragher
Department of Maths & Stats.
Post Office Box 4400
Fredericton
N.B.
Canada E3B 5A3

Marjorie C. Carss
Dept. of Education
University of Queensland
St. Lucia Q4067
Australia

Dr. J. A. Charlesworth
School of Mathematics
Robert Gordon Institute of Technology
St. Andrew's Street
Aberdeen AB9 1FR

Dr. Bernard K. S. Cheung
Dept. of Mathematical Studies
Hong Kong Polytechnic
Hung Hom
Kowloon
Hong Kong

Mr. A. W. Clayton
120 Park Drive
Sittingbourne Kent

Dr. R. Clements
Dept. of Engineering Maths
University of Bristol
Queens Building
Bristol BS8 1TR

Mr. G. N. Copley
4 Manor Close
Templecombe
Somerset BA8 OLA

Dr. Mark Cross
Thames Polytechnic
Wellington Street
Wpplwich
London SE18 6PF

Mr. E. Cumberbatch
Chairman, Dept. of Mathematics
Claremont Graduate School
Claremont, California 91711

Mr. A. J. Davies
Hatfield Polytechnic
P.O. Box 109
College Lane
Hatfield Herts AL10 9AB

Mr. H. B. Davies
Department of Maths & Stats.
Portsmouth Polytechnic
Mercantile House
Hampshire Terrace
Portsmouth PO1 2EG Hants.

Mrs. Joy Davis
10 Stone Road
Broadstairs Kent

J. de Lange
Tiberdreef 4
3561 66 Utrecht Holland

Dr. D. Edwards
Thames Polytechnic
Wellington Street
Woolwich
London SE18 6PF

Dr. Judith Ekins
19 Bingham Road
Sherwood
Nottingham

Dr. D. J. Ellis
Department of Eng. & Computing

Salisbury College of Technology
Salisbury Wilts SP1 2LW

Mr. R. J. Eyre
High Beech
Elkstone
Cheltenham Glos. GL53 9PA

Mr. B. Faithfull
Mathematics Department
Brighton Polytechnic
Moulsecoomb
Brighton BN2 4GJ

Mr. S. L. Saveri Fontana
Istituto Matematica Dini
Viale Morgag 67A
50134 Firenze Italy

Fulvia M. Furinghetti
University of Genova
Via L.B. Alberti 4
16132 Genova
Italy

Dr. A. M. Gadian
Maths. Department
Teesside Polytechnic
Borough Road
Middlesborough TS1 3BA

Girish Gupta
B-1 Inderpuri
New Delhi 110012

Mr. C. R. Haines
Dept. of Mathematics
The City University
Northampton Square
London EC1V 0HB

Dr. W. D. Halford
Department of Maths and Stats.
Massey University
Palmerston North
New Zealand

Dr. K. Haliste
Umea University
Department of Mathematics
S-90187 Umea Sweden

Professor G. G. Hall
University of Nottingham
Department of Mathematics
Nottingham

Mr. M. J. Hamson
School of Mathematics
Thames Polytechnic
Woolwich
London SE18

Mr. D. L. Haynes
Maths Department
Liverpool Polytechnic
Byrom Street Liverpool L3 3AF

Mr. P. J. Haysman
Royal Military College of Science
Shrivenham

Mr. J. W. Hearne
University of Natal
P.O. Box 375
Pietermaritzburg 3200

Mr. A. Henderson
12 Hermitage Park
Edinburgh EH6 8HB

Dr. Brian Henderson-Sellers
Dept. of Civil Engineering
University of Salford
Salford M5 4WT

Dr. M. H. Hendriks
De Dreijen 8
6703 BC Wageningen
The Netherlands

Mr. Innes C. Hendry
Robert Gordon Institute of Technology
School of Mathematics
St. Andrew Street
Aberdeen

B. Henkelon
P.O. Box 58
Gronesigen Netherlands

Dr. Wilfried Herget
Technische Universität Clausthal

Institut für Mathematik
D-3392 Clausthal-Zellerfeld FRG

Mr. M. J. Herring
22 St. Albans Close
Warden Hill
Cheltenham GL51 5DP

Dr. F. Hickman
Polytechnic of the South Bank
54 Inchmery Road
London SE6 2NE

Professor A. F. Horadan
The New South Wales Inst. of
Technology
School of Mathematical Sciences
P.O. Box 123
Broadway
NSW 2007 Australia

Dr. S. K. Houston
School of Mathematics
Ulster Polytechnic
Newtown Abbey BT37 0QB

Dr. P. C. Hudson
Dept. Maths & Stats.
Teesside Polytechnic
Borough Road
Middlesbrough
Cleveland TS1 3BA

Dr. I. Huntley
Department of Math. Statistics and OR
Sheffield City Polytechnic
Pond Street
Sheffield S1 1WB

Dr. G. Ivanus
I.I.T.P.I.C.
Petrochemical Department
56–58 Caderea
Bstiliei St.
71.139 Bucharest Romania

Dr. G. James
Department of Mathematics
Coventry Polytechnic
Priory Street
Coventry

Mrs. B. E. Jarvis
The Open University
Walton Hall
Milton Keynes MK7 6AA

John Jaworski
BBC Open University Production Centre
Walton Hall
Milton Keynes MK7 6BH

Mr. H. Johansson
University of Lulea
Department of Mathematics
S-951 87
Lulea
Sweden

Mr. Alun Jones
Polytechnic of the South Bank
Borough Road
London SE1 0AA

Mrs. Gabriele Kaiser
Heinrich-Plett Strasse 40
3500 Kassel FRG

Mr. Kevin J. Kelly
Cork Regional Technical College
Cork Ireland

Mr. Kindt
Tiberdreef 4
3561 Utrecht Netherlands

Dr. P. Landers
Dept. of Electrical Engineering
Dundee College of Technology
Bell Street
Dundee DD1 1HG

Mr. Yitzak Lavi
Tel-Aviv University
P.O. Box 419
Kiryat-Ata 28103
Israel

Dr. E. G. Lawrence
University of Surrey
Department of Mathematics
Guildford Surrey

Professor D. H. Lee
School of Maths & Computer Studies
P.O. Box 1
Ingle Farm
South Australia 5098

Mr. D. W. Le Masurier
Mathematics Department
Brighton Polytechnic
Moulsecoomb
Brighton BN2 4CJ

Dr. Henry C. Levy
University of Otago
(14 Ribblesdale House
West End Lane
London N.W.6)

Mr. D. N. Lipscomb
Portsmouth Polytechnic
Mercantile House
Hampshire Terrace
Portsmouth

Mr. A. W. C. Lun
Dept. of Mathematical Studies
Hong Kong Polytechnic
Hung Hom
Hong Kong

Dr. George Luna
Mathematics Department
California Polytechnic State University
San Luis obispo
CA 93407

Mr. David Lynam
Airfield
Glen Road
Belfast BT11 8SQ

Mr. I. G. Mackenzie
Robert Gordon Institute of Technology
School of Mathematics
St. Andrew Street
Aberdeen AB1 1HG

Dr. V. P. Madan
Red Deer College
P.O. Box 5005
Red Deer
Alberta Canada

Mrs. Marianne Marme
Zentralblatt für Didaktik der Mathematik
Fachinformationszentrum Energie Physik Mathematik GmbH
D-7514
Eggenstein-Leopoldshafen 2

Dr. John H. Mason
Maths Faculty
The Open University
Walton Hall
Milton Keynes MK7 6AA

Dr. R. R. McLone
Faculty of Mathematical Studies
University of Southampton
Southampton S09 5NH

Dr. Daphne G. Medley
School of Mathematics
University of Leeds
Leeds LS2 9JT

Boleslaw Mikolajczak
Institute of Control Engineering
Technical University of Poznan
60-965 Poznan Poland

Mr. Patrick Moloney
35 Fairmile Avenue
Cobham Surrey

Mr. A. Moscardini*
Sunderland Polytechnic
Langham Tower
Ryhope Road
Sunderland SR2 7EE

Mr. Thomas Naylor
29 Spencer Road
Wigan WN1 2QR
Greater Manchester

Dr. M. J. O'Carroll
Teesside Polytechnic
Borough Road
Middlesbrough
Cleveland TS1 3BA

Mr. K. H. Oke
Dept. of Mathematical Science & Computing
Polytechnic of the South Bank
Borough Road
London SE1 OAA

Professor M. Olinick
Middlebury College
Middlebury
Vermont 05743

Mr. B. P. C. Patel
Dept. of Maths. Stats. and Computing
Plymouth Polytechnic
Plymouth PL4 8AA

Chris J. Patterson
Theoretical & Applied Mechanics
Private Bag
Auckland New Zealand

Dr. H. O. Pollak
Bell Laboratories
600 Mountain Avenue
Murray Hill
New Jersey 07974

Mr. D. E. Prior
Sunderland Polytechnic
Dept. of Maths. & Comp. Studies
Chester Road
Sunderland SR1 3SD

Dr. G. F. Raggett
Dept. of Maths, Stats. & O.R.
Sheffield City Polytechnic
Sheffield S1 1WB

Dr. Stephen C. Ryrie
Dept. of Computer Studies & Mathematics
Bristol Polytechnic
Coldharbour Lane
Frenchay
Bristol

Professor Eero Matti Salonen
Helsinki University of Technology
SF-02150 Espoo 15
Finland

Dr. Lino Sant
Dept. of Maths & Science
University of Malta
Tal QrooQ Malta

Dr. Jeffrey W. Searle
West Denton High School
West Denton Way
Newcastle Upon Tyne NE5 2SZ

Dr. S. Sheel
Department of Electrical Engineering
M. N. R. Eng. College
Allahabad 211004 India

Mr. M. G. Simpson
Maths Faculty
The Open University
Walton Hall
Milton Keynes MK7 6AA

Professor D. K. Sinha
Jadavpur University
Calcutta 700032 India

Dr. P. P. Siwak
Institute of Control Eng.
Technical University of Poznan
Piotrowo 3 a
60–965 Poznan Poland

Dr. Gillian Slater
Dept. Maths, Stats & O.R.
Sheffield City Polytechnic
Pond Street
Sheffield S1 1WB

Mr. N. C. Steele
Dept. of Mathematics
Coventry (Lanchester) Polytechnic
Priory Street
Coventry CV1 5FB

Dr. Sheila S. Stone
The Open University
Walton Hall
Milton Keynes MK7 6AA

Mr. Peter Strachan
School of Maths
Robert Gordon Institute of Technology

St. Andrew Street
Aberdeen

Dr. D. Sullivan
University of New Brunswick
Fredericton
New Brunswick
Canada E3B 5A3

Mr. P. B. Taylor
Polytechnic of the South Bank
Dept. of Mathematical Sciences & Computing
Borough Road
London SE1 0AA

Mr. D. Thompson
Dept. of Maths, Science & Computing
College of St. Paul and St. Mary
The Park
Cheltenham GL50 2RH

Mr. Miha Tomsic
Institute Joseph Stefan
Janova 39
61111 Ljublana Yugoslavia

Mr. M. S. Townend
Department of Mathematics
Liverpool Polytechnic
Byrom Street
Liverpool L3 3AF

Dr. W. R. Tunnicliffe
Faculty of Maths
The Open University
Walton Hall
Milton Keynes MK7 6AA

Dr. J. R. Usher
Robert Gordon Institute of Technology
School of Mathematics
St. Andrew Street
Aberdeen

Mr. D. J. Waddams
51 Southend Crescent
Eltham
London SE9 2SD

Dr. Costa Wahde
Dept. of Mathematics
Chalmers University of Technology
S-41269 Goteborg Sweden

Mr. L. J. Walker
Dept. of Maths & Computing
Paisley College
High Street
Paisley Renfrewshire

Mr. B. J. Wilson
Mirembe
Mews Lane
Winchester SO22 4PS

Mr. David B. Wilson
7 Ray Road
Romford Essex

Dr. D. T. Wilton
Dept. of Maths, Stats. & Computing
Plymouth Polytechnic
Plymouth